Petroleum Science and Technology

Petroleum Science and Technology

Edited by Natalie Mitchell

STATES
ACADEMIC PRESS
www.statesacademicpress.com

Published by States Academic Press,
109 South 5th Street,
Brooklyn, NY 11249, USA
www.statesacademicpress.com

Petroleum Science and Technology
Edited by Natalie Mitchell

© 2022 States Academic Press

International Standard Book Number: 978-1-63989-410-9 (Hardback)

This book contains information obtained from authentic and highly regarded sources. Copyright for all individual chapters remain with the respective authors as indicated. All chapters are published with permission under the Creative Commons Attribution License or equivalent. A wide variety of references are listed. Permission and sources are indicated; for detailed attributions, please refer to the permissions page and list of contributors. Reasonable efforts have been made to publish reliable data and information, but the authors, editors and publisher cannot assume any responsibility for the validity of all materials or the consequences of their use.

Trademark Notice: Registered trademark of products or corporate names are used only for explanation and identification without intent to infringe.

Cataloging-in-Publication Data

Petroleum science and technology / edited by Natalie Mitchell.
p. cm.
Includes bibliographical references and index.
ISBN 978-1-63989-410-9
1. Petroleum. 2. Petroleum engineering. 3. Petroleum--Geology. I. Mitchell, Natalie.
TN870 .P48 2022
553.28--dc23

TABLE OF CONTENTS

PREFACE

The naturally occurring oil which is yellowish-black in color, found in the body of rocks beneath the surface of the earth is termed as petroleum. It is also known as crude oil. Petroleum science and technology is an area of study which deals with the production of hydrocarbons, crude oil and natural gas. Composition of petroleum consists of crude oil, and solid, liquid and gaseous hydrocarbons. The two main subdisciplines in petroleum engineering are oil and gas industry. They focus on maximizing the economic recovery of hydrocarbons from subsurface reservoirs. Oil and gas industry is broadly classified into upstream, midstream and downstream sectors. The exploration and production of petroleum fall under the umbrella of the upstream sector of the oil and gas industry. This book unravels the recent studies in the field of petroleum science and technology. The various advancements in this area of study are glanced at and their applications as well as ramifications are looked at in detail. For all those who are interested in this field, this book can prove to be an essential guide.

Significant researches are present in this book. Intensive efforts have been employed by authors to make this book an outstanding discourse. This book contains the enlightening chapters which have been written on the basis of significant researches done by the experts.

Finally, I would also like to thank all the members involved in this book for being a team and meeting all the deadlines for the submission of their respective works. I would also like to thank my friends and family for being supportive in my efforts.

Editor

1

Using an Elastic, Expandable Sealant System for Zonal Isolation of Maroon Wells

Hamid Soltanian[1]*, Mohammad Ataei[1], Reza Khalokakaie[1], and Ezzatollah Kazemzadeh[2]

[1] School of Mining, Petroleum and Geophysics Engineering, University of Shahrood, Iran
[2] Petroleum Engineering Department, Research Institute of Petroleum Industry (RIPI), Tehran, Iran

ABSTRACT

An oil and gas well cementing in Gachsaran formation, where sustained annular pressure has been reported in many wells, presents a big challenge in Maroon field. The main challenges are preventing gas migration and achieving zonal isolation using a competent cement sealant system which is able to withstand downhole stresses and high temperatures during production cycles. Unlike conventional cement systems, properties, such as, high Poisson's ratio and low Young's modulus compared to that of the rock were optimized in the new system to achieve mechanical resistance and durability. The use of elastic-expandable additives to solve problems in oil well cementing has been investigated in recent years by several research groups in the petroleum industry. This study includes the laboratory examination of the effect of an elastic-expandable additive on the physical properties of a new cement sealant system. In the research process, a candidate well was selected and the properties of the used cement slurry in a problematic section of the well were evaluated in the laboratory. Then, the elastic-expandable additive was added as an elastic agent and the improvements in the cement slurry and stone properties were studied. This article discusses the problems associated with the conventional cement used in the candidate well and gives the detail of the improvements in cement properties obtained by adding the elastic-expandable additive to the cement slurry formulation as an elastic agent. The elastic-expandable additive increases the Poisson's ratio and expansion set cement, but it decreases the Young's modulus and fluid loss of the cement slurry. In addition, to prevent gas migration and achieve zonal isolation, there is an optimum concentration of the elastic-expandable additive at which the maximum compressive strength is reached. The results of this study can be used to optimize the cement slurry design in any given set of conditions.

Keywords: Expandable Additive, Compressive Strength, Rheological Properties, Young's Modulus, Poisson's Ratio

INTRODUCTION

Interzonal communication in a wellbore may lead to loss of reserves, the contamination of zones, the production of unwanted fluids, or safety and environmental issues. Remedial solutions are available to solve the problems, but for technical or economic reasons, the well may be shut in or abandoned To improve the lifetime of the well, the

***Corresponding author**
Hamid Soltanian
Email: soltanianh@ripi.ir

cement sheath must be chemically and mechanically durable [1-2]. Sealants resistant to aggressive formation fluids are designed when required. In the same way, sealants should be designed to withstand the stresses experienced during production and well operations for instance casing pressure tests, stimulation treatments or temperature changes during production cycles-throughout the well life. To achieve this, a better understanding of the mechanical behavior of different sealants under downhole conditions is required to design fit-for-purpose materials [3-5].

Several papers have been written on the subject. According to Thiercelin et al., three changes in downhole conditions can cause mechanical damage to the cemented annulus (mechanical failure or the creation of microannuli) that may lead to a loss of zonal isolation [6]. The key conclusion of that paper was that, instead of considering the strength of the sealant as the primary property, one should rather look at the complete mechanical system formed by the steel casing, the cemented annulus, and the formation. Indeed, the increase of pressure and/or temperature in the wellbore initially expands the inner steel casing, which instantly imposes this deformation on the neighboring cement sheath [7-8]. As a consequence, imposed displacements rather than imposed stresses are applied to the cement inner diameter (ID). At a greater time scale (the lifetime of the well), the cement sheath must withstand multiple displacement cycles. Numerical models have been proposed by several authors to simulate the sealant mechanical behavior and predict the initiation of failures, according to known mechanical properties of the complete system (steel, cement, and rock) [9-10]. In the past, the perception for judging the mechanical properties of cement was to look at the compressive strength of the set cement. However, the recent lessons learned from the wells experiencing sustained annular pressure showed the limitation of this traditional approach [11]. Changes in downhole conditions in terms of temperature and pressure can induce sufficient stresses to destroy the integrity of the cement sheath, which will cause long term gas migration and sustained annular pressure. Hence, the set cement mechanical properties have to be carefully designed in order to withstand the downhole stresses. Cement mechanical failure is caused by stresses induced by variations in downhole conditions such as [12-13]:

- Change in mud weight after cement placement;
- Pressure increase due to gas production;
- Pressure integrity test;
- Stimulation treatments;
- Temperature Changes;

Two scenarios need to be distinguished; the above stresses can cause cement failure either in traction or compression depending on downhole conditions [14-15]. Traction or compression failure is caused by the pressure or temperature increment. If the cement is placed across a soft formation and subjected to a tangential stress which exceeds the cement tensile strength, this will cause radial cracks, and thereby cement traction failure. Cement compression failure may occur if the cement is placed between the two casings or across a hard formation and the radial stresses exceed the cement rupture strength [16]. Another important cause of the loss of zonal isolation is the formation of microannulus, which is caused by temperature or/and pressure decrement. This paper summarizes the implementation and evaluation of a new sealant system designed to overcome this problem by preventing the formation of a microannulus across the gas zone [17-18].

Elastic, Expandable Sealant Cement System

The unique performance factors of this new sealant cement system are the mechanical parameters of the set cement. These are controlled by the system design and may be affected by the bottom hole static temperature. The mechanical parameters can be controlled independently of slurry density and properties. Based on the optimized, tri-modal, particle-composition theory, it encompasses all the advantages of an engineered particle-size distribution blend and the additional benefits of specific particles which can adjust the flexible properties and expansion capabilities of the set cement [19]. During the production life of a well, the cement sheath is exposed to varying stress fields as pressure and temperature fluctuate in the wellbore. These stresses can compromise the integrity of the cement sheath and affect its ability to maintain hydraulic isolation. Studies have shown that cement with a low Young's modulus and good expansion properties during the hardening phase is best for mechanical durability and resistance to stresses [20-21].

This system is adaptable to pressure and temperature changes, and the ability to expand improves sealing at the casing/formation interface. The addition of flexible material to the optimized blend plays an important role in increasing the flexibility of the set cement by decreasing the Young's modulus. The system applications focus on the need for zonal isolation in wells with severe changes in the downhole temperature and pressure, gas wells, high-pressure and high- temperature wells, plug-and abandon-applications, multilateral wells, and tectonically active environments or areas, where subsidence or compaction causes casing failure. There are four key cement mechanical parameters that must be determined to allow modeling the cement sheath behavior. These parameters are as follows:

- Compressive strength;
- Young's modulus;
- Poisson's ratio;
- Expansion capability;

EXPERIMENTAL PROCEDURES

This study is based on laboratory testing of the improvements in the cement slurry and stone by using an elastic-expandable additive. Figure 1 shows the SEM images of the elastic-expandable additive. The specifications of the elastic-expandable additive are presented in Table 1. This section provides a detailed description of the key sample preparation and testing methods required for the determination of the mechanical behavior of the set cement.

Figure 1: SEM images of the elastic-expandable additive.

Table 1: Physical and chemical properties the elastic-expandable additive.

Physical and Chemical Properties				
Size (micron)	**Specific Gravity**	**Humidity (%)**	**Color**	**Phase**
75	1.1	4-5	dark brown	solid

Sample Preparation

It is very important to prepare samples that are representative of the systems which are to be pumped. The systems must have zero free-fluid, which show no settling, and be free of air bubbles. The systems should be cured under downhole conditions. In many cases this is the bottomhole static temperature in the zone of interest, but in some cases, e.g. for steam injection wells, the curing may be done in two steps and at two different temperatures to simulate the parameters before and after steam injection. Ideally, a pressurized curing chamber should be used. However, if one is not available, then the slurry should be degassed under vacuum prior to cure to ensure that no air can enter through the sample. The curing period should be sufficiently long to enable the system to attain equilibrium. The systems to be tested are cured in molds of an appropriate size to allow correctly-sized samples to be prepared (Figure 2). For compressive strength, Young's modulus, and Poisson's ratio, cylindrical samples are required with a length to diameter ratio greater than 2.5. This length to the diameter ratio provides uniaxial compression conditions. For samples with smaller length to diameter ratios, the ultrasonic cement analyzer (UCA) would be overestimated, and correction factors are required. For length to diameter ratios of 1 (e.g., cube samples), the UCA can be overestimated by 15%. Typically, the samples which their diameters are equal to 68 mm, and their length is equal to 210 mm (length/diameter=3) have been used.

For tensile test measurements (Brazilian test), a length to the diameter ratio of 0.5 is required. Samples are first cured, and then the ends are cut perpendicular to the cylinder length using a rotating saw. In both operations, a continuous stream of tap water is used to cool the cutting surfaces and remove debris. The saw has two parallel blades and a special core holder which ensures that the ends are both parallel to each other and perpendicular to the cylinder axis. Throughout the preparation process, the samples must be kept saturated with water to avoid cracking related to drying shrinkage.

Figure 2: Cylindrical rubber mold ready for placement in a pressure curing chamber.

Figure 3: Cement samples ready for the test.

Equipment

The main requirement for the measurements is a load frame, with controllable load and displacement rates, equipped with a force transducer. The press that was used in the current work was a computer-controlled electromechanical press which can be fitted with different load cells depending on the strength and the geometry of the materials to be tested. The axial and radial displacements were measured with linear variable-displacement transducers (LVDT) and a strain gauge cantilever respectively (Figure 4). The LVDT's directly measure the displacement of the loading platens to eliminate deformation of the load frame assembly, which may occur at high loads. The cantilever is placed at the middle of the sample height under test and measures the radial displacement in three directions. The output from the 3 LVDT's is averaged to give the radial displacement, and the output from the cantilever is axial displacement. The LVDT's and cantilever were calibrated against the displacement of the press crosshead which was calibrated annually against a standard displacement transducer. The amount of the set cement expansion can only be measured after being cured in special rectangular cylinders called expansion cell under downhole conditions (Figure 6), and it is then tested in an expansion apparatus (see Figure 7).

Study Method of the New Cement System

The key advantage of the new cement is its ability to provide elasticity that can be adapted based on the well conditions. Another important advantage of the new system is its ability to provide flexibility characteristics along with the expansion. This study is based on taking the laboratory tests of the improvements in the cement slurry and stone by using the elastic-expandable additive at five concentrations. The conventional and new cement system slurries are mixed in the laboratory, and their properties including weight, rheology, free water, fluid loss and thickening time are measured according to API RP 10B-2. The compressive strength of cement stones is measured by UCA. The elastic properties are measured on cylindrical samples by a uniaxial compressive strength tester equipped with axial and radial strain gauges (Figure 4). The elastic properties of the cement stone are determined based on the ISRM (International Society for Rock Mechanics) and ASTM (American Society for Testing and Materials) rock mechanic standards. The cylindrical cement stone samples are prepared by curing the cement slurry for 48 hours at 210 °F and 3000 psi in a rubber mold with a diameter of 68 mm and a length of 210 mm in a curing chamber (Figure 5). After the ends of the sample are cut flat, they are cooled down and kept saturated by immersing in the saturated water, and the loading rate is 4000 psi/min, which is the maximum loading rate of the available equipment. Table 2 shows the conventional cement recipe and 4 other recipes modified by different amounts of the elastic-expandable additive. The elastic-expandable additive is added to the cement slurry composition at four different concentrations (3.3, 6.6, 9.9, and 13.2% by weight of cement). Tables 3 and 4 show the measured properties of the cement slurries and stones, according to API RP 10B-2. Table 5 shows the elastic properties measured for the 5 samples.

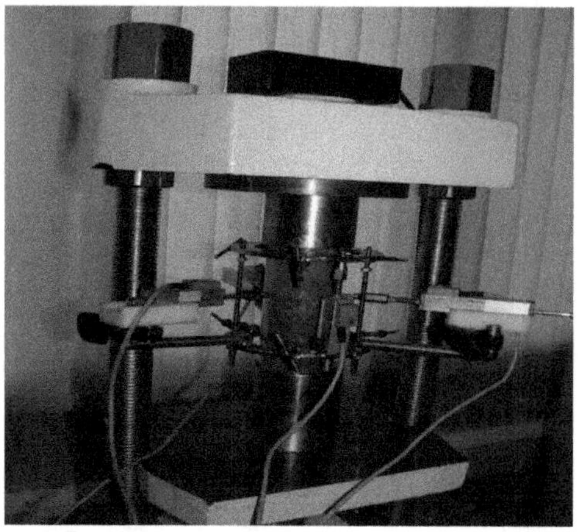

Figure 4: Detailed view showing the LVDT's and cantilever.

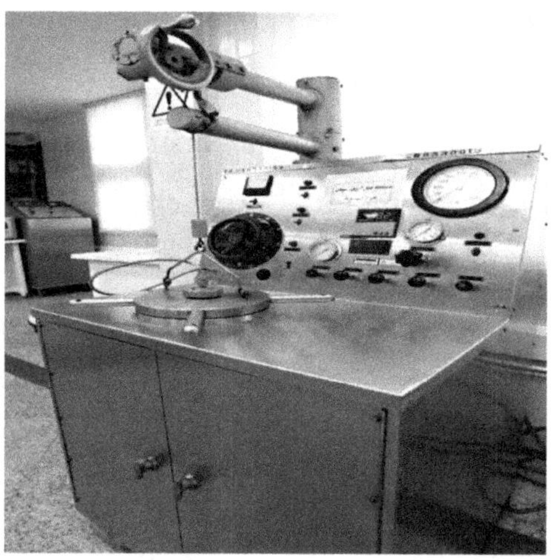

Figure 5: API Cement pressure curing chamber.

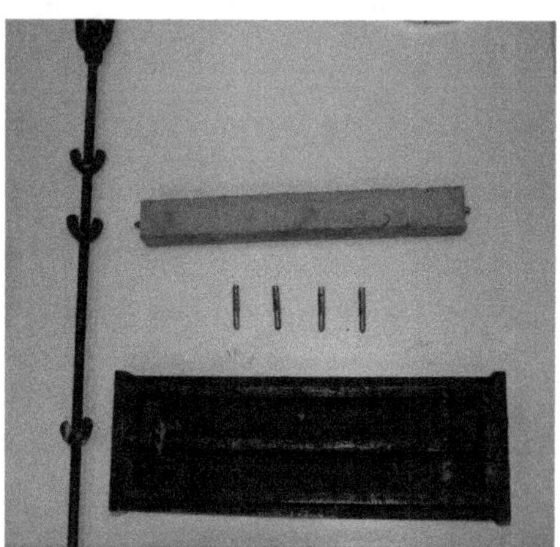

Figure 6: Cement expansion cell.

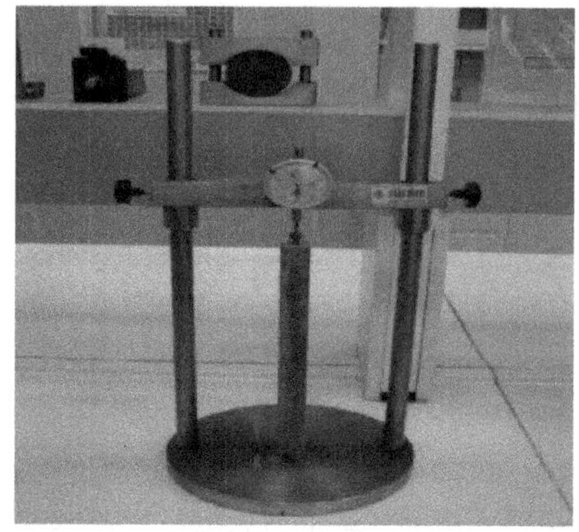

Figure7: Cement expansion equipment.

Expansion Capability

Expansion is an increase in the bulk volume of the initial hardened cement. One of the main causes of the loss of zonal isolation is the formation of microannuli. An inner microannulus between the casing and the cement can be created, for instance, by the radial displacement of the casing, resulting from wellbore temperature or/and pressure decrement, especially when the wellbore pressure is decreased (change in drilling fluid density for example), once the cement has set. An outer microannulus between the cement and the casing can be generated by cement bulk shrinkage (the bulk volume shrinkage of conventional cement system is 3 to 4% of total volume). The formation of microannuli can be avoided by the use of expanding cement systems, which have been specifically designed to compensate for the inward radial displacement of the casing. This problem has been overcome after adding an expansion agent to the new cement system to seal off possible microannuli and tighten the cement against the casing and the formation. Bonding between the cement/formation and cement/casing increases with time due to the fact that expansion occurs after the cement has set. The concentration of the expansion additive required for a certain value depends on several factors. These include bottomhole static temperature (BHST), the nature of the surrounding formation, cement type, and slurry additives. Table 6 shows the expansion measurements made at 210°F for 5 samples.

New Cement Properties

The new cement system was designed to be gastight to prevent gas migration, which could ultimately lead to an undesired sustained annular pressure. The fluid migration analyzer (FMA) test was performed to measure the tightness of cement slurry to gas or fluid migration during its transition from liquid to solid under downhole conditions. Figures 8 and 9 show the results of the gas migration test through the cement column (Slurry No. 5). The data from the static gel strength test has been used to calculate the pattern of cement slurry pore pressure changes, which was the input into the calculations used in the fluid migration analyzer apparatus. All realistic conditions in a wellbore; such as, curing cement slurry at downhole condition, fluid loss, slurry pore pressure decline, and finally the formation fluid migration through the cement column have been simulated in this apparatus. While the cement pore pressure declines (the (a) curve) and becomes close to gas-bearing formation pressure (Figure 8), the process of injecting gas to the cement is started. gradually eliminated due to the compressive and gel strength development in the slurry. In this test, the total gas volume migrated through cement has been equal to 80 ml per 5 hrs test period (Figure 9, (b) curve); thus the gas migration rate was nearly 2.67 cc/min. In accordance with the literature, the maximum gas migration rate should not exceed the 30 ml/min limit.

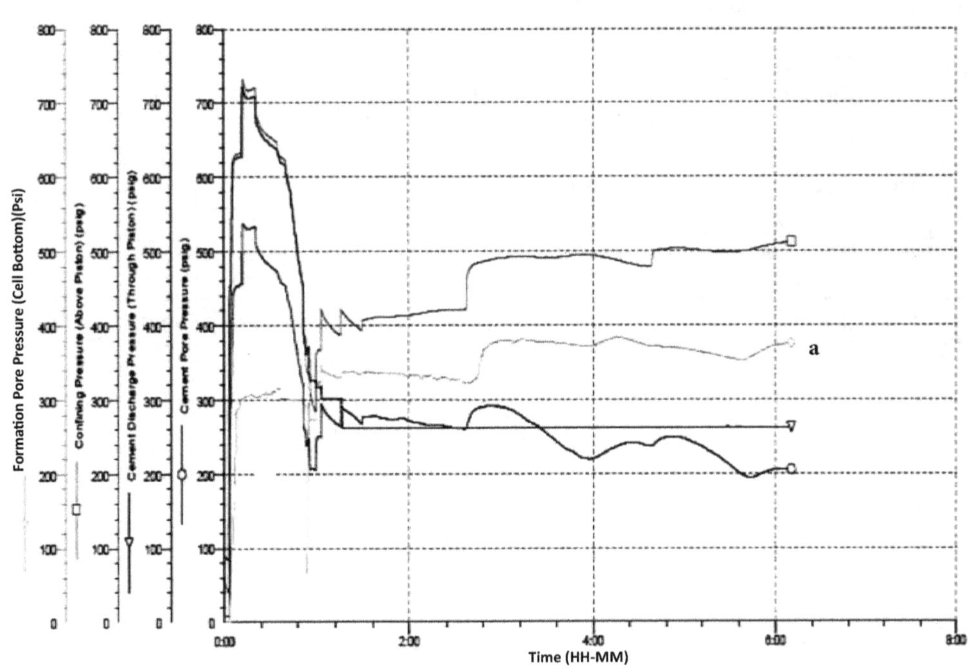

Figure 8: The fluid migration test shows that there is no indication of gas migration in cement slurry No. 5.

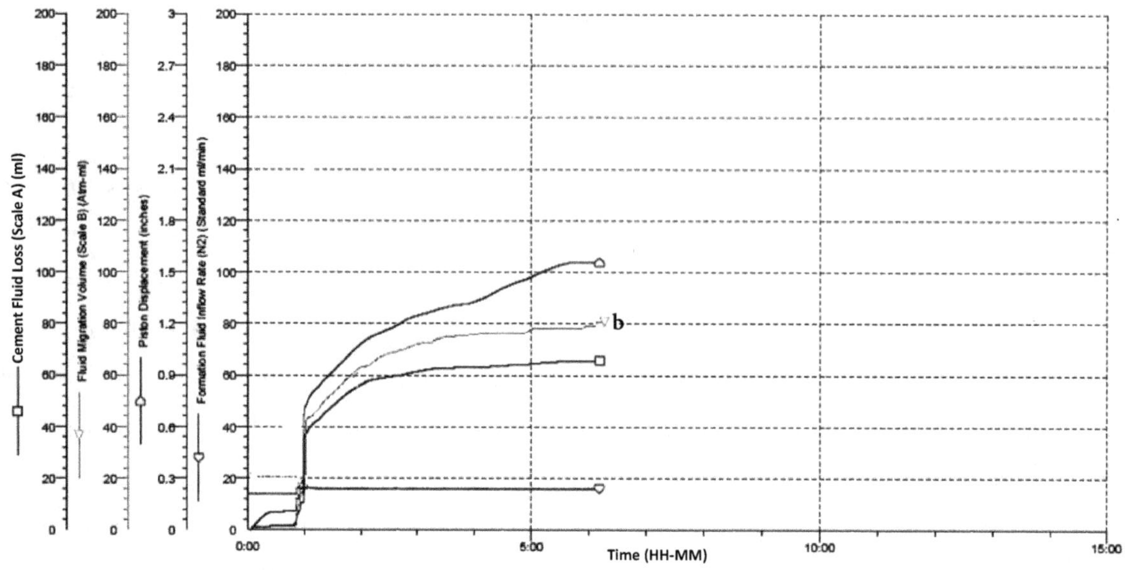

Figure 9: The fluid migration volume and cement fluid loss changes in cement slurry No. 5.

Table 2: Cement slurry compositions.

Component	Unit	Slurry 1	Slurry 2	Slurry 3	Slurry 4	Slurry 5
Cement class E	lb.	110	110	110	110	110
Elastic-expandable additive	lb./sk	-	3.3	6.6	9.9	13.2
Salt	lb./sk	15.51	15.51	15.51	15.51	15.51
Hidense	lb./sk	60.5	60.5	60.5	60.5	60.5
Boric acid	lb./sk	0.33	0.33	0.33	0.33	0.33
Fluid loss control	lb./sk	-	0.33	0.33	-	-
Cement friction reducer	lb./sk	-	0.55	0.55	0.11	1.32
Antifoam	gal/sk	0.02	0.02	0.02	0.02	0.02
Water	ft^3	0.6696	0.6696	0.6696	0.6696	0.6696

Table 3: Cement slurry properties.

Slurry number	Slurry weight (PCF)	Free Water (cc)	Fluid loss (ml) (T=180 °F, P=1100 psi)	Thickening Time(min) (T=180 °F, P=9000 psi)	Compressive strength 24 hrs(psi) (T=210 °F, P=3000 psi)
Slurry 1	149	1.5	30	145	3650
Slurry 2	147	1	27	160	3100
Slurry 3	145	1	30	180	2770
Slurry 4	143	0.5	31.5	193	2150
Slurry 5	141	0	33	203	1900

Table 4: cement slurries rheological properties.

Slurry number	(θ_{600})	(θ_{300})	(θ_{200})	(θ_{100})	(θ_6)	(θ_3)	PV (cP)	YP (lb/100ft²)
Slurry 1	>300	273	190	107	15	13	249	24
Slurry 2	>300	250	177	98	20	16	228	22
Slurry 3	>300	286	181	100	19	15	249	17
Slurry 4	>300	276	165	107	17	13	253.5	22.5
Slurry 5	>300	248	177	98	29	19	225	23

Table 5: Cement stones elasticity properties.

Sample	Elasticity Properties	
	Young's Modulus (GPa)	Poisson's Ratio
1	2.86	0.04
2	2.76	0.06
3	2.48	0.16
4	1.94	0.25
5	1.76	0.31

Table 6: Expansion set cement stones.

Sample	Expansion (%)	
	24 hr	3 Day
1	0.0	0.0
2	0.04	0.09
3	0.10	0.19
4	0.15	0.25
5	0.16	0.30

CONCLUSIONS

The stresses induced in the cement matrix through the deformation of the cemented casing can cause damage due to the variations in the downhole conditions. Appropriate mechanical properties of cement can reduce potential cement failure and debonding. High compressive cement is not always the best solution; however, the elastic cement is the appropriate approach. The new cement system has been examined in the lab and the following conclusions have been obtained:

- The new cement system is more ductile than the regular cement. As a result, the new cement system enhances cement resistance to stress cycling.
- The new cement system expands after placement in order to enhance the bonding between the casing and the formation.
- The new cement system has a low fluid loss, free water, and compressive strength compared to conventional systems.
- Increasing the concentration of elastic-expandable material in the new cement system reduces Young's modulus.
- The new cement system prevents gas migration and achieves zonal isolation.
- Increasing the concentration of the elastic-expandable material in the new cement system increases the Poisson's ratio and the expansion of the set cement.
- The lifetime of a well may be greatly improved by the selection of the most appropriate sealant.
- The key advantage of the new cement is its ability to provide the elasticity which can be adapted based on the well conditions.
- Another important advantage of the new system is its ability to provide flexibility characteristics along with the expansion.

REFERENCES

1. Bonett A. and Pafitis D., "Getting to the Root of Gas Migration," *Oil Field Review Spring*, **1996**, 36-49.
2. Rogers M. J. and Dillenbeck R. L., "Transition Time of Cement Slurries, Definitions and Misconceptions Related to Annular Fluid Migration," SPE 90829, **2004**, 26-29 September.
3. Webster W. W. A and Eikerts J. V., "Flow after Cementing: Field and Laboratory Study," SPE 8259, **1979**, September, 23-26.
4. Buraik K. A., Abdulqer K. A., and Basibes R., "Prevention of Shallow Gas Migration through Cement," SPE 47745, **1998**.
5. Robert M., "Gas Flow in Cements," SPE 11207, **1985**.
6. Boumgarte C., Thiercelin M., and Klaus D., "Case Studies of Expanding Cement to Prevent Microannular Formation," SPE 56535, **1999**.
7. Stewart R. B. and Schouten F. C., "Gas Invasion and Migration in Cemented Annuli," SPE Drilling Engineering, **1988**, 77-82.
8. Jennings S. S. and Ansari A. A., "Gas Migration after Cementing Greatly Reduced," SPE 11209, **2003**.
9. "API Recommended Practice for Testing Well Cements," RP 10B API, Washington, **2010**.
10. Boukhelifa L., Moroni N., James S.G., Le Roy-Delage S., et al., "Evaluation of Cement Systems for Oil and Gas Well Zonal Isolation in a Full-scale Annular Geometry," SPE 87195-MS, **2004**.
11. Destrade M. et al., "Slight Compressibility and Sensitivity to Changes in Poisson's Ratio," *International Journal for Numerical Methods in Engineering*, **2011**, *90*, 403-411.
12. El-Hassan H., Sultan M., Saeed A. A., Johnson C., et al., "Using a Flexible, Expandable Sealant System to Prevent Microannulus Formation in a Gas Well: A Case History," SPE 92361, SPE Middle East Oil & Gas Conference, Bahrain, **2005**, 12–15 March.
13. Le Roy-Delage S., Baumgarte C., Thiercelin M., and Vidick B., "New Cement Systems for Durable Zonal Isolation," SPE 59132-MS, IADC/SPE Drilling Conference, New Orleans, Louisiana, USA, **2000**, 23-25 February.
14. Mack J. and Dillenbeck L. R., "Cement: How Tough is Tough Enough, A Laboratory and Field Study," SPE 78712-MS, SPE Eastern Regional Meeting, Lexington, Kentucky, USA, **2002**, 23-26 October.
15. Morris W., Criado M. A., Robles J., and Bianchi G., "Design of High Toughness Cement for Effective Long Lasting Well Isolations," SPE 81001-MS, SPE Latin American and Caribbean Petroleum Engineering Conference, Port-of-Spain, Trinidad and Tobago, **2003**, 27-30 April.
16. Mueller D. T., "Producing Stress-resistant High-temperature/High-pressure Cement Formulations through Microstructural Optimization," SPE 84562-MS, SPE Annual Technical Conference and Exhibition, Denver, Colorado, USA, **2003**, 5-8 October.
17. Ravi K., Bosma M., and Gastebled O., "Safe and Economic Gas Wells through Cement Design for Life of the Well," SPE 75700-MS, SPE Gas Technology Symposium, Calgary, Alberta, Canada, **2002**, 30 April-2 May.
18. Wray B., Bedford D., Leotaud L., and Hunter B., "The Application of High-density Elastic Cements to Solve HP/HT Challenges in South Texas: the Success Story," SPE 122762-MS, SPE Annual Technical Conference and Exhibition, New Orleans, Louisiana, **2009**, USA, 4-7 October.
19. The University of Texas at Austin-Petroleum Extension services; "Well Cementing (Oil and Gas production)," **1983**.
20. Ravi K. et al., "Optimizing the Cement Sheath Design in HPHT Shearwater Field," SPE 79905-MS, **2003**.
21. Wray B. et al., "The Application of High-density Elastic Cements to Solve HP/HT Challenges in South Texas: the Success Story," SPE 122762-MS, **2009**.
22. Yuan Z., "HPHT Gas Well Cementing Complications and its Effect on Casing Collapse Resistance," SPE 153986-MS, **2012**.

2

Identification of Gas Accumulation Adjacent to the Gas Hydrate Bearing Zone by Inversion Utilizing AVO Attribute

Effat Behboudi[1]*, Nasser Keshavarz[2], and Asghar Naderi[2]

[1]Department of Petroleum Engineering, Amirkabir University of Technology, Tehran, Iran

[2]Exploration and Production Research Center, Research Institute of Petroleum Industry (RIPI), Tehran, Iran

ABSTRACT

The presence of gas hydrate and associated gas in the Makran Accretionary Prism, offshore Iran, has been confirmed by various seismic indicators, including the bottom-simulating reflector (BSR), amplitude blanking, and bright spot. The pre-stack analysis and the study of amplitude-variation-with-offset (AVO) attributes are the most reliable tools for semi-quantitative study of gas hydrate and free gas resources. Due to the lack of well data in the most regions of gas hydrate accumulations in deep oceans, the study of AVO attributes and application of seismic inversion techniques are solutions to evaluate both gas hydrate and the free gas trapped beneath it. Conventional seismic stacked data increases the signal-to-noise ratio; however, amplitude stretching in far offsets resulted in normal move-out (NMO), and Poisson's error in velocity analysis decreases the frequency content of stack section. Moreover a stacked seismic data do not present a real zero-offset seismic data. In this study, the AVO intercept section was used to remove far offset effects and produce a more realistic zero-offset section with a high frequency content. Consequently, the acoustic impedance produced from the inverted intercept section shows acoustic property of the sediment in a better accuracy. According to the inverted section from AVO intercept data, free gas zone, and double bottom-simulating reflector, which identified about 50 ms below the first BSR, were observed with high vertical resolution.

Keywords: Amplitude-variation-with-offset (AVO), Intercept Attribute, Gas Hydrate, Free Gas, Double Bottom-simulating Reflector (DBSR)

INTRODUCTION

Natural gas hydrates are crystalline solids of gas (mostly methane) and water formed in low temperature and high pressure thermodynamic conditions[1]. Studying natural gas hydrates as a potential future energy resource has been increased during the past decades. The total amount of methane trapped in gas hydrates or accumulated as free gas beneath the gas hydrate stability zone is quite large [2]. Main methods of recovering gas from gas hydrates resources include; depressurization method: the reservoir

***Corresponding author**

Effat Behboudi

Email: effatbehboodi@aut.ac.ir

pressure is lowered to adjacent zones of solid hydrate pressure due to the production well in the gas hydrate reservoir [3,4] and pressure reduction frees the methane molecules from hydrate; thermal stimulation method: the reservoir temperature is increased above the hydration temperature by hot water injection, steam stimulation or circulating hot fluids downhole[5]; chemical inhibitors injection: the injection of chemical inhibitors (such as salts and alcohols) shifts the pressure-temperature equilibrium conditions to lead a rapid dissociation of gas hydrate [6]; and finally carbon dioxide displacement method: this method produces methane via the exchange between CO_2 and CH_4 in this method [7]. In practice, a combination of these techniques has also been proposed. Depressurization method is considered as a relatively feasible method and the most efficient and economical recovery technique to produce methane from seabed gas hydrates [8].

A comprehensive gas hydrate exploration technology is integration of 3-D seismic data (such as multicomponent ocean bottom cables (OBC)), Marine controlled-source-electromagnetic (CSEM), marine sediment sampling, and gas geochemical analysis to reduce drilling risk. OBC technology provides a direct measurement of shear wave to estimate elastic modules and to provide a more reliable image of structure associated with gas hydrate reservoirs than primary wave data [8]. CSEM technique provides information on rock and fluid electrical properties due to the electrical resistivity contrast of gas hydrate and free gas bearing zones with water-saturated rocks[9]. Numerous seismic techniques such as seismic inversion, rock physics templates and modeling, pre-stack seismic attribute study, and seismic velocity analysis have been developed to identify gas hydrates in marine sediments and the trapped free gas beneath them [10-15]. Most gas hydrate resources which is generally found in deep marine areas confront lack of well data, subsequently, exploration approach is heavily depends on seismic data for further evaluating these resources [16,17]. The dominant seismic indicator of gas hydrates accumulation is bottom-simulating reflector (BSR) which is sub-parallel to the seafloor and has a reverse polarity compared to the seabed on a seismic section [18]. This reflector corresponds as the base of gas-hydrate stability zone, and most of the seismic analyses have been focused on this reflector to characterize the above hydrates associated with its possible underneath free gas. BSR generally depicts the interface between higher P-wave velocity and lower P-wave velocity since a small amount of free gas below the gas hydrate stability zone decreases P-wave velocity considerably without affecting the S-wave velocity [19]. Thus, important information on gas hydrate and free gas accumulation across the base of the gas hydrate stability zone is provided by amplitude-variation-with-offset (AVO) characteristics of the BSR, which depends on P-wave and S-wave velocity changes [20-22].

In this study, the main target is investigation of the regions which likely contain highly concentrated gas hydrate and free gas in the sediments across the BSR located in the Iranian part of Makran accretionary prism. AVO analysis and post-stack inversion of the AVO attributes are utilized to reduce ambiguity and increase reliability to identify a potential target of gas hydrate accumulation.

EXPERIMENTAL PROCEDURES

Methods

AVO inversion

The AVO technique has been considered as an effective tool for reservoir characterization including porosity, density, lithology identification, and fluid content due to the fact that seismic amplitudes at layer boundaries are affected by the variations of the physical properties just above and below the boundaries [20,23]. Theoretical equations which completely determine the exact coefficients of the transmitted and reflected plane waves and its mode converted waves when the plane wave incident occurs at an interface between two semi-infinite isotropic and homogeneous layers have been developed by Zoeppritz in 1919 [24]. The Zoeppritz equation was simplified by making some approximations that describe the variation of P-wave reflection coefficients with the angle of incidence as a function of the P-wave velocity, the S-wave velocity, and the densities above and below the interface [25]. A series of linear approximations to the Zoeppritz equations has been developed for AVO analysis technique to predict the presence of natural gas. A useful approximation of the Zoeppritz equations has been proposed by Shuey in 1985 [26]; where AVO is estimated by performing a least-squares fit of the amplitudes from each time sample of the gather. This equation contains three terms of the intercept (A), gradient (B), and curvature (C):

$$R(\theta_i) \approx A + B\sin^2\theta_i + C(\tan^2\theta_i - \sin^2\theta_i) \quad (1)$$

where

$$A \approx \frac{1}{2}\left(\frac{\Delta V_P}{V_P} + \frac{\Delta\rho}{\rho}\right) \quad (2)$$

$$B = \frac{1}{2}\left(\frac{\Delta V_P}{V_P}\right) - 4\left(\frac{V_P}{V_S}\right)^2\left(\frac{\Delta V_S}{V_S}\right) - 2\left(\frac{V_S}{V_P}\right)^2\left(\frac{\Delta\rho}{\rho}\right), \quad (3)$$

and

$$C = \frac{1}{2}\frac{\Delta V_P}{V_P}. \quad (4)$$

θ can be estimated by equation 5 that relates the offset to the angle in the presence of an appropriate estimation of velocity [27]:

$$\theta = \arcsin\left(\sqrt{\left(\frac{x^2 v_i^2}{v_r^2\left(v_r^2 t_0^2 + x^2\right)}\right)}\right), \quad (5)$$

where x is the source-receiver offset, v_i is the interval velocity of a particular layer, v_r is the RMS velocity, and t_0 is the zero offset two-way time (TWT). The values V_P, V_S, and ρ are the P-wave velocity, S-wave velocity, and density respectively. The first term (A) denotes the intercept which measures the band-limited reflection amplitudes at normal incidence or zero offset. The second term (B) is the gradient that predominates at the intermediate angles Gradient measures the rate of change of P-wave amplitude with offset and it is related to Poisson's ratio. The gradient section is often a good indicator of gas reservoirs since Poisson's ratio is greatly affected by gas in porous media [20]. The third term (C) is the curvature which controls the behavior of the reflection variation at far angles, and it is related to P-wave velocity.

For the low incidence angles (up to ~ 30°) based on the Shuey's approximation of the Zoeppritz equation, the third term can be ignored and the intercept and gradient are estimated by a linear fit of P-wave reflection amplitudes to the $\sin^2\theta$ for each time samples (Figure 1). Equation 1 can be expressed as [26]:

$$R(\theta_i) \approx A + B\sin^2\theta_i \quad (6)$$

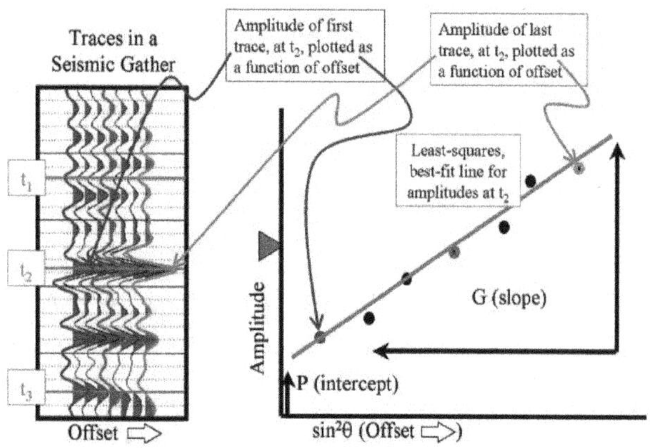

Figure 1: Intercept and Gradient calculation based on least-squares fit of the amplitudes of all traces at each time sample of the CMP-gather [28].

For a conventional stacked trace, the amplitude value for a time sample is the average of the amplitudes of all offsets. This average does not consider the relative amplitude changes based on different offsets while an AVO intercept trace gives normal incidence amplitude considering the amplitude value of each time sample depending on the related offset. Therefore, AVO intercept section can be considered as a better zero-offset section than a conventional stack section.

Impedance Inversion

Good –quality impedance inversion provides the detail of the subsurface due to the fact that acoustic impedance is a layer property (e.g. porosity and lithology) rather than interface property. The acoustic impedance is the result of the integrating of seismic data with well logs or velocity data. AVO analysis in conjunction with acoustic impedance improves interpretation and discrimination of lithology, and fluids. Subsequent AVO attribute analysis followed by AVO inversion was carried out to verify initial interpretation as well as to assess the possibility of distinguishing between different lithologies and fluids.

The inversion workflow utilized in this research constitutes of two steps. The first step is reprocessing of seismic data to recover true amplitude. This procedure includes geometrical spreading, transmission loss, attenuation, receiver directivity corrections, conversion of amplitude to reflection coefficient, and trim static correction [29]. The second stage is estimation of AVO attributes followed by post-stack inversion in order to estimate P-wave impedances. This workflow is depicted in Figure 2.

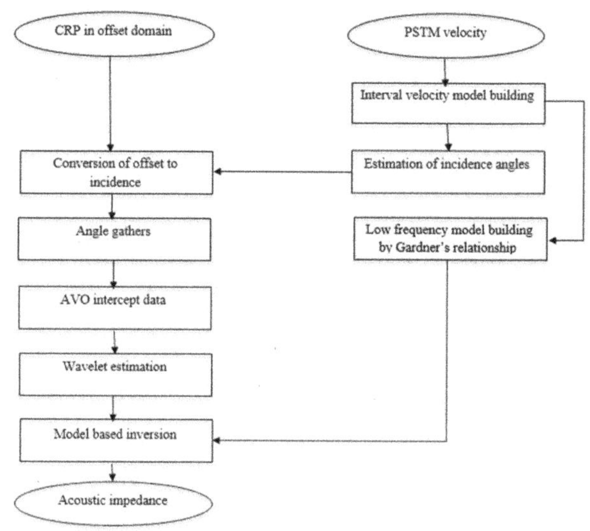

Figure 2: Workflow for calculating the acoustic impedance section from seismic data using AVO and post-stack inversion.

Seismic inversion requires initial low-frequency impedance model to constrain the inversion results. Seismic data generally lack low-frequency components; thus, inversion is limited to the frequency content of the seismic data. The low-pass filtered impedance log data is used as a background model to overcome this shortage of low frequencies. Usually, these background trends (p-wave impedance and density profile) are extracted from well data but due to the lack of well information in the area of study, these values were derived from the velocity model obtained during Pre-stack time migration process. Furthermore, the density profile was derived using Gardner's relationship [30].(Gardner et al., 1974). Figure 3 shows the initial impedance model in which the acoustic impedance is continuously increasing with depth while no gas layer could be detected in pre-stack time migration (PSTM) velocity model nor as stated earlier on the resultant acoustic impedance initial model.

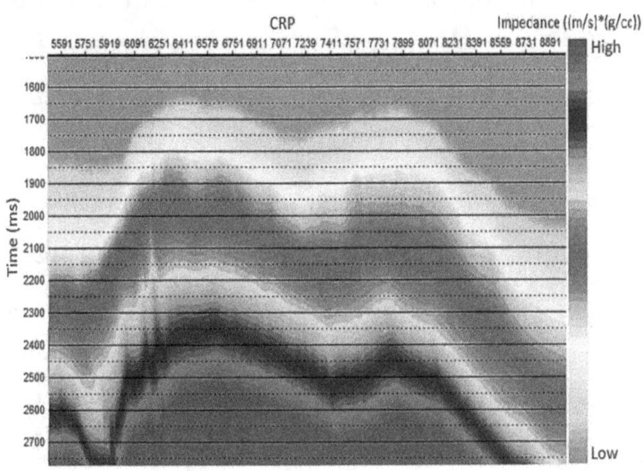

Figure 3: The initial impedance model on which the acoustic impedance steadily increases with depth while no gas layer is visible.

Zero-phase wavelet was statistically derived from the de-phased seismic data (Figure 4). Post-stack AVO intercept inversion was first computed at five selected control point (pseudo wells) locations to test the input parameters for robustness before computing the inverted P-impedance section. Afterward, a model based inversion (generalize linear inversion) which was constrained not to vary by more than 50% from the input model was applied on the background model to update acoustic impedance iteratively by minimizing the difference between synthetic seismic trace and real seismic data.

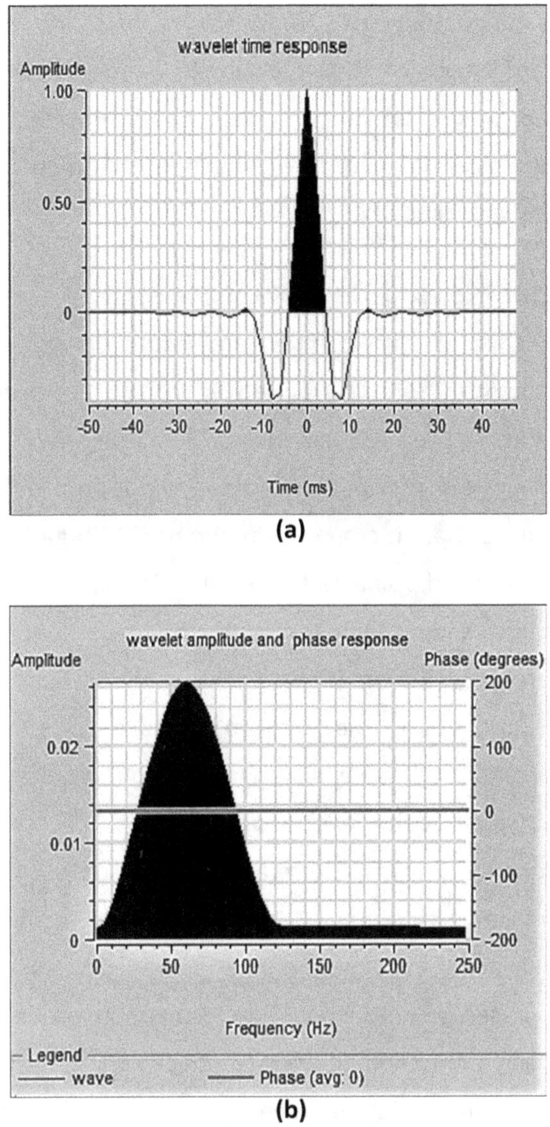

Figure 4: The statistical wavelet was extracted from the dephased seismic data. (a) The extracted zero-phase wavelet in time domain. (b) Amplitude and phase spectrum of the extracted wavelet in frequency domain.

Field Data

The data used in this study includes: 2-D pre-stack seismic data, pre-stack time migration velocity, and 2-D post-stack seismic data. These data have been acquired with 6100 m streamers length that consisting of 480 hydrophone group with 12.5 m receiver interval at a depth of 6 m. The energy source constitutes 2842 cubic inch airgun array at a depth of 5 m with a 25 m long shot interval providing foldage of 120. Sample rate is 2 ms with record length of 7000 ms. Dominant frequency of the data is 50 Hz. The data were kindly provided by Exploration Directorate of National Iranian Oil Company (NIOC).

Datasets and Data Reprocessing

Pre-stack amplitude inversion should be applied to common reflection point (CRP) gathers in order to derive the AVO attributes. Analyzing AVO in CRP gathers after pre-stack time migration is more accurate than in common-midpoint (CMP) gathers on steeply dipping reflectors and in complicated structures due to the effects of CMP smearing. Therefore, Kirchhoff pre-stack time migration was applied to the data to produce reflection coefficients at image locations and provide suitable data required for AVO analysis.

The raw seismic data were also reprocessed in order to maintain true amplitudes [31]. For AVO analysis, the relative amplitudes of the seismic data should be preserved throughout the analysis in order to extract information on primary reflectivity. Therefore, all steps in the data reprocessing sequence could be fulfilled through the spherical divergence correction according to Ursin method [32], the transmission loss correction [33], the attenuation correction for the layers overlying top of gas accumulation (BSR) as Q=191 that was considered for the P-wave of this seismic line [31], receiver directivity correction was accomplished using Sheriff and Geldart equation [34], amplitude conversion to reflection coefficient using Warner function [35], and Trim static correction to accurately preserve the relative amplitudes to remove misalignments of seismic events which have not been corrected after normal move-out (NMO) corrections and the computation of angle gathers.

RESULTS AND DISCUSSIONS

Acoustic impedance inversion was applied to the Makran 2D seismic data to determine the extension of the gas-hydrate and free gas saturated zones. In this study, the seismic inversion has produced a data set of the acoustic impedance of different layers associated with the BSR by matching a reflectivity model to the seismic data.

Figure 5 shows the AVO intercept section of the selected line between CRPs 5000 to 9000 in the Iranian part of Makran accretionary prism. The pre-stack time migration RMS velocity was converted to the interval velocity using Dix formula [36] to convert offset gathers to angle gathers. All CRP traces were transformed from the offset domain to the angle domain by estimating the incidence angles for each offset and time sample using Equation 5. Then, for each time samples the intercept and gradient were estimated by equation 6. In the intercept section, the BSR is clearly characterized by reversed polarity with respect to the seafloor reflection at about 2000 ms TWT. In Figure 5, the seismic reflections above the BSR within the hydrate zone are weak compared to the seafloor and BSR reflectivity. It can be explained by the presence of hydrates that fill the sediment pore spaces and decreasing the impedance contrast between sedimentary layers. Also, below the BSR, a weak reflector is observed that is subparallel to the seafloor topography and has a reverse polarity with respect to it. This seismic event which is identified about 50 ms below the first BSR is interpreted as a second or a double BSR (DBSR) (Figure 6) while it is less visible on the AVO intercept and the conventional post-stack seismic sections. The presence of another BSR have been reported in Victoria Land Basin [37], Barents Sea and Norwegian continental margin [38], Nankai Trough off central Japan [39,40] and Danube deep sea fan in the Black Sea [41]. Therefore, the presence of double BSR and the gas below BSR were investigated through post-stack inversion.

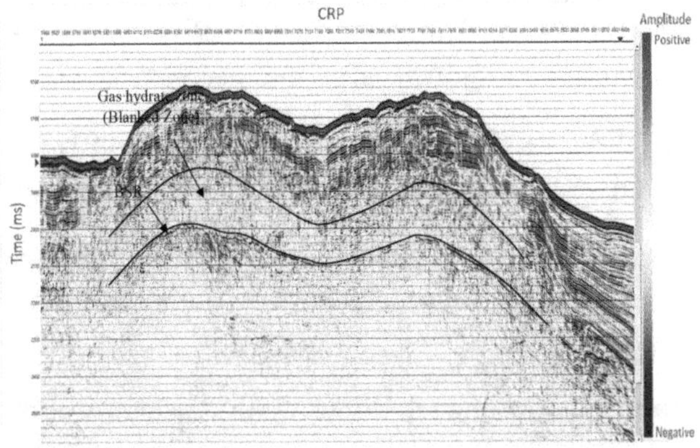

Figure 5: AVO intercept section sample over Makran accretionary prism shows weak reflections due to the presence of hydrates in sediments and a negative and strong BSR that is produced by the accumulation of free gas at the base of the hydrate layer.

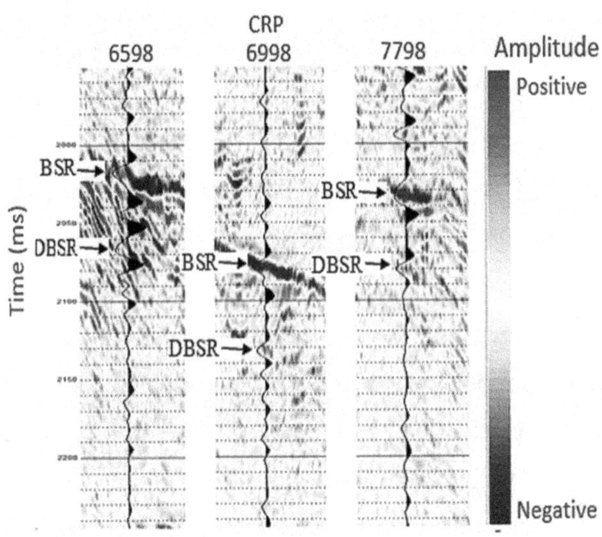

Figure 6: The wiggle traces show the seismic characters of the BSR and double BSR in CRPs 6598, 6998, and 7798. BSR is characterized by a negative polarity corresponding to the seafloor reflection due to decreasing of acoustic impedance of the underlying layer which is related to the gas presence. The double BSR is observed by amplitude lowering of the BSR and by a negative polarity with respect to the seafloor.

Figure 7 shows the acoustic impedance section provided by inverting AVO intercept data. The P-impedance contrasts at each subsurface position were estimated by applying a model-based inversionmethod to the AVO intercept section. Also, Figure 8 shows impedance section by inverting the conventional post-stack seismic data. In both mentioned figures, the low acoustic impedance zone containing free gas positioned underneath of the gas hydrate stability zone, which has been observed as BSR due to the decrease in P-wave velocity at the transition zone from hydrate to gas sediments while it was not included in the initial acoustic model. Comparing Figures 7 and 8 demonstrates that P-wave impedance of AVO intercept inversion has a high vertical resolution which helps thin layers such as DBSR be clearly observed while it is barely visible in the conventional post-stack inversion. In Figure 7, the high acoustic impedance between CRPs 6000-7000 could be due to the high concentration of gas hydrate (cementing effect) in the deepest part of the gas hydrate stability zone. The potential cemented gas hydrate reservoir at the base of the gas hydrate stability zone and existence of free gas beneath BSR would place these locations as a suitable candidate as well as an economical target for gas hydrate exploration in the future.

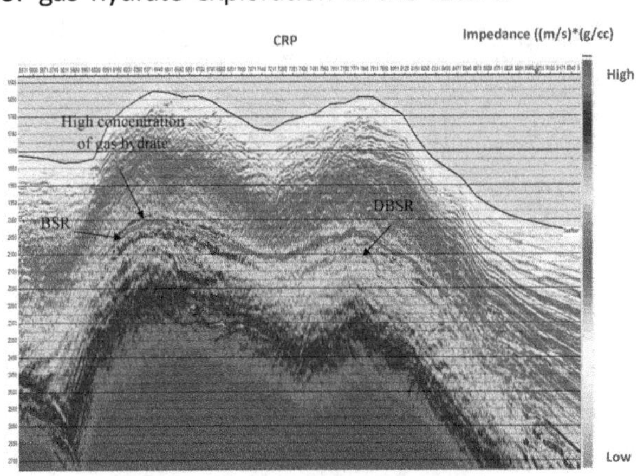

Figure 7: P-wave impedance from post-stack inversion of AVO intercept data. BSR and DBSR are clearly seen by a transition from high acoustic impedance containing gas hydrate to scattered reflectors with low acoustic impedance including free gas. Also, highly concentrated hydrate zone at bottom part of gas hydrate zone is shown.

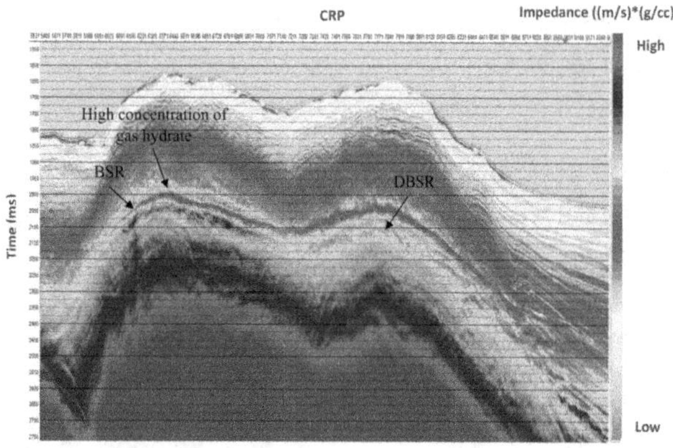

Figure 8: P-wave impedance from conventional post-stack inversion. BSR is clearly seen by a transition from high acoustic impedance containing gas hydrate to a low acoustic impedance including free gas while high concentration of gas hydrate at bottom of hydrate stability zone and DBSR are barely visible.

Several hypothesis for a DBSR could be considered: (1) the presence of sediments partially saturated with gas hydrates containing a different gas chemical composition which produces a BSR with a different stability zone [42,38,15], (2) regional uplifting of the bottom of the hydrate stability field from a previous depth to the current depth due to the compressional tectonic [39], (3) migration of free gas through competent or concrete layers of fractured sediments or in response to a recent sea bottom warming related to Global Warming effect [39,43]; and (4) diagenetic reflections that are produced by the opalA/opal-CT transition show normal polarity to the seafloor [39]. Referring to [44], considering the thickness modeling of gas hydrate stability zone with different gas compositions, geothermal gradient, water depth, pore pressure gradient, seabed temperature, and salinity revealed that the source of gas hydrate in some areas in Iranian part of Oman Sea includes thermogenic gases (containing 95% methane). Therefore, the formation mechanism for negative and weak DBSR in the studied area could be referred to migration of gas through unconsolidated sediments.

CONCLUSIONS

The AVO intercept data provided a better zero-offset section than a conventional stack section due to taking into consideration of the relative amplitude changes in different offsets. Therefore, AVO intercept inversion was undertaken to identify regions where are likely to contain gas under the BSR. Comparing inverted conventional stack seismic data with acoustic impedance derived from intercept section, revealed the efficiency of the proposed approach in depicting better distribution of free gas and gas hydrate zones.

The amplitude blanking, high acoustic impedance, and strong reflections of BSR identified from the 2D seismic profiles below seabed have been affected by the presence of gas hydrate. High acoustic impedance above the BSR is related to the presence of gas hydrates while a sharp drop in acoustic impedance below the BSR (sometimes appears as scattered layers) indicates the presence of free gas. Blanking zone could be related to high acoustic impedance which is associated with low reflectivity. A second BSR, which mimics the seafloor topography representing reverse polarity corresponding to this one, has appeared about 50 ms below the BSR and could be considered as a second gas hydrate resource containing thermogenic gas composition.

ACKNOWLEDGMENTS

The authors thank the Geophysics Department of National Iranian Oil Company (NIOC) Exploration Directorate for providing seismic data, their kind consent to publish this work and their financial support. The associate editor and three anonymous reviewers are also appreciated for their useful comments and suggestions.

NOMENCLATURES

BSR	: Bottom-simulating Reflector
CMP	: Common-MidPoint
NMO	: Normal Move-Out
AVO	: Amplitude-variation-with-offset
DBSR	: Double Bottom-simulating Reflector
OBC	: Multicomponent Ocean Bottom Cables
CSEM	: Marine Controlled-Source-Electromagnetic
TWT	: Two-way Time
NIOC	: National Iranian Oil Company
PSTM	: Pre-Stack Time Migration
CRP	: Common Reflection Point

REFERENCES

1. Hardage B. A. and Roberts H. H., "Gas Hydrate in the Gulf of Mexico: What and Where Is the Seismic Target?," *Leading Edge*, **2006**, *25*, 566 -571.
2. Kvenvolden K. A. and Lorenson T. D., "Global Occurrences of Gas Hydrate," In *Proceedings of the Eleventh International Offshore and Polar Engineering Conference*, **2001**, *1*, 462 -467.
3. Su Z., Moridis G. J., Zhang K., and Wu N., "A Huff-and-puff Production of Gas Hydrate Deposits in Shenhu Area of South China Sea through a vertical well," *Journal of Petroleum Science Engineering*, **2012**, *86–87*, 54–61.
4. Moridis G. J. and Reagan M. T., "Strategies for Gas Production from Oceanic Class 3 Hydrate Accumulations," In *Proceedings of the Offshore Technology Conference*, Houston, TX, USA, **2007**.
5. Tan Z., Pan G., and Liu P., "Focus on the Development of Natural Gas Hydrate in China," *Sustainability*, **2016**, *8*, 520- 539.
6. Holder G. D. and Lee S. Y., "Methane Hydrates Potential as a Future Energy Source," *Fuel Processing Technology*, **2001**, *71*, 181-186.
7. White M. D. and McGrail B. P., "Numerical Simulation of Methane Hydrate Production from Geological Formations via Carbon Dioxide Injection," *Proceedings of the Offshore Technology Conference*, Houston, TX, USA, **2008.**
8. Backus M. M., Murry P. E., Hardage B. A., and Graebner R., "High-resolution Multicomponent Seismic Imaging of Deep-water Gas Hydrate System," *The Leading Edge*, **2006**, *25*,578-596.
9. Edwards N., "Marine Controlled Source Electromagnetics: Principles, Methodologies, Future Commercial Applications," *Surveys in Geophysics*, **2005**, *26*, 675-700.
10. Fohrmann M. and Pecher I. A., "Analyzing Sand-dominated Channel Systems for Potential Gas-hydrate-reservoirs Using an AVO Seismic Inversion Technique on the Southern Hikurangi Margin, New Zealand," *Marine and Petroleum Geology*, **2012**, *38*, 19-34.
11. Ghosh R. and Sain K., "Effective Medium Modeling to Assess Gas Hydrate and Free Gas Evident from the Velocity Structure in the Makran Accretionary Prism, Offshore Pakistan," *Marine Geophysical Researches,* **2008**, *29*, 267-274.
12. Xia G., Sen M. K., and Stoffa P. L., "Mapping of Elastic Properties of Gas Hydrates in the Carolina trough by Waveform Inversion," *Geophysics*, **2000**, *65*, 735-744.
13. Riedel M. and Shankar U., "Combining Impedance Inversion and Seismic Similarity for Robust Gas Hydrate Concentration Assessments-a Case Study from the Krishna-Godavari basin, East Coast of India," *Marine and Petroleum Geology*, **2012**, *36*, 35-49.
14. Ojha M. and Sain K., "Seismic Velocities and Quantification of Gas-hydrates from AVA

Modeling in the Western Continental Margin of India," *Marine Geophysical Researches*, **2007**, *28*, 101-107.

15. Lu S. and McMechan G. A., "Elastic Impedance Inversion of Multichannel Seismic Data from Unconsolidated Sediments Containing Gas Hydrate and Free Gas," *Geophysics*, **2000**, *69*, 164-179.
16. Pevzner R. L., Volkonskaya A. L., Bouriak S. V., Bocharova A. A., and et al., "Prediction of Properties of Gas Hydrate Bearing Sediments from the Dynamic of Reflected Waves without Borehole Data," *70th EAGE Conference and Exhibition*, Italy, Rome, **2008**.
17. Shelander D., Dai J., and Bunge G., "Predicting Saturation of Gas Hydrates Using Pre-stack Seismic Data, Gulf of Mexico," *Marine Geophysical Research*, **2010**, *31*, 39-57.
18. Shipley T. H., Houston M. H., Buffler R. T., Shaub F. J., and et al., "Seismic Evidence for Widespread Possible Gas Hydrate Horizons on Continental Slopes and Rises," *AAPG Bulletin,* **1979**, *63*, 2204-2213.
19. Lee M. W. and Dillon W. P., "Amplitude Blanking Related to the Pore-filling of Gas Hydrate in Sediments," *Marine Geophysical Researches,* **2001**, *22*, 101-109.
20. Ostrander W. and Gassaway G. S., "The Use of Offset Dependent Reflectivity in Exploration," *SEG*, **1983**, 637–638.
21. Ecker C., Dvorkin J., and Nur A., "Sediments with Gas Hydrates: Internal Structure from Seismic AVO," *Geophysics,* **1998**, *63*, 1659-1669.
22. Chen M. A., Riedel M., Hyndman R. D., and Dosso S. E., "AVO Inversion of BSRs in Marine Gas Hydrate Studies," *Geophysics*, **2007**, *72*, 31-43.
23. Fatti J. L., Smith G. C., Vail P. J., Strauss P. J., and et al., "Detection of Gas in Sandstone Reservoirs Using AVO Analysis: a 3-D Seismic Case History Using the Geostack Technique," *Geophysics*, **1994**, *59*, 1362-1376.
24. Zoeppritz K., "Erdbebenwellen VIII B, on the Reflection and Penetration of Seismic Waves through Unstable Layers," *Goettinger Nachr,* **1919**, 66-84 (in German).
25. Smith G. C. and Gidlow P. M., "Weighted Stacking for Rock Property Estimation and Detection of Gas," *Geophysics Prospecting*, **1987**, *35*, 993-1014.
26. Shuey R. T., "A Simplification of Zoeppritz Equations," *Geophysics*, **1985**, *50*, 609-614.
27. Walden A. T., "Making AVO Sections More Robust," *Geophysical Prospecting*, **1991**, *39*, 915-942.
28. Young R. A. and LoPiccolo R. D., "A Comprehensive AVO Classification," *The Leading Edge,* **2003**, *22*, 1030-1037.
29. Salehi E., Javaherian A., Ataee Pour M., Keshavarz Farajkhah N., and et al., "Quantitative Seismic Pre-stack Analysis of Potential Gas-hydrate Resources in the Makran Accretionary Prism: offshore Iran," *Marine and Petroleum Geology*, **2013**, *48*, 160-170.
30. Gardner G. H. F., Gardner L. W., and Gregory A. R., "Formation Velocity and Density-the Diagnostic Basics for Stratigraphic Traps," *Geophysics*, **1974**, *39*, 770-780.
31. Hosseini Shoar B., Javaherian A., Keshavarz Farajkhah N., and Seddigh Arabani M., "Reflectivity Template, a Quantitative Intercept-gradient AV O Analysis to Study Gas Hydrate Resources-A Case Study of Iranian Deep Sea Sediments," *Marine and Petroleum Geology*, **2014**, *51*, 184 -196.
32. Ursin B., "Offset-dependent Geometrical

Spreading in Layered Medium," *Geophysics,* **1990**, *55*, 492-496.

33. Fink C. R. and Spence G. D., "Hydrate Distribution off Vancouver Island from Multi Frequency Single-channel Seismic Reflection Data," *Journal of Geophysical Research*, **1999**, *104,* 2909-2922.
34. Sheriff R. E. and Geldart, L. P., "Exploration Seismology," Cambridge University Press, **1995**.
35. Warner M., "Absolute Reflection Coefficients from Deep Seismic Reflections," *Tectonophysics,* **1990**, *173*, 15-23.
36. Dix C. H., "Seismic Velocities from Surface Measurements," *Geophysics,* **1955**, *20*, 68-86.
37. Geletti R. and Busetti M., "A Double Bottom Simulating Reflector in the Western Ross Sea, Antarctica," *Journal of Geophysical Research*, **2011**, *116*, 1-15.
38. Andreassen K. J., Miniert P. Bryan, and Singh S. C., "A Double Gas-hydrate Related Bottom Simulating Reflector at the Norwegian Continental Margin, in Gas Hydrates: Challenges for the Future," *Annals of the New York Academy of Sciences*, **2000**, *912*, 126–135.
39. Foucher J. P., Nouzè H., and Henry P., "Observation and Tentative Interpretation of a Double BSR on the Nankai Slope," *Marine Geology,* **2002**, *187*, 161-175.
40. Baba K. and Yamada Y., "BSRs and Associated Reflections as an Indicator of Gas Hydrate and Free Gas Accumulation: an Example of Accretionary Prism and Forearc Basin System along the Nankai Trough, Off Central Japan," *Resource Geology*, **2004**, *54*, 11-24.
41. Popescu I., Batist M. D., Lericolaisc G., Nouzéc H., and et al., "Multiple Bottom-simulating Reflections in the Black Sea: Potential Proxies of Past Climate Conditions," *Marine Geology,* **2006**, *227*, 163-176.
42. Posewang J. and Mienert J., "The Enigma of Double BSRs: Indicators for Changes in the Hydrate Stability Field," *Geo-Marine Letters,* **1999**, *19*, 157-163.
43. Rajput S., Muller T. M., Clennell M. B., Rao P. P., and et al., "Constraints on Seismic Reflections and Mode Conversions at Bottom Simulating Reflectors Associated with Gas Hydrates," *Journal of Petroleum Science and Engineering,* **2012**, *88*, 48-60.
44. Afazeli E., Gerami S., and Badakhshan A., "Sensitivity Analysis on Parameters of Affecting the Thickness of Gas Hydrate Zone in the Gulf of Oman," *the SPE Middle East Unconventional Gas Conference and Exhibition,* Oman, Muscat, **2013.**

3

Pore Characteristics of the Upper Carboniferous Taiyuan Shale in Liaohe Depression

Jian Xiaofei[1, 2*], Liu Renhe[2, 3, 4], Tang Shuheng[1], Lin Wen[3], Zhang Qin[3], and Jia Lingxiao[1]

[1] China University of Geosciences, Beijing 100083, China

[2] National Research and Development (Experimental) Center for Shale Gas, LangFang 065007, China

[3] Research Institute of Petroleum Exploration and Development-LangFang, LangFang 065007, China

[4] China University of Petroleum, Beijing, 102249, China

ABSTRACT

High pressure mercury, nitrogen adsorption, nano-CT, and scanning electron microscope with energy spectrum analysis were conducted on core shale samples for studying the characteristics of Taiyuan formation in the eastern uplift of Liaohe depression. The research results show that the shale gas reservoir pores are mainly open pores such as the wedge-shape pores and parallel-plate pores. By a genetic type, pores mainly include organic pore, pyrite crystal particle pore, illite intragranular pore, illite-smectite mixed layer intragranular pore, and feldspar dissolved pore. The micropore and mesopore play an important role in shale gas reservoir, and their surface area and pore volume are 9.56 m^2/g, 0.0414 mL/g, 97.3%, and 68.8% respectively. The pores diameter presents a bimodal distribution with two main peaks at 43 nm and 6.35 μm. Based on the nano-CT, the porosity is 4.36% and the permeability is 204 nD. The brittle minerals played a supportive and protective role for the pores and controlled their spatial distribution.

Keywords: Liaohe, Marine-continental Transitional Facies, Taiyuan Formation, Shale, Pore

INTRODUCTION

Shale gas refers to the natural gas exploited from organic-rich shale. It is accumulated continuously in shale and is characterized by "self-generation and self-preservation" [1-2]. The successful development of the shale gas in North America proves that it is an important type of alternative clean energy [3,4]. The shale of China is characterized by an old age, multiple phases of tectonic reworking, and high evolution degree compared with that of the North America [1-5]. Previous experts (Liu Honglin et al., 2012; Wang Libo et al., 2013; Pu Boling et al., 2014) have conducted a detailed study on the hydrocarbon enrichment pattern, reservoir space, adsorption features, and storage conditions of the marine shale gas reservoirs in the Southern China, and significant breakthrough has been acquired [5-7]. The shale gas of Chongqing Fuling was the first to put into commercial development. Moreover, the demonstration area in Southern Sichuan and Northern Guizhou has also made a significant progress [8, 9]. Compared with marine shale, continental shale is characterized by a thin thickness, a low content of brittle minerals, a high

***Corresponding author**

Jian Xiaofei

Email: 872411052@qq.com

content of clay minerals, and a low pressure [10-13]; however, the productivity construction of continental shale gas demonstration areas is still steadily advancing through adopting new technologies such as fracturing by CO_2 [14].

At present, studies on the marine-continental transitional facies shale are poor. There is relatively large controversy about the resources potential of the shale of marine-continental transitional facies since it is characterized by a thin reservoir thickness and abruptly changed lithofacies. According to the US Energy Information Administration (EIA) (2011), the shale of marine-continental transitional facies and continental facies in China has no conditions for forming shale gas reservoirs. Assessment of Ministry of Land and Resources (2012) shows the shale gas resources of marine-continental transitional facies are 8.9 trillion square kilometers more than that of continental facies (7.9 trillion square kilometers); therefore, the study on the marine-continental transitional facies shale in the Eastern Uplift, Liaohe Depression is of an important practical significance.

Currently, the kerogen of the marine shale of North American and Southern China as well as the continental shale of the Ordos Basin is predominantly type I and II [15, 16], and studies on shale with kerogen III are lacking. There are two sets of shale of marine-continental transitional facies present in the Eastern Uplift, Liaohe Depression, which were from Upper Paleozoic Carboniferous Taiyuan Formation and Permian Shanxi Formation respectively. Kerogen type of rocks in Taiyuan and Shanxi Formation is type III; therefore, this study on them is of practical and theoretical significance. At present, the shale gas in the area of interest is still in the exploration phase. Previous experts (Ge Mingna et al., 2012; Mao Junli et al., 2012; Ren Zhunlin et al., 2013; Yu Peng, 2013) have studied the geochemical features, gas generation capacity, gas-bearing property as well as target area selection and the evaluation of shale gas reservoirs [17-21]; however, the economic production of shale gas needs new technology to transform the reservoirs into "artificial reservoirs" with good porosity and permeability. Therefore, the study on reservoir features is critical for the efficient development of shale gas [3, 22-24].

In order to gain an insight into the marine-continental transitional facies shale with kerogen III and to characterize the shale reservoir, this paper quantifies the pores features of the Upper Carboniferous Taiyuan formation shale. Also, the paper provides a reference for the economic development of the marine-continental transitional facies shale in the future.

EXPERIMENTAL PROCEDURES

Geological Setting

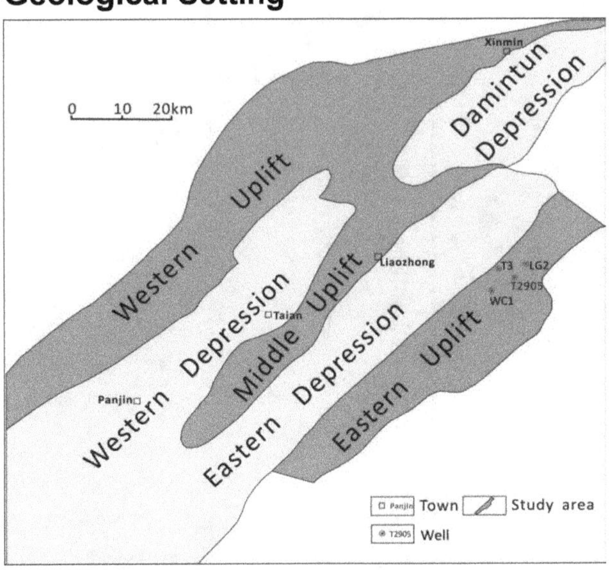

Figure 1: Geological map of eastern uplift in Liaohe depression (according to reference [21]).

Liaohe depression is located in the eastern part of North China plate, adjacent to Liaodong fold belt to the east and Yanshan subsidence zone to the west. It is a Cenozoic rift basin in the central and northern part of the Bohai Bay Basin, with an area of 1.0×104 km². Liaohe depression presents a tectonic framework of "three uplifts and three sags," including Western Uplift, Western Sag, Damintun Sag, Central Uplift, Eastern Sag, and Eastern Uplift [19, 21] (Figure 1). The Eastern Uplift is the first-order structural unit in Liaohe depression and is located in the east part, with N-S length of about 80 km, E-W width of about 20 km and an area of 1600 km². Kerogen of Taiyuan formation shale in Liaohe depression is type III and typical humic organic matter, which predominantly generated gas. Its microscopic components were dominated by vitrinite (accounting for 77%), followed by inertinite (accounting for 41%), exinite was the least. The organic matter of shale had an average TOC of 3% and maturity of 2.1-2.7%. Its Ro value is high, indicating it is in a dry gas phase (Ro> 1.4%). Taiyuan formation of the Upper Carboniferous contains shale, sand, and coal in Liaohe depression. Taiyuan formation shale is the object of this study (Figure 2).

Existing drilling data revealed that the formation thickness of Taiyuan formation was 122.9 to 171.3 m in the Eastern Uplift of Liaohe depression, among which, 48 to 87.4 m was shale interval (Table 1). Eight core samples of shale were all selected from the target shale interval of Taiyuan formation in Well T2905 (Figure 2). Shale samples depth and other related data are in Table 2.

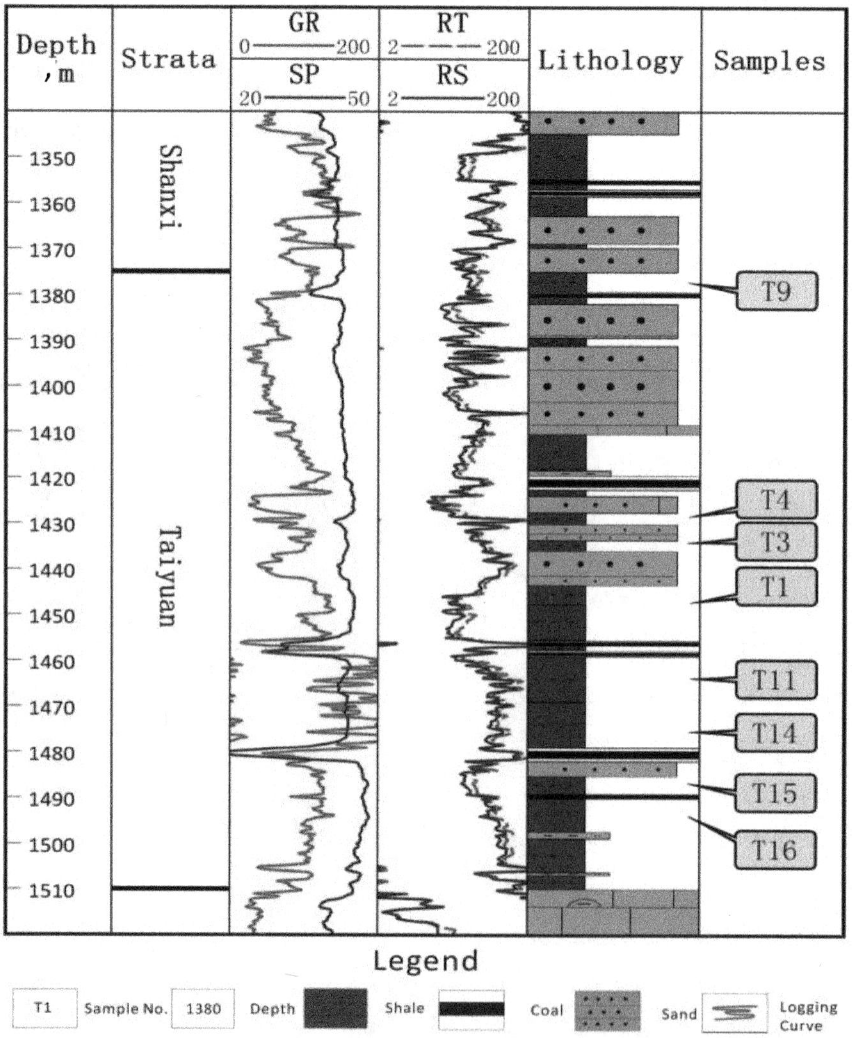

Figure 2: Taiyuan formation comprehensive stratigraphic column.

Table 1: Thickness of Shanxi (P1s) and Taiyuan (C3t) formation in the Upper Carboniferous.

Well	Burial Depth (m) P1s/C3t	Total (THK, m) P1s/C3t	Shale (THK, m) P1s/C3t	Sand (THK, m) P1s/C3t	Coal (THK, m) P1s/C3t
WC1	1754.48/1914.04	159.5/123.1	31.6/60.3	115.2/48.8	12.7/14
LG2	844.5/1015.8	171.4/140	85.3/67.1	77.3/61.1	8.8/11.9
T3	1941.05/2070.95	129.9/122.9	49/87.4	68.5/19.1	12.4/16.4
T2905	1266/1375.06	109/134.8	40.6/48.1	65.4/78.3	3/8.4

Experiments like scanning electron microscopy, CT, nitrogen adsorption, and high pressure mercury injection had obtained some achievements in the characterizing shale pores [25-29], but all of them had their limitations; for example, scanning electron microscopy and CT could not accurately quantify pore features, and nitrogen adsorption method was only limited to characterize the pores with size of 2 to 50 nm; high pressure mercury injection method could not provide adsorption capacity evaluation. In this study, the experiments mentioned above were comprehensively utilized in order to characterize the shale reservoirs of marine-continental transitional facies in the Eastern Uplift of Liaohe depression from multi-scale and multi-view field accurately, qualitatively, and quantitatively. After polished by argon ions, the samples were observed by high-resolution scanning electron microscope, based on which, the pore type, structure, and size were qualitatively analyzed; combined with high pressure mercury injection method and nitrogen adsorption method, the pores with different sizes were quantitatively and accurately characterized. Advanced nano-CT technology had achieved non-destructive test and has been used to exhibit the 3D spatial distribution of the pores, organic matters, and mineral components of shale.

RESULTS AND DISCUSSION

Argon ion polishing, high resolution field emission scanning electron microscopy, energy spectrum analysis, high pressure mercury injection, nitrogen adsorption, and nano-CT experimental test were performed in order to finely characterize the pore morphology, pore size, pore distribution and porosity, and the permeability of shale reservoirs (Figure 3). The specification of the shale samples for argon ion polishing was 10 mm×10 mm×4 mm. Specific preparation process was as follows: After a small piece of shale sample was knocked off, it was firstly polished using coarse sandpaper; then, it was finely ground using 2000 mesh fine sandpaper in order to make its upper and lower side parallel to each other and vertical to one of its narrow facade. After the sample was processed, its "edge" should be polished by argon ion in order to form a polished triangular surface at the intersection between upper surface and narrow facade. Argon ion polishing experiment was performed by the Key Laboratory of Unconventional Oil and Gas Research Institute of Petroleum Exploration and Development LangFang Branch with the instrument of GATAN693(HAC-200) and gas source of high-purity argon (99.9996%).

After the sample was polished by argon ions, scanning electron microscopy and energy spectrum analysis experiment were performed in Beijing Normal University Testing Center with S-4800(BNUATC-023) field emission scanning electron microscopy and EDAX spectrometer. 8.0 kV voltage was adopted for energy spectrum analysis, whereas, 20.0 kV voltage was adopted for high-resolution scanning electron microscopy analysis. The imaging resolution had reached nanoscale.

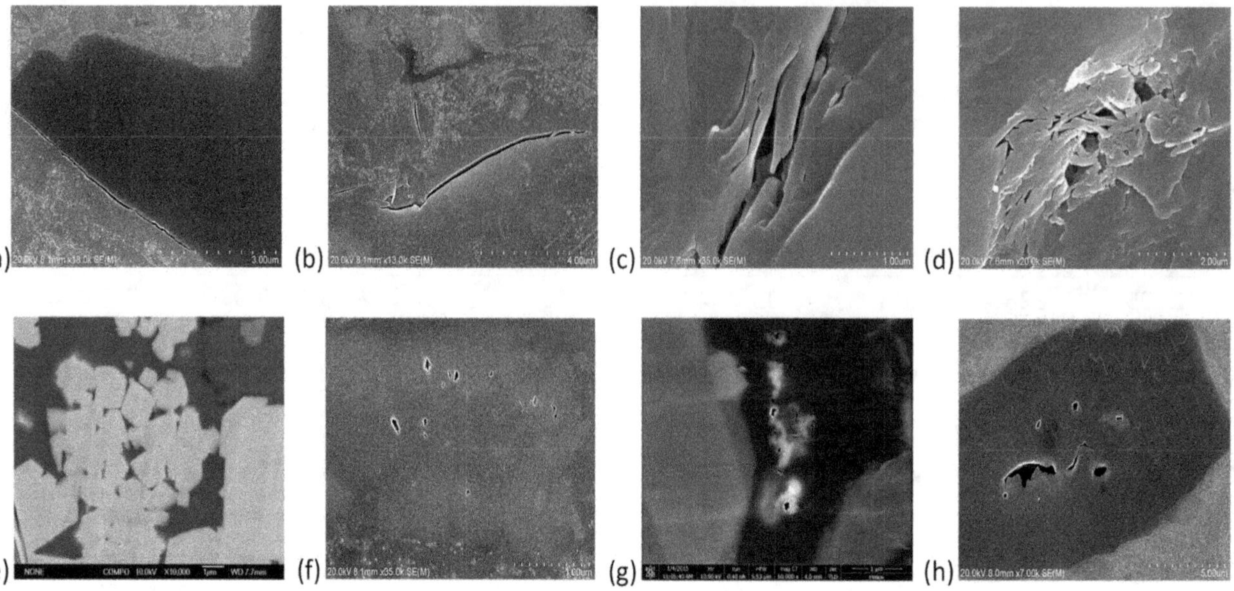

Figure 3: SEM energy spectrum characteristics of minerals and pores in shale samples.
(a): T15 organic matter and clay mineral marginal fracture; (b): T15 fractures within illite/smectite mixed layer; (c): T9 illite intercrystalline pore; (d): T9 kaolinite intercrystalline pores; (e): T9 pyrite intercrystalline pore; (f): T15 feldspar surfaces dissolved pore; (g): T1 organic matter pore; (h): T16 organic matter pore.

A high pressure mercury injection experiment was performed (Table 2) using AutoPore IV 9520 (YQ-10007) full-automatic mercury injection apparatus with pore size measuring range of 1.8 nm-200 μm, maximum mercury injection pressure of 400 MPa, and mercury injection and ejection volume accuracy of less than 0.1 μL. A high pressure mercury injection experiment was performed by Wuxi Experimental Research Center, Research Institute of Petroleum Exploration and Development, SINOPEC according to the petroleum industry standard SY / T5346-2005.

Table 2: High pressure mercury injection characteristic parameters of Taiyuan formation samples.

Parameters Samples	Burial depth (m)	Brittle minerals (%)	Displacement pressure (MPa)	Maximum connection radius (μm)	Median pressure (MPa)	Median radius (μm)	Residual saturation (%)	Ejection efficiency (%)
T1	1449	55.3	53.25669	0.013801	123.3005	0.005961	45.44427	54.55573
T3	1435.7	56.9	24.07091	0.030535	99.5688	0.007382	45.3444	54.6556
T4	1431.8	59.7	37.79729	0.019446	137.3208	0.005352	38.46313	61.53687
T9	1376.5	51.4	53.27746	0.013796	127.7273	0.005754	31.03269	68.96731
T11	1463.5	55.6	53.26795	0.013798	125.5626	0.005854	31.25196	68.74804
T14	1477.6	62.4	53.26465	0.013799	154.8728	0.004746	22.08751	77.91249
T15	1485.4	55.1	48.06381	0.015292	87.69488	0.008381	42.42424	57.57576
T16	1493.2	47.1	53.28251	0.013794	120.1871	0.006115	32.39362	67.60638
Average	/	55.44	47.04	0.017	122.03	0.0062	36.06	63.94

Nitrogen adsorption specific surface analysis and test were performed (Table 3) after the sample was grinded to 200 meshes or less. The experiment was undertaken by the Key Laboratory of Unconventional Oil and Gas, Research Institute of Petroleum Exploration and Development LangFang Branch. According to the petroleum industry standard SY/T 6154-1995, the test was conducted at a temperature of 120 °C, at a vacuum degree of 1.0×10^{-3} Pa for 5 hrs.

Table 3: Pore size distribution of shale samples tested by N_2 adsorption.

Pore size / Samples	Average diameter (nm)	Specific surface (m^2/g)	Pore volume (mL/g)	Pore volume proportion (%)			Specific surface proportion (%)		
				<2nm	2-50nm	50-150nm	<2nm	2-50nm	50-150nm
T1	59.3454	12.5473	0.0527	13.699	55.842	31.206	56.993	44.543	2.746
T3	46.7672	8.5802	0.0466	13.322	58.415	28.325	51.855	46.76	2.529
T4	30.1151	8.1101	0.0311	17.872	66.297	16.052	58.105	41.53	1.445
T9	60.1385	10.7762	0.0474	13.362	57.838	29.561	56.131	45.4	2.79
T11	54.0789	10.7667	0.0463	11.943	60.73	27.791	52.966	47.191	2.689
T14	68.7560	6.5099	0.0394	10.377	61.086	28.609	48.271	49.078	3.091
T15	38.9129	11.1711	0.0428	16.879	62.665	21.244	58.595	43.33	1.885
T16	27.1690	11.2335	0.0425	14.704	59.013	26.294	55.038	43.015	1.964
Average	48.16	9.96	0.0436	14.02	60.24	26.14	54.74	45.11	2.39

Nano-CT scanning experiment was performed (Figure 4) using the nano-CT scanner UltraXRM-L200 produced by Xradia Company. The constant voltage of nano-CT X-ray source was 40 kV and the photon energy of X-ray was 8 keV. In a high-resolution mode, the viewing area was 15 μm and resolution was 50 nm.

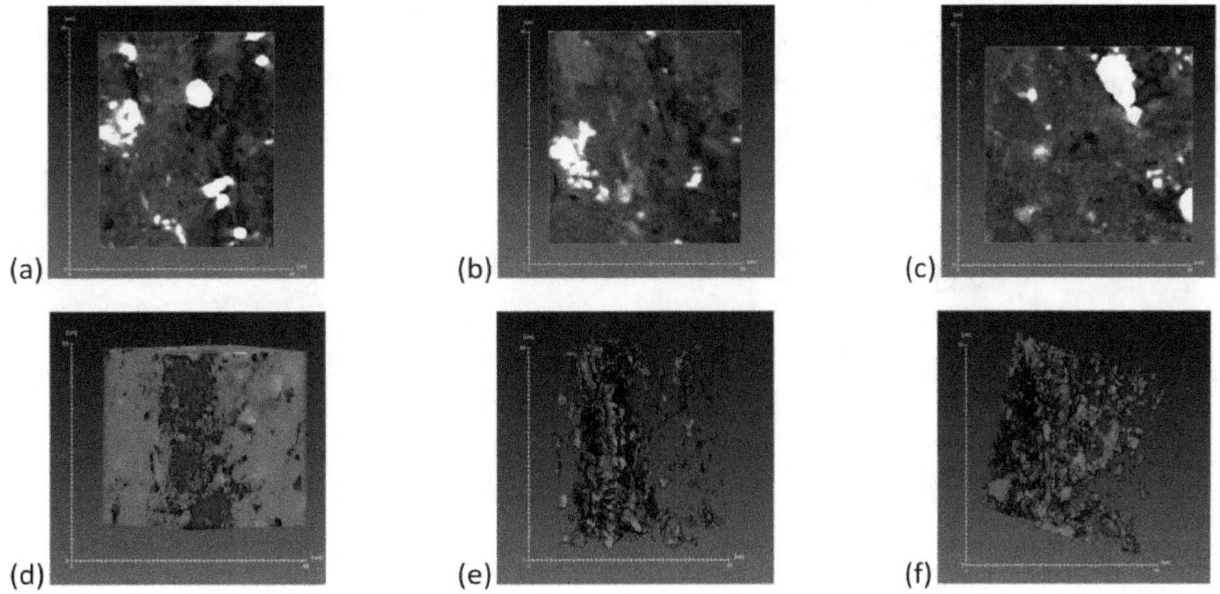

Figure 4: CT images of shale sample T1 from the eastern uplift of Liaohe depression; (a): Cross-section of x-y direction; (b): Cross-section of x-z direction; (c): Cross-section of y-z direction; (d): Segmentation stereographic map; (e), (f): Pore framework.

Pore Morphology

Most pores of shale reservoirs were micron and nano-scaled pores and their development was related to the size of shale reservoir space [4, 30, 31]. Zou Caineng et al. [32] had divided the pores of shale reservoirs into millimeter-scaled (>1 mm), micron-scaled (1 to 1000 μm), and nano-scaled (<1 μm) according to their sizes. International Union of Pure and Applied Chemistry Association (IUPCU) had divided the micron-scaled pores into macro-pore (> 50 nm), meso-pore (2 to 50 nm), and micro-pore (<2 nm) [33]. They also had divided desorption curves into four types (H1, H2, H3, and H4). The adsorption and desorption branch of H1 were perpendicular to pressure axis and parallel to each other within a wide desorption range, while the adsorption and desorption branch of H4 were horizontal and parallel to each other within a wide pressure range. Samples with the regular pores often presented the loop of type H1, and the samples with micro-pores often presented the loop of type H4; however, the samples with irregular pores often presented the loop of type H2 and H3 [34]. Irregular pores could be divided into open pore (including four-side opened slit-shaped pore and two-side opened cylindrical pore), closed pore (including one-end closed cylindrical pore, parallel plate-like pore and conical pore), and ink bottle pore. Open pore could generate desorption curve, but closed pore could not generate adsorption curve; however, ink bottle pore could generate desorption curve although it was closed at one end. Its desorption curve had a sharply declined inflection point [35].

The nitrogen desorption curves of the experimental samples (Figure 5) were steep in the vicinity of saturated vapor pressure and gentle around moderate pressure, which was closed to the H3 curve recommended by IUPAC and had the features of H4 curve at the same time. Moreover, the loops exhibited a superposition of a number of standard loops and were a comprehensive reflection of the sample morphology, indicating the pores of the shale gas reservoirs of Shanxi formation and Taiyuan formation in Eastern Uplift Liaohe depression were dominated by the nano-pore with an irregular (amorphous) structure. The studied samples had generated desorption curve, indicating open pores existed in the shale reservoirs of Taiyuan formation (four-side opened slit-shaped pore and two-side opened cylindrical pore); part pores could not generate desorption curve, indicating closed pores existed (one-end closed cylindrical pore, parallel plate-like pore and conical pore); the inflection point of desorption curve was unobvious, indicating ink bottle pores were scarce.

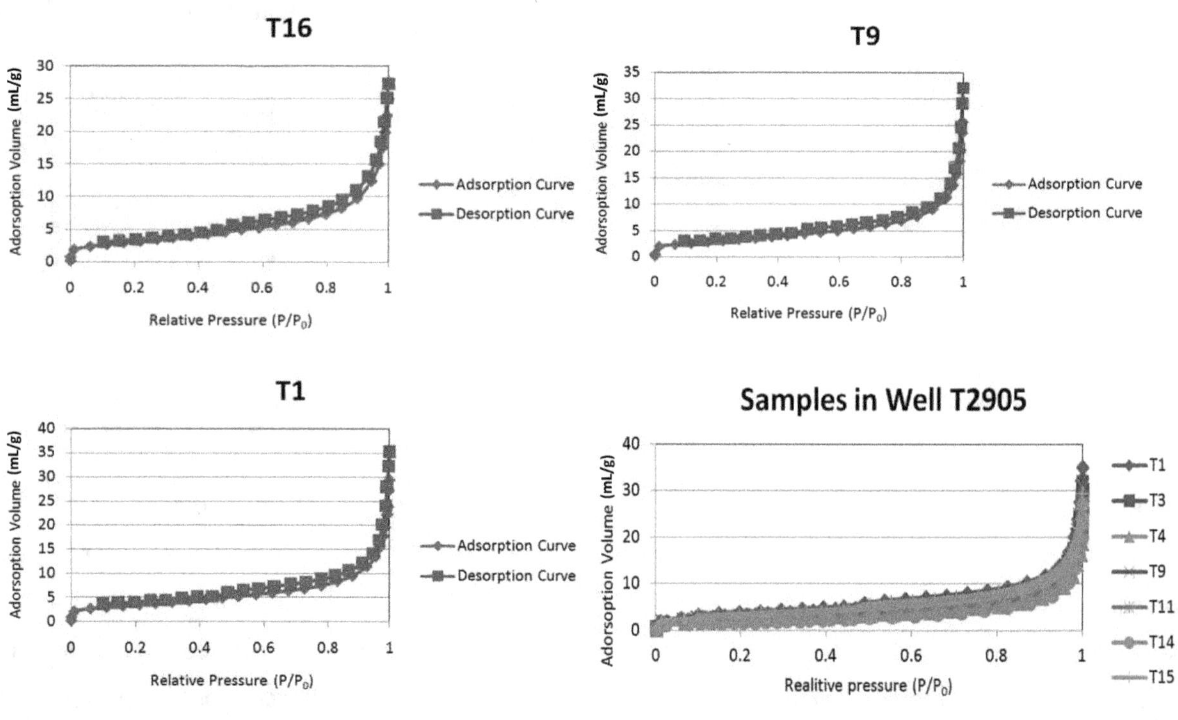

Figure 5: Adsorption isotherms of shale samples from eastern uplift of Liaohe depression.

The measured nitrogen adsorption curves of different experimental samples were typical type II curves, all of which presented an «S» shape. In low-pressure section, the first half of the curve raised slowly and presented a convex shape, indicating the transition stage from adsorbed monolayer into multilayer; in the middle section of the curve, the adsorption amount gradually increased with increasing pressure, indicating a multilayer adsorption stage; in the last half of the curve, the adsorption curve raised sharply; at the end of the curve, adsorption saturation phenomenon did not occur although the pressure was closed to saturation vapor pressure; this indicates that capillary condensation took place during adsorption process, resulting in the macro-pores filled; therefore, a small amount of meso-pores and macro-pores were existed.

The pore throat of the shale reservoirs of the Eastern Uplift, Liaohe depression was measured by high pressure mercury injection method (Figure 6). The capillary force curve of the shale samples in the area of interest was divided into three sections: initial ascending section, middle gentle section, and end ascending section.

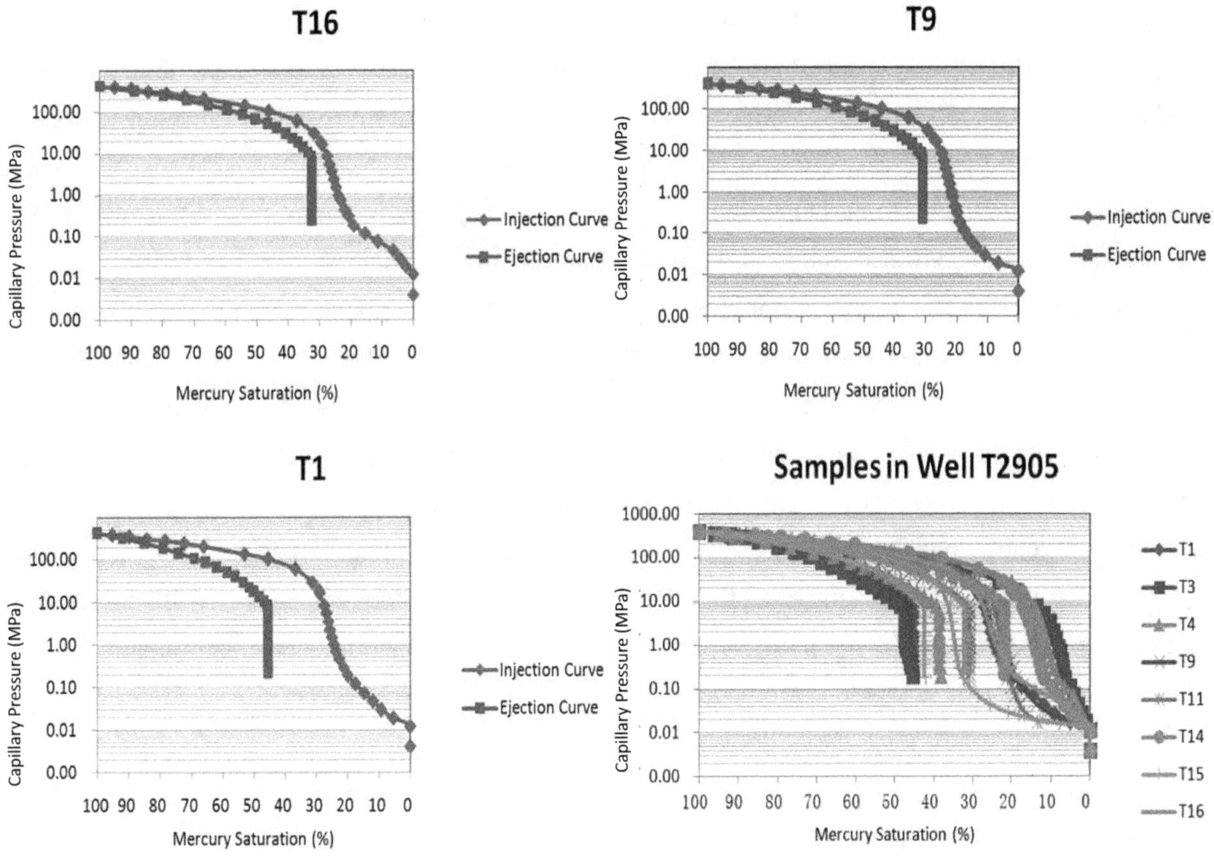

Figure 6: Capillary pressure curve of shale samples from eastern uplift of Liaohe depression.

In initial ascending section, with increasing pressure, mercury saturation increased; non-wetting mercury began to gradually enter into shale, particularly into the pits on shale surface or dissected macro-pore; in the middle gentle section, with further increasing pressure, mercury saturation gradually increased and the number of the macro-pores in shale decreased; at the end of ascending section, with further increasing pressure, mercury saturation increased sharply, indicating mercury began to enter into shale pore in a large amount. The major mercury intrusion section was gentle and long, indicating the pores in the reservoirs were well sorted. The mercury intrusion pressure of the end gentle section ranged between 8 to 378 MPa, indicating the pores with a radius of 2 to 89 nm were extensive in shale. The saturation median radius was between 4.7 to 8.3 nm, with an average of 6.2 nm. The capillary pressure curves were quantitatively studied (Table 2); the results revealed that the averaged displacement pressure of the samples was 47.04 MPa (displacement pressure represented the minimum pressure when mercury started entering into shale, corresponding to the capillary pressure of the largest pore of shale), which was obtained from the intersection point between the extensional line of end ascending section and the y-axis with mercury saturation of 0; this indicates that the maximum connected radius of shale pores was 16.8 nm and the pore connectivity was poor. Moreover, the phenomenon that capillary force curve deviated to upper right, and fine skewness indicated that the pore throat was small in size, and that the pore connectivity was poor. The averaged mercury ejectional efficiency was calculated as 64.85% through residual mercury saturation. The mercury ejectional efficiency was high, indicating most pores were four-side opened slit-shaped pore and two-side opened cylindrical pore, which were advantageous for gas migration.

Pore Type

The micro-pores in shale were semi-quantitatively observed under scanning electron microscope, which were divided into micron-scaled pore and nano-scaled pore. Based on the microscopic components of shale determined from energy spectrum analysis, the pores were divided into micro-fracture, mineral pore, and organic matter pore according to their occurrence.

1) Micro-fracture

Micro-fracture was not apparent in shale, but it was of great significance for the petrophysical property of shale reservoir. It could not only serve as the channel for gas seepage and migration, but also connect micro-pore and macro-fracture, which was of great significance for development [23]. A small amount of micron-scaled micro-fractures were observed in shale, which occurred at the margin of organic matter clumps and minerals (Figure 3-a) or within the clay minerals (Figure 3-b). Organic matter margin fractures occurred at the margin of organic matter and their morphology was related with organic matter clumps. The fractures at the two ends of clay minerals expressed blended saw toothed shape, with a length of about 6 μm, while the fractures within clay minerals were straight and unfilled by minerals, with a length of about 8 μm. The nano-scaled pores and fractures generated by clay minerals had a great contribution to the methane adsorption capacity of shale; this phenomenon was also discovered by other scholars [36, 37].

2) Mineral pore

Matrix pores were extensive in shale in addition to a small amount of micro-fractures. The development of matrix pores was predominately related with the minerals and organic matters in shale, corresponding to mineral pore and organic matter pore. Mineral pore included intergranular pore and intragranular pore.

Intergranular pore predominated in shale reservoirs and occurred between the particles of clay minerals such as illite and kaolinite, and brittle minerals like pyrite. The intergranular pore observed in the shale samples included illite intercrystalline pore (Figure 3-c), kaolinite intercrystalline pore (Figure 3-d), and pyrite intercrystalline pore (Figure 3-e). Intragranular pore included the dissolved pore on feldspar surface (Figure 3-f). Illite intercrystalline pore (Figure 3-c) was located between illite lamella and represented long and narrow tabulate shape. It was formed due the directional arrangement of illite lamella, with a length of 1 to 2 μm and a width of 50 to 300 nm in general.

Kaolinite intercrystalline pores (Figure 3-d) represented two forms; one was the short plate-shaped pore occurred at the margin of kaolinite, part of which had triangular cross-section. It was formed due to the vertical superimposition and arrangement of kaolinite particles, with a width of about 100 nm and a length of no more than 1 μm; the other was the kaolinite intercrystalline pore that represented "honeycomb". It was formed due to the planner superimposition of hexagonal kaolinite particles. Its diameter was up to 400 nm, representing a good reservoir property.

Pyrite was formed in reducing environment and often associated with organic matter clumps (Figure 3-e) or present in the clay minerals (Figure 3-f). Pyrite crystals generally represented a regular cluster shape, between which an intercrystalline pore was generally formed. Moreover, affected by the crystal shape, the cross section of pyrite intercrystalline pore often exhibited triangular or other polygonal shapes; furthermore, halted by pyrite crystals, the pores were generally unconnected, with a pore size of less than 1 μm.

The dissolved pores on feldspar surface (Figure 3-g) were isolated and scattered and generally exhibited irregular cross section such as circular and ellipse, with a pore size of generally less than 200 nm.

3) Organic matter pore

Organic matter pore was one type of matrix pore. A large amount of nano-pore was existed within the organic matter particles of shale. Organic matter pore was observed during the study (Figure 3-g and Figure 3-h). Organic matter pore exhibited an irregular shape such as ellipse and was unconnected with each other, with a smooth edge and clear edge sharpness. The pore size of organic matter pore was generally less than 1 μm, with the maximum size of 5 μm. Organic matter pore was closely related with the specific surface area and pore volume of shale and played a positive role in shale adsorbability.

Figure 3 showed that clay mineral pores and pyrite intercrystalline pores were extensive, but organic matter pores were rare, which was resulted from the high content of clay minerals in the samples. The clay mineral content of sample T1, T9, and T15 was 44.7%, 48.6%, and 44.9% respectively, with an average of 46.1% confirming a unimodal distribution in the specific good marine shale reservoir (the clay mineral content of good marine shale reservoir was < 30%). The existing pores shrinkage even disappeared due to the high plasticity of clay minerals, whereas the organic matter pores formed during hydrocarbon generation and expulsion at the late stage was difficult to be preserved since the support and protection of brittle minerals are not present. Similar to the successful development of shale (which also contains high clay minerals more than 40%) in Erdos basin, the shale reservoir of the Eastern Uplift in Liaohe depression still had a development potential.

Pore Size and Distribution

The study results of pore morphology showed that the pores in the shale of Taiyuan formation in the Eastern Uplift, Liaohe depression were dominated by nano-pores, with a certain amount of macro-pore. The pore size and distribution of the nano-pores in shale were characterized from different scales through the combined method of nitrogen adsorption method and high pressure mercury injection method [38].

According to the results measured by nitrogen adsorption experiments (Table 3), the average pore size of shale was 48.16 nm. BJH pore volume showed that the pore volume of shale reservoir exhibited a bimodal distribution, with two peaks at 2 nm and 100 nm (Figure 7); BET specific surface area showed that the specific surface area of all the samples represented a unimodal distribution, with a peak at 2 nm. Compared with the pores of the marine shale in Southern China surface area and pore volume, the pores of shale reservoirs in the Eastern Uplift, Liaohe depression expressed a bimodal distribution in pore volume and a unimodal distribution in specific surface area, indicating the polarization of the pores in the shale of marine-continental transitional facies. Meso-pore and macro-pore contributed significantly to pore volume, whereas meso-pore and micro-pore contributed significantly to specific surface area.

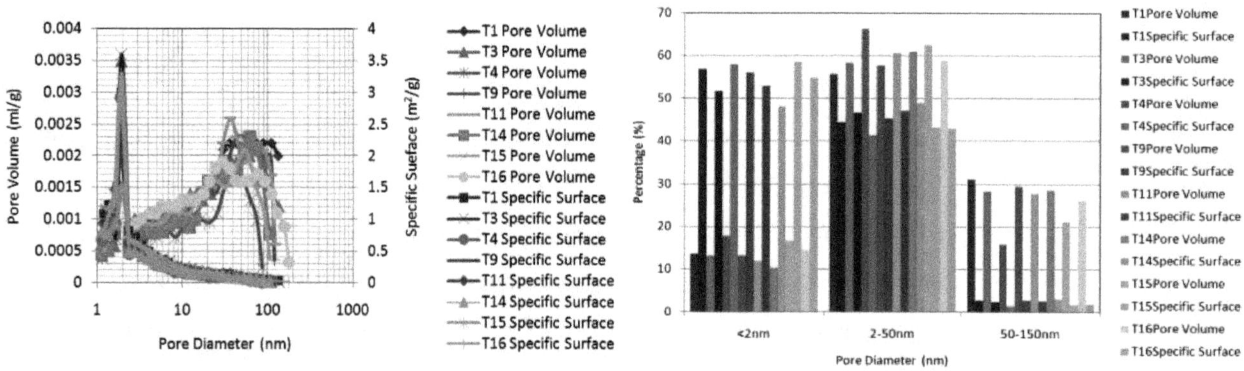

Figure 7: Pore size distribution of shale samples tested by N_2 adsorption.

Quantitative analysis results (Table 3) also revealed a bimodal distribution for pores. The BET specific surface area of shale was 6.37 m^2/g to 12.55 m^2/g, with an average of 9.56 m^2/g; the specific surface area of micro-pore, meso-pore, and macro-pore accounted for 14.6%, 62.4%, and 23.5% respectively, among which the specific surface area of meso-pore and micro-pore accounted for 97.86%. BJH pore volume was 0.0234 mL/g to 0.0527 mL/g, with an average of 0.0414 mL/g. The pore volume of micro-pore, meso-pore, and macro-pore accounted for 14.6%, 62.4%, 23.5% respectively, among which, the pore volume of both meso-pore and macro-pore accounted for 85.4%; therefore, for the pores in the shale reservoir in the Eastern Uplift Liaohe depression, the meso-pore and micro-pore had significantly contributed to the specific surface area, which had provided the primary place for methane adsorption; moreover, the smaller the pore size was, the stronger the adsorption capacity was (adsorption potential was inversely proportional to pore size); however, meso-pore and macro-pore had significantly contributed to the pore volume, which was also the primary place for the occurrence of free gas.

The distribution of macro-pore was characterized by a high pressure mercury injection experiment since nitrogen adsorption method would create a large testing error for the pores with a size greater than 100 nm. Figure 8 showed the mercury saturation corresponding to different pore sizes, which represented the contribution of different pore sizes on pore volume. High pressure mercury injection results indicated that the peak of the macro-pore size was 6 μm and the pore volume of macro-pore accounted for 48.6%.

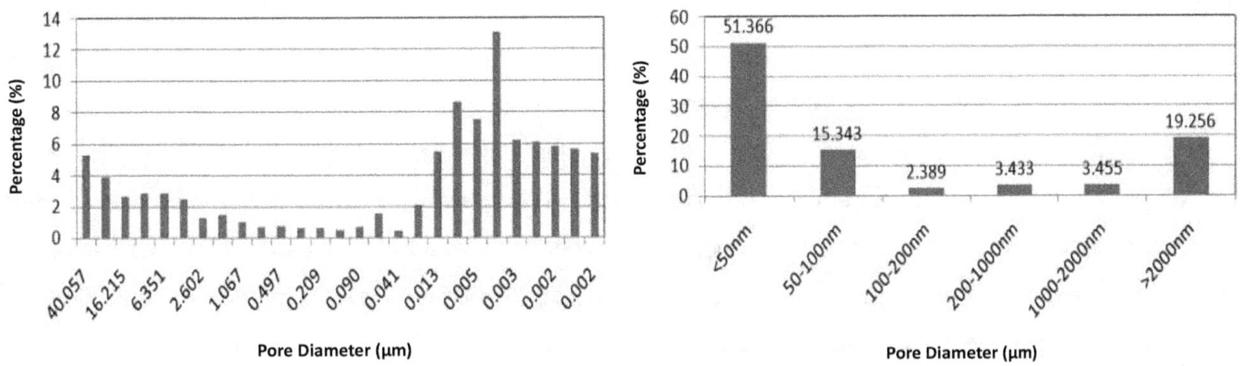

Figure 8: Pore size distribution of shale sample T1 tested by high pressure mercury injection.

Nitrogen adsorption method and high pressure mercury injection method had different measurement results for the pores with different scales. The optimal result segments of each method were selected to be analyzed: for the pores of 2 to 50 nm, the test data of nitrogen adsorption method was selected, while for the macro-pore (> 50 nm), the test data of high-pressure mercury intrusion method was selected. Given that the high pressure mercury intrusion test had a wide applicable range and good macroscopic property, the overall weight of the testing ranges of the two methods were allocated according to the proportion of high pressure mercury injection test (51:49). The comprehensive analysis of the two testing methods is shown in Figure 9. The results of accurate quantitative analysis showed that the two peak values of the bimodal distribution of shale occurred at the point of 43 nm and 6.35 μm respectively.

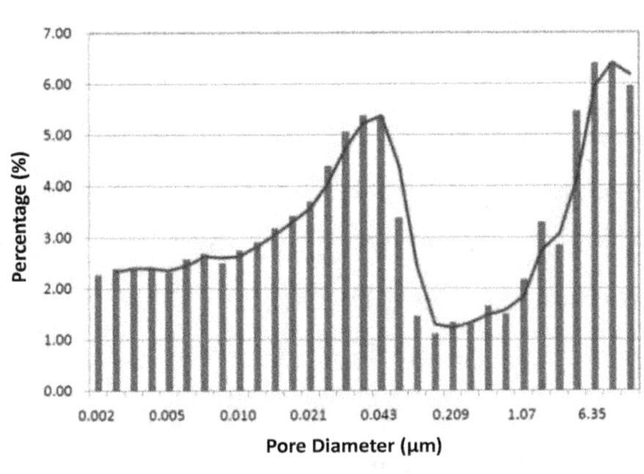

Figure 9: Bimodal distribution of pores in sampleT1.

The bimodal distribution and polarization of the micro-pore and macro-pore of shale reservoirs made the pore space have a "clear responsibility" advantage in preserving adsorbed gas and free gas. The two types of pores played different roles in different exploitation phases after fracturing; in the early exploitation phase, free gas predominantly migrated into wellbore and produced productivity, but in late exploitation phase, the adsorbed gas continued desorbing and migrating into meso-pores and macro-pores and provided the major productivity for shale gas. The bimodal distribution of the pores in the shale of marine-continental transitional facies was conductive to the sustainable development of shale gas.

Pore Distribution

CT test was a new technology for charactering pore variation and micro-structure [39]. The pore space distribution of shale reservoir could be quantitatively and semi-quantitatively characterized by CT scanning experiment. The average core porosity measured by CT scanning technology was closed to that measured by He, which exhibited high reliability. The nano-CT experiment of this time obtained 2D and 3D images by taking T1 sample as an example (Figure 4).

The methane molecules needed the pore throat with a diameter greater than 0.38 nm to migrate since their diameter was only 0.38 nm [40]. The measured CT shale porosity was 4.36% and permeability was 204 nD; under the conditions that the diameter of shale pores was much larger than that of methane molecules the physical property of shale reservoirs could still meet the conditions for shale gas development. CT scanning images showed different colors for the components with different densities, which could be used to observe the distribution of the different minerals and pores in shale intuitively (Figure 4). Moreover, the brittle minerals such as quartz and feldspar exhibited light color due to high density; clay minerals exhibited gray color due to moderate density; organic matter represented black color due to small density.

After the pores marked blue, it could be found from the thin sections (the blue part in Figures 4a, 4b, 4c), 3D segmentation (the blue part in Figure 4d), and pore stripping framework map (the blue part in Figures 4e and 4f) from different directions that pores represented a continuous and isolated spatial distribution pattern. It could be found from Figure 4a and 4c that the distribution of pores (blue part) and the distribution of brittle minerals were closely related; around the area where brittle minerals were abundant, pores were extensive and continuous; in the area where brittle minerals were scarce, pores were scarce and exhibited an isolated and scattered distribution.

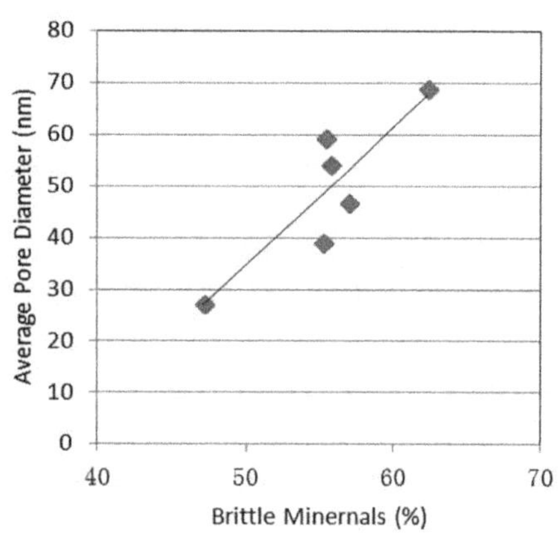

(a) Brittle mineral content and pore size represented positive correlation

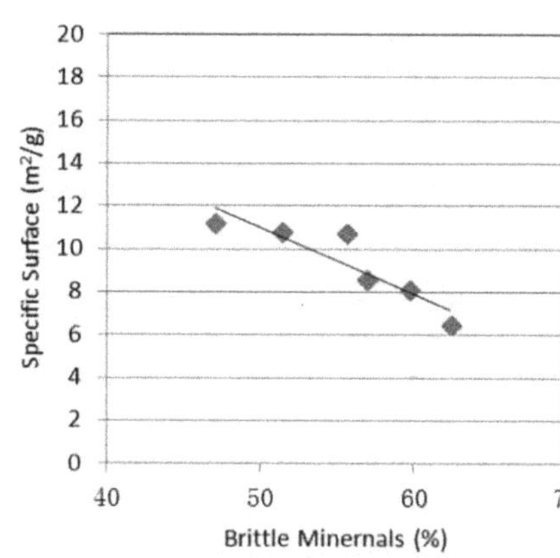

(b) Brittle minerals and pore specific surface area represented negative correlation

Figure 10: The relationship between brittle minerals and pores.

Based on the nano-CT qualitative and semi-quantitative analysis, total rock and clay minerals X-ray diffraction experiments were performed in order to analyze the relationship between brittle minerals and pores (Figure 10). Moreover, the results could be used to verify the analysis results of nano-CT experiment. Figure 4a showed that the brittle mineral content was positively correlated with pore size, and that with increasing brittle mineral content, the average pore size increased from 30 nm to 70 nm; Figure 4b showed that the brittle mineral content was negatively correlated with specific surface area, and with increasing brittle mineral content, the pore specific surface area decreased from 12 m^2/g to 6 m^2/g. The former revealed the more the brittle minerals were, the larger the pore size was (mostly macro-pore), indicating brittle minerals could support and protect pores; therefore, the pores with a large pore size were formed; the latter revealed that the less the brittle minerals were, the larger the specific surface area was; Table 2 shows that the meso-pore and micro-pore provides 97.86% of the total specific surface area, indicating the specific surface area had reflected the number of meso-pore and micro-pore; thus, the less the brittle minerals were, the more the meso-pore and micro-pore were; this indicates that the evolved micro-pores even disappear under compaction if they are not supported and protected by brittle minerals. Figure 4a and 4c show that the brittle minerals had a regional distribution (white part), indicating that the pore size also had a regional distribution (macro-pore took dominance around brittle minerals, but micro-pore took dominance in the area where brittle minerals were rare); this verified that the pores in the shale of marine-continental transitional facies had a bimodal distribution from the other side.

The pore spatial distribution of shale was clarified through quantitative-semi-quantitative analysis; the continuous and isolated distribution of pores were mainly controlled by brittle minerals; brittle minerals played a supportive and protective role for pores; where brittle minerals were abundant, pores mostly occurred as large pores with a continuous distribution, but in the area where brittle minerals were not abundant, pores were scarce and mostly occurred as meso-pores and micro-pores with an isolated and scattered distribution.

CONCLUSIONS

Argon ion polishing, scanning electron microscope plus energy spectrum analysis, nitrogen adsorption, high pressure mercury injection, and nano-CT experiments were conducted on the shale reservoirs of marine-continental transitional facies in Taiyuan formation, Eastern Uplift, Liaohe depression, and the following conclusions were achieved:

(1) Most pores were four-side opened slit-shaped pores, two-side opened cylindrical pores, and closed pores, without ink bottle pore in morphology, which included pyrite intercrystalline pore, organic matter pore, feldspar surface dissolved pore, organic matter and clay mineral marginal pore, illite/smectite mixed layer intragranular tabulate pore, and illite intercrystalline pore.

(2) The BJH pore volume of shale was 0.0234 mL/g to 0.0527 mL/g, with an average of 0.0414 mL/g. Meso-pores and macro-pores were the major occurrence place for free gas, accounting for 85% of the total volume. The BET specific surface area of the shale was 6.37 m^2/g to 12.55 m^2/g, with an average of 9.56 m^2/g. Meso-pores and micro-pores had enhanced the adsorption capacity of the shale, accounting for 97.86% of the total specific surface area. The pore size of shale presented a bimodal distribution, with two peaks distributed at 43 nm and 6.35 μm, which was conductive for the shale gas development.

(3) The shale reservoir had CT porosity of 4.36% and permeability of 204 nD; the pores were continuous and concentrated. Brittle minerals controlled the spatial distribution of the pores in shale and played a supportive and protective role. In the area where brittle minerals were abundant, pores mostly occurred as large pores with a continuous distribution, but in the area where brittle minerals were not abundant, pores were scarce and mostly occurred as meso-pores and micro-pores with an isolated and scattered distribution.

Because of the thin shale and small tight pores in it, many researchers consider that the marine-continental transitional facies shale has no gas potential for development. Also, no company has interest in marine-continental facies transitional shale of Liaohe depression. According to above three conclusions, it is beneficial to produce gas from the shale in Liaohe depression as it has different types; moreover, a bimodal distribution and brittle minerals supported pores.

REFERENCES

1. Jinchuan Z., Zhijun J., and Mingsheng Y., "Reservoiring Mechanism of Shale Gas and its Distribution," *Natural Gas Industry*, **2004**, *24*(7), 15-18.
2. Caineng Z., Zhi Y., Guosheng Z., et al, "Conventional and Unconventional Petroleum "Orderly Accumulation": Concept and Practical Significance," *Petroleum Exploration and Development*, **2014**, *41*(1), 14-27.
3. Currtis J.B., "Fracture Shale-gas Systems," *AAPG Bulletin*, **2002**, *86*(11), 1921-1928.
4. Loucks R. G., Reed R. M., Ruppel S. C., et al., "Spectrum of Pore Types and Networks in Mud Rocks and a Descriptive Classification for Matrix-related Mud Rocks Pores," *AAPG Bulletin*, **2012**, *96*(6), 1071-1098.
5. Honglin L. and Hongyan W, "Adsorptivity and Influential Factors of Marine Shales in South China," *Natural Gas Industry*, **2012**, *32*(9), 5–9.
6. Libo W., Kai J., Weite Z., et al., "Characteristics of Lower Cambrian Marine Black Shales and Evaluation of Shale gas Prospective Area in Qianbei Area, Upper Yangtze Region," *Acta Petrologica Sinica*, **2013**, *29*(9), 3263–3278.
7. Boling P., Dazhong D., Songtao W., et al., "Microscopic Space Types of Lower Paleozoic Marine Shale in Southern Sichuan Basin," *Journal of China University of Petroleum* (Edition of Natural Science), **2014**, *38*(4), 19–25.
8. Tong-lou G. and Ruo-bing L., "Implications from Marine Gas Exploration Breakthrough in Complicated Structural Area at High Thermal Stage: Taking Longmaxi Formation in Well JY1 as an Example," *Natural Gas Geoscience*, **2013**, *24*(4), 643–651.
9. Zhigang W., "Breakthrough of Fuling Shale Gas Exploration and Development and its Inspiration," *Oil & Gas Geology*, **2015**, *36*(2), 1–6.
10. Peng L and Liming J, "Reservoir Characteristics and Potential Evaluation of Continental Shale Gas," *Natural Gas Geoscience*, **2013**, *24*(5), 1060–1068.
11. Xiangzeng W., Shengli G., and Chao G., "Geological Features of Mesozoic Continental Shale gas inSouth of Ordos Basin, NW China," *Petroleum Exploration and Development*, **2014**, *41*(3), 294–304.
12. Guoheng L., Zhilong H, Zhenxue J., et al, "The Characteristics and Reservoir Significance of Lamina in Shale from Yanchang Formation of Ordos Basin," *Natural Gas Geoscience*, **2015**, *26*(3), 408–417.
13. Wei Y., Guojun C., Chengfu L., et al., "Micropore Characteristics of the Organic Rich Shale in the 7th Member of the Yanchang Formation in the Southeast of Ordos Basin," *Natural Gas Geoscience*, **2015**, *26*(3), 418–426.
14. Xiangzeng W., Jinqiao W., and Juntao Z., "Application of Fracturing Technology for Terrestrial Shale Gas Reservoirs," *Natural Gas Industry*, **2014**, *34*(1), 64–67.
15. Xin-jing L., Su-yun H., and Cheng Ke-ming C., "Suggestions from the Development of Fractured Shale Gas in North America," *Petroleum Exploration and Development,* **2007**, *34*(4), 392-400.
16. Xian-ming X., Zhi-guang S., Yan-ming Z., et al., "Summary of Shale Gas Research in North American and Revelations to Shale Gas Exploration of Lower Paleozoic Strata in China South Area" *Journal of China Coal Society*, **2013**, *38*(5), 721-727.
17. Ming-na G, Jin-chuan Z., Xiao-guang L., et al., "Gas-bearing Property Analysis on Upper Paleozoic Shale in Eastern Uplift of Liaohe Basin," *Fault-Block Oil Gas Field*, **2012**, *19*(6), 722-726 (in Chinese with English abstract).

18. Ming-na G, Jin-chuan Z, Xiao-guang L., and et al., "Evaluation on Neopaleozoic Shale Gas Resource Potential in the Eastern Salient of the Liaohe Depression," *Natural Gas Industry*, **2012**, *32*(9), 28-32.
19. Junli M., Xiaoguang L., Yansheng S., et al., "Shale Gas Accumulation Conditions of Eastern Region of Liaohe Depression," *Earth Science Frontiers*, **2012**, *19*(5), 348-355.
20. Zhunlin R and Xiaoguang L., "Enrichment Condition of Lishugou Formation Shale Gas and Favorable Area Selection in Eastern Area of Liaohe," *Fault-block Oil & Gas Field*, **2013**, *20*(6), 704-708.
21. Peng Y., "Application of Widely-covered Hydrocarbon Generation Concept to Hydrocarbon Accumulation in Eastern Uplift, Liaohe Depression," *Journal of Xi'an Shiyou University (Natural Science Edition)*, **2014**, *29*(4), 24-37.
22. Dazhong D., Caineng Z., Jianzhong L., et al., "Resource Potential, Exploration and Development Prospect of Shale Gas in the Whole World," *Geological Bulletin of China*, **2011**, *30*(2/3), 324-326.
23. Hongyan W., Yuzhang L., Dazhong D., et al., "Scientific Issues on Effective Development of Marine Shale Gas in Southern China," *Petroleum Exploration and Development*, **2013**, *40*(5), 574-579.
24. Caineng Z., Dazhong D., Shejiao W., et al., "Geological Characteristics, Formation Mechanism and Resource Potential of Shale Gas in China," *Petroleum Exploration and Development*, **2010**, *37*(6), 641-643.
25. Shujing J., Hui H., Qingping W., et al., "Scanning Electron Microscope Analysis of Porosity in Shale," *Journal of Chinese Electron Microscopy Society*, **2012**, *31*(5), 432-436.
26. Gengshe Y. and Hui L., "Study on the Rock Damage Characteristics Based on the Technique of CT Image Processing," *Journal of China Coal Society*, **2007**, *32*(5), 463-468.
27. Bin B., Rukai Z., Songtao W., et al, "Multi-scale Method of Nano(Micro)-CT Study on Microscopic Pore Structure of Tight Sandstone of Yanchang Formation, Ordos Basin," *Petroleum Exploration and Development*, **2013**, *40*(3), 329-333.
28. Wang Jialu W., Jian G., and Li L., "Porosity Characteristics of Sandstone by X-ray CT Scanning System," *Acta Petrolei Sinica*, **2009**, *30*(6), 887-893.
29. Xiaoxia S., Yuegang T., Wei L., et al., "Advanced Characterization of Seepage Pores in Deformed Coals Based on Micro-CT," *Journal of China Coal Society*, **2013**, *38*(3), 435-440.
30. Haikuan N., Ruikang B., Peixian Z., et al., "Micro-types and Characteristics of Shale Reservoir of the Lower Paleozoic in Southeast Sichuan Basin, and their Effects on the Gas Content," *Earth Science Frontiers*, **2014**, *21*(4), 331-343.
31. Boling P., Dazhong D., Chuang E., et al., "Favorable Reservoir Characteristics of the Longmaxi Shale in the Southern Sichuan Basin and their Influencing Factors," *Natural Gas Industry*, **2013**, *33*(12), 41-47.
32. Caineng Z., Rukai Z., Bin B., et al., "First Discovery of Nano-pore Throat in Oil and Gas Reservoir in China and its Scientific Value," *Acta Petrologica Sinica*, **2011**, *27*(6), 1857-1864.
33. Rouquerol J., Avnir D., Fairbridge C. W., et al., "Recommendations for the Characterization of Porous Solids," *Pure & Applied Chemistry*, **1994**, *66*(8), 1739-1785.
34. Hui L., Shaohua W., Xiumin J., et al., "The

Configuration Analysis of the Adsorption Isotherm of Nitrogen in Low Temperature with the Lignite Char Produced under Fast Pyrolysis," *Journal of China Coal Society*, **2005**, *30*(4), 507-510.

35. Ping C. and Xiuyi T., "The Research on the Adsorption of Nitrogen in Low Temperature and Micro-pore Properties in Coal," *Journal of China Coal Society,* **2001**, *26*(5), 552-556.
36. Ross D. J. and Bustin R. M., "Shale Gas Potential of the Lower Jurassic Gordondale Member, Northeastern British Columbia, Canada," *AAPG Bulletin*, **2007**, *55*(1), 51-75.
37. Shuheng T. and Erping F., "Methane Adsorption Characteristics of Clay Minerals in Organic-rich Shales," *Journal of China Coal Society*, **2014**, *39*(8), 1700-1706.
38. Feng Y., Zhengfu N., Detao K., et al., "Pore Structure of Shales from High Pressure Mercury Injection and Nitrogen Method," *Natural Gas Geoscience*, **2013**, *24*(3), 450-455.
39. Jialu W., Jian G., and Li L., "Porosity Characteristics of Sandstone by X-ray CT Scanning System," *Acta Petrolei Sinica*, **2009**, *30*(6), 887-893.
40. Nelson P H., "Pore-throat Sizes in Sandstones, Tight Sandstones, and Shales," *AAPG Bulletin*, **2009**, *93*(3), 329-340.

4

A Mechanical Model and its Experimental Verification for a Water Injection String in a Highly Deviated Well

Liu Jun[1]*, Wang Wei[2], Liu Qingyou[1], Xiao Guohua[3], Zhang Jianzhong[3], and Sang Guo[2]

[1] School of Mechatronic Engineering, Southwest Petroleum University, Cheng Du Si Chuan 610500, China
[2] School of Civil and Architecture Engineering, Southwest Petroleum University, Cheng Du Si Chuan 610500, China
[3] Drilling Technology Research Institute of Jidong Oilfield Company, Tang Shan He Bei 063000, China

ABSTRACT

Water injection strings in highly deviated wells are subjected to complex forces on the string bore. In this work, a mechanical model is developed for these forces and for those on downhole tools. On the basis of this model, and taking account of the characteristics of the string in different working conditions, a temperature field model and a pressure field model are introduced, and a statically indeterminate structural calculation method is adopted. A force tester for highly deviated wells is developed and used in eight tests on strings in the Jidong Nanpu oilfield making a comparison between the measurements from the tests and the results of calculations using the mechanical model indicates that the mechanical model exhibits high computational accuracy, with the errors at the wellhead being no more than 10%, and that the forces on packers calculated from the model are in accordance with those measured in the tests. Thus, the mechanical model developed in this work is suitable for analyzing the forces on water injection strings and downhole tools in highly deviated wells.

Keywords: Highly Deviated Well, Water Injection String, Packer, Mechanical Model, Mechanics Experiment

INTRODUCTION

In many offshore oilfields, the restrictions imposed by surface and underground conditions are such that it is necessary to use directional and horizontal wells to enable exploitation. In addition, high-angle directional wells account for a high percentage of injection wells [1,2]. The use of both multiple pipe segments and high angles both increase the effect of friction on pipe columns, causing difficulties in their lowering and hoisting. Additionally, the higher intensity of pipe deformation and creep can be ascribed to the large differential forces caused by pipe setting, water injection, inverse well-flushing, and other work conditions [3,5]. Thus, the permanent helical buckling of columns, plastic failure, fracture failure, early failure of the packer, and other detrimental phenomena are prone to appear during the operation [6]. These problems, in particular with regard to the safety of the pipe string and early failure of the packer, are becoming increasingly prominent, constraining the development of oilfield water injection and directly affecting normal

***Corresponding author**
Liu Jun
Email: wwswpu@qq.com

production, with consequent economic losses [7,8]. Meanwhile, the water injection technique is being extended to zonal approaches and to deep wells [9–11], and the higher water injection pressures required lead to an increase in pressure differences between layers and to more diverse and complex working conditions; this results in further challenges to the reliability and safety of high-angle well water injection strings [12].

Although many theoretical and experimental studies have been carried out on various aspects of pipe string mechanics [13–20], there has been little research on high-angle water injection pipe columns, and no mature theoretical description of such systems is available. Three deficiencies in particular should be mentioned: (i) the existing mechanical models of high-angle water injection wells have not taken comprehensive account of the influence of temperature and pressure fields, frictional resistance, and downhole tools on the force and deformation of the pipe string; (ii) there have been no thorough examinations of the force characteristics of the pipe string; (iii) making a comparison between the measured data and theoretical results has rarely been carried out, owing to a lack of measurements of the force state of the pipe string, so the existing models have not been calibrated and validated satisfactorily, and their accuracy and practicability are limited. Therefore, this contribution aims to improve the theoretical mechanical model of water injection pipe strings in high-angle wells to, verify it using field test data, and to provide theoretical support for the design and construction of water injection pipe columns for high-angle wells.

EXPERIMENTAL PROCEDURES

Mechanical Model for Water Injection in a Highly Deviated Well

Three Dimensional "Rigid Rope" Model

In this model, the effects of the water injection string stiffness and the curvature of the wellbore are taken fully into account in the force analysis of the curved portion. It is assumed that (i) the interaction between the well wall and the pipe string is a rigid continuous contact, (ii) the axis of the string coincides with the axis of the borehole, and (iii) the computational elements of the bent pipe in the sloping plane are iso-curvature segments of a circular arc. A three-dimensional "rigid rope" mechanical model is then established.

A rectangular coordinate system is established as shown in Figure 1, and a natural system of coordinates on the string axis is also established. The equations describing the force on the curved portion of the string are then as follows:

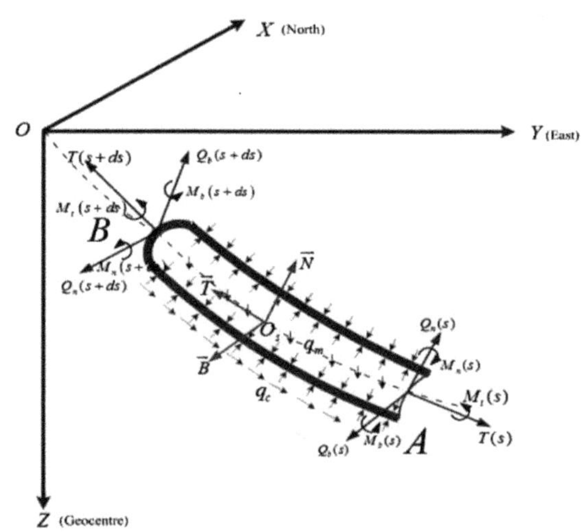

Figure 1: an infinitesimal section force diagram.

$$\begin{cases} \frac{dT_t - dP_i A_i + dP_o A_o}{ds} + K \cdot EI \frac{dK}{ds} \pm \mu_\alpha N + f_\lambda - q_m \cos\alpha = 0 \\ \frac{dM_t}{ds} = \mu_t RN + 2\pi R^3 \omega \left[\frac{\tau_f}{\sqrt{v^2 + (R\omega)^2}} + \frac{2\mu}{D_w - 2R} \right] \\ -EI \frac{d^2K}{ds^2} + K \cdot T + N_n + \mu_t N_b + q_m \sin\alpha \frac{K_\alpha}{K} = 0 \\ -K \frac{dM_t}{ds} + \mu_t N_n - N_b - q_m \sin^2\alpha \frac{K_\varphi}{K} = 0 \\ N^2 = N_n^2 + N_b^2 \end{cases} \quad (1)$$

where, α is the deviation angle of the borehole (rad), and φ is the azimuthal angle (rad); K_α is the rate of change of the deviation angle (rad/m), and K_φ is the rate of change of the azimuthal angle (rad/m); $K_f = 1 - \rho_m / \rho_s$ is the buoyancy factor (where ρ_m is the fluid density in the wellbore, and ρ_s is the density of the string), and q is the unit weight of the string in air (kN/m); q_m is its unit weight in the fluid (kN/m); N_n is the normal pressure on the string in the normal direction (kN), and N_b is the normal pressure on the string in the binormal direction (kN), μ_t is the friction coefficient in the circumferential direction; μ_α is the friction coefficient in the axial direction, and A_i is the inner chamber cross-sectional area of the string (m^2); A_o is the inner chamber cross-sectional area of the casing (m^2), and P_o represents the pressure in the pipe (MPa); τ_f is the fluid force on unit length pipe due to the motion of the structure (N/m), and μ is the kinetic viscosity of the fluid (N·s/m^2), ω stands for the rotational speed of the string (rad/s), D_w is the wellbore diameter (m), R is the outside radius of the string (m), and v is the fluid velocity (m/s); E is the Young's modulus of elasticity (kN/m^3), and I is the moment of inertia of the string (m^4).

Under the effects of axial friction and interlayer pressure differences during the operation, once the load at the end of the string reaches a certain threshold, pipe string buckling instability will occur. This will generate additional positive contact pressure and thus additional friction. To provide a more accurate description, the three-dimensional "rigid rope" mechanical model is amended by including a critical load from sinusoidal or spiral buckling as follows:

$$\begin{cases} T_{crs} = \dfrac{4EI}{Rr_c} & sinusoidal\ bending \\ T_{crh} = 7.56\dfrac{EI}{Rr_c} & spiral\ bending \end{cases} \tag{2}$$

When there is the axial force, there is no buckling, and there is sinusoidal buckling; moreover, when $T_z \geq T_{crh}$, there is spiral buckling. The corresponding additional reaction forces are:

$$\begin{cases} N_s = \dfrac{rT_z^{\ 2}}{8EI} & sinusoidal\ bending \\ N_h = \dfrac{rT_z^{\ 2}}{4EI} & spiral\ bending \end{cases} \tag{3}$$

When sinusoidal buckling occurs, the positive pressure on the string N becomes $N + N_s$; when spiral buckling occurs, it becomes $N + N_h$.

Three-dimensional "Soft Rope" Mechanical Model

In this model, the effects of water injection string stiffness and the curvature of the wellbore are not taken into account in the force analysis. Thus, in comparison with the "rigid rope" mechanical model, the following simplifying assumptions are made: (i) the effects of stiffness on the force of the string are neglected; (ii) the shear force on the string section is neglected. The resulting three-dimensional "soft rope" mechanical model is then given by:

$$\begin{cases} \dfrac{dT_t - dP_iA_i + dP_oA_o}{ds} \pm \mu_a N + f_\lambda - q_m \cos\alpha = 0 \\ \dfrac{dM_t}{ds} = \mu_t RN + 2\pi R^3\omega\left[\dfrac{\tau_f}{\sqrt{v^2 + (R\omega)^2}} + \dfrac{2\mu}{D_w - 2R}\right] \\ K \cdot T + N_n + \mu_t N_b + q_m \sin\alpha \dfrac{k_\alpha}{k} = 0 \\ -K\dfrac{dM_t}{ds} + \mu_t N_n - N_b - q_m \sin^2\alpha \dfrac{K_\varphi}{k} = 0 \\ N^2 = N_n^2 + N_b^2 \end{cases} \tag{4}$$

These are nonlinear differential equations, which can be solved by a quasi-Newton method.

Temperature and Pressure Field Models for a Water Injection String

Changes in temperature and pressure are significant factors in the mechanical analysis of strings. As the water is injected and flows down from the top, the forces on the string will vary considerably, with the string elongating or shortening because of changes in temperature and pressure, while its motion is simultaneously restricted, owing to the presence of packers and other downhole tools.

Temperature Field Model

The following assumptions are made in the numerical simulation models: (i) the stiffness of the string is such that it is in rigid and continuous contact with the borehole walls; (ii) the axis of the string coincides with the axis of the borehole, with the hole trajectory of the highly deviated well being described by the Frenet formula; (iii) the computational element of the pipe string is a circular arc on the sloping plane; (iv) before water injection, the pipe string, the fluid in the wellbore, and the ground are in thermal equilibrium. The initial temperature distribution is exhibited as the table geothermal gradient. After water injection, heat transfer in the radial and vertical directions of the wellbore due to mass transfer and thermal conduction is taken into account; (v) when discredited in the spatial domain, the material properties of each micro body domain such as the heat transfer coefficient and the specific heat are relatively stable; (vi) the water injection rate, water injection temperature, annual average subsurface temperature, and the geothermal gradient are constant; (vii) sufficiently far from the wellbore axis, $r \geq R_{max}$, the rock formation temperature is the original ground temperature. At the surface of the earth, $Z=Z_m$, the temperature Tm does not change with time; below the surface, $Z>Z_m$, the original rock formation temperature satisfies a linear relationship: $T_{dz}=T_m+\alpha z$, where α is the rock formation temperature gradient; (viii) given the variety of conditions occurring in oil production operations, the influence of the last process on the next process must be considered in the calculation of wellbore temperature field.

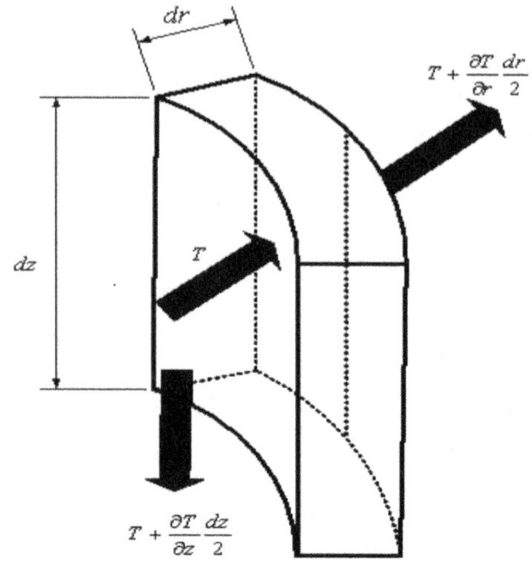

Figure 2: Infinitesimal body diagram.

On the basis of the above assumptions, in light of the first law of thermodynamics and the basic principles of heat transfer, considering the liquid in the pipe string, the string wall, and the liquid in the annulus, and with the stratigraphic unit taken as a control volume (see Figure 2), the mathematical model of the temperature field is given by the following energy balance equations:

$$
\begin{cases}
-c_f \rho_f r_{ti} v_f \frac{\partial T_f}{\partial z} + \frac{2}{r_{ti}} \alpha_{tf} \left(T_t - T_f\right) + \frac{1}{4 r_{ti}} \lambda_f \rho_f v_f^{\,3} = c_f \rho_f \frac{\partial T_f}{\partial t} \\
k_t \frac{\partial^2 T_t}{\partial z^2} - 2\alpha_{tf} \frac{r_{ti}}{r_{to}^2 - r_{ti}^2} \left(T_t - T_f\right) + 4k_t \frac{r_{to}}{\left(r_{to}^2 - r_{ti}^2\right)\left(r_{to} - r_{ti}\right)} \left(T_a - T_t\right) = c_t \rho_t \frac{\partial T_t}{\partial t} \\
k_a \frac{\partial^2 T_a}{\partial z^2} - 4k_t \frac{r_{to}}{\left(r_{ci}^2 - r_{to}^2\right)\left(r_{to} - r_{ti}\right)} \left(T_a - T_t\right) + 4k_a \frac{r_{ci}}{\left(r_{ci}^2 - r_{to}^2\right)\left(r_{ci} - r_{to}\right)} \left(T_c - T_a\right) = c_a \rho_a \frac{\partial T_a}{\partial t} \\
k_c \frac{\partial^2 T_c}{\partial z^2} - 4k_a \frac{r_{ci}}{\left(r_{co}^2 - r_{ci}^2\right)\left(r_{ci} - r_{to}\right)} \left(T_c - T_a\right) + 4k_c \frac{r_{co}}{\left(r_{co}^2 - r_{ci}^2\right)\left(r_{co} - r_{ci}\right)} \left(T_e - T_c\right) = c_c \rho_c \frac{\partial T_c}{\partial t} \\
k_e \frac{\partial^2 T_e}{\partial z^2} - 4k_c \frac{r_{co}}{\left(r_{ce}^2 - r_{co}^2\right)\left(r_{co} - r_{ci}\right)} \left(T_e - T_c\right) + 4k_e \frac{r_{ce}}{\left(r_{ce}^2 - r_{co}^2\right)\left(r_{ce} - r_{co}\right)} \left(T_w - T_e\right) = c_e \rho_e \frac{\partial T_e}{\partial t} \\
\frac{\partial^2 T_w}{\partial z^2} + \frac{\partial^2 T_w}{\partial r^2} + \frac{1}{r} \frac{\partial T_w}{\partial r} = \frac{c_w \rho_w}{k_w} \frac{\partial T_w}{\partial t}
\end{cases}
\qquad \textbf{(5)}
$$

where, C_f is the specific heat of the fluid in the pipe string (J/(kg·°C)), and α_{tf} is the coefficient of convective heat transfer between the fluid in the string and the pipe wall (W/(m²·°C)); λ_f is the friction coefficient of the fluid in the string (which is related to the fluid Reynolds number and the flow pattern), and ρ_f is the fluid density in the string (kg/m³); v_f is the speed of fluid motion (m/s). k_t , k_a , k_c , k_e, and k_w are respectively the heat conductivity coefficients of the injected fluid, the cylinder wall, the fluid within the annulus, the rock formation, and the original liquid mixing unit; c_t , c_a, c_c, c_e,and c_w (J/(kg·°C)) are the respective specific heat capacities; and ρ_t , ρ_a , ρ_c , ρ_e , and ρ_w (kg/m³) are the respective densities. r_{to}, r_{co}, r_{ci}, r_{ti}, and r_{ce} (m) are respectively the radius of the outer wall of the pipe string, the radius of the outer wall of the casing, the string inner radius, the radius of the inner wall of the casing, and the cement ring radius. The spatial domain is discredited using high-precision isoperimetric elements, and an optional implicit finite difference time domain with variable steps is used in a loop iteration, which allows the efficient and accurate calculation of the whole temperature field [21].

Pressure Field Model

The momentum equation for one-dimensional flow of the fluid under pressure is:

$$\frac{dp}{dz}=\frac{dP_f}{dz}+\frac{dP_a}{dz}+\frac{dP_g}{dz} \tag{6}$$

The variation in pressure due to changes in frictional resistance is given by:

$$\frac{dp_f}{dz}=-\tau \tag{7}$$

The variation due to changes in kinetic energy is given:

$$\frac{dP_a}{dz}=\frac{U_m^2}{A^2}\frac{d(1/\rho)}{dz} \tag{8}$$

Also, the variation due to changes in gravity and the hole deviation angle is given:

$$\frac{dP_g}{dz}=g\rho\cos(\theta) \tag{9}$$

Then, neglecting changes in fluid density, one way obtain:

$$\frac{dP}{dz}=\frac{dP_f}{dz}+\frac{dP_g}{dz}=-\tau+g\rho\cos(\theta) \tag{10}$$

The pressure distribution along the well depth is found by integration, taking account of the boundary conditions:

$$P(z)=\int_0^z[-\tau(z)+g\rho\cos(\theta)]dz+P_0 \tag{11}$$

When the coefficient of frictional resistance τ is constant, Equation 11 can be simplified:

$$P(z)=-\tau\times z+g\rho\times z\cos(\theta)+P_0 \tag{12}$$

Mechanical Models for Pipe String and Downhole Tools

As shown in Figure 3, the packers are numbered from top to bottom, 1, ... n, the pipe anchoring tools as M_1, M_2, ..., M_{n+1}, the corresponding forces on the packers as F_1, F_2,..., F_n, the annulus pressures under the packers as P_1, P_2,..., P_n, and the lengths of the string between the packers as L_1, L_2,..., L_n. The general force equations for an n-time statically indeterminate structure are then given by the following equations:

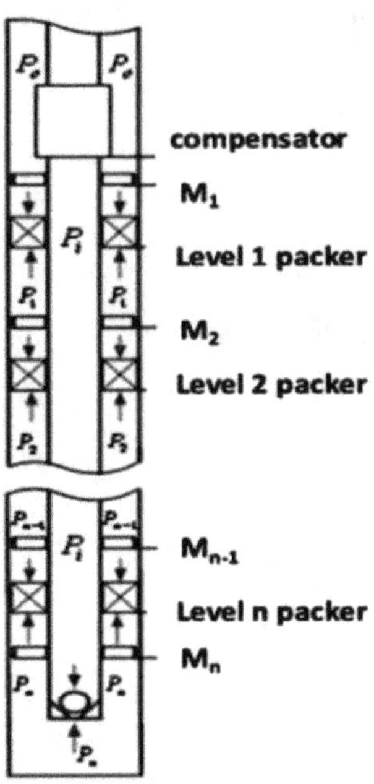

Figure 3: Force diagrams of pipe string and downhole tools.

$$\begin{cases}\delta_{11}X_1+\delta_{12}X_2+\cdots+\delta_{1n}X_n+\Delta_{1p}+\Delta_{1t}=0\\ \delta_{21}X_1+\delta_{22}X_2+\cdots+\delta_{2n}X_n+\Delta_{2p}+\Delta_{2t}=0\\ \cdots\cdots\cdots\\ \delta_{n1}X_1+\delta_{n2}X_2+\cdots+\delta_{nn}X_n+\Delta_{np}+\Delta_{nt}=0\end{cases} \tag{13}$$

where, Δ_{ip} and Δ_{it} are the displacements in the X_i direction due to differential pressure and temperature; δ_{ij} is the displacement in the X_i direction due to unit force $X_j = 1$, and $\overline{F_i}$ is the internal force at the level *i* packer caused by a dummy unit load; F_{ip} and F_{it} are the internal forces due to differential pressure and temperature respectively.

The internal forces on the statically indeterminate structure are given by the superposition principle:

$$F_i = \overline{F_1}X_1 + \overline{F_2}X_2 + \cdots + \overline{F_n}X_n + F_{ip} + F_{it} \tag{14}$$

Strength Calculations for a Water Injection String in a Highly Deviated Well

To ensure the safety of the water injection string, its tensile, compressive, and collapsing strengths are first calculated. The force intensity is then obtained using the distortion energy theory. Finally, the synthesis safety factor is calculated.

The tensile-strength safety factor is given by the following equations:

$$K_{rd} = \frac{F_{rd}}{q_e \times L} \tag{15}$$

$$q_e = q + \rho_i g A_i - \rho_o g A_o \tag{16}$$

Brst-strength safety factor is given by the following equation:

$$K_{rpi} = \frac{P_{ri}}{Max(p_i - p_o)} \qquad (p_i > p_o) \tag{17}$$

The safety coefficient for resistance to external pressure is given by the following equation:

$$K_{rpo} = \frac{P_{ro}}{Max(p_o - p_i)} \qquad (p_o > p_i) \tag{18}$$

From the distortion energy theory, the tri-axial stress is given by the following equation:

$$\sigma_{xd} = \sqrt{\frac{1}{2}\left[(\sigma_r - \sigma_\theta)^2 + (\sigma_\theta - \sigma_z)^2 + (\sigma_z - \sigma_r)^2\right]} \tag{19}$$

The synthesis safety factor is also given by the following equation:

$$K = \mathrm{Min}(K_{rd}, K_{rpi}, K_{rpo}, K_{xd}) \tag{20}$$

Force Measurements for a High-angle Pipe String

To verify the correctness of the model, a force tester, the composition of a resistance strain gauge transducer, a data acquisition circuit board, a power module, and a force test structure have been developed. The force test structure is composed of an inner barrel and an outer barrel, as shown in Figure 4. The data acquisition circuit board is contained within an electronic equipment enclosure, which is positioned eccentrically between the inner and outer barrels. The resistance strain gauge transducer is fixed to the inner ring groove. The complete force test substructure is shown in Figure 5. This force tester was used in site tests on a string in the Jidong Nanpu Oilfield, as described in the next section.

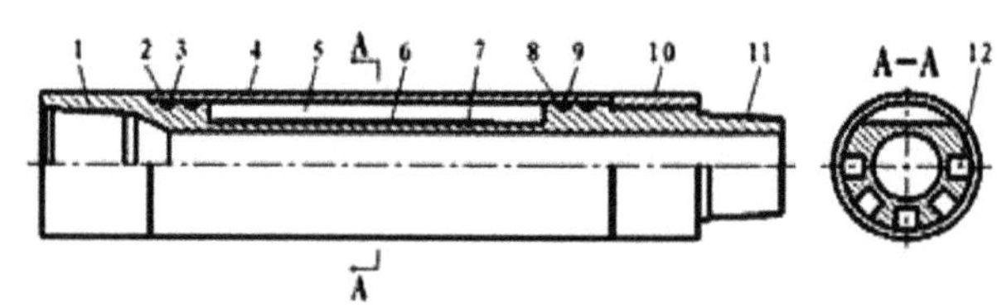

Figure 4: Mechanical structure of the test instrument. 1-screwed joint; 2,9-seal ring; 3,8-seal retainer; 4-outer barrel; 5-electronic equipment; 6-acquisition of circuit board; 7-inner barrel; 10-baffling barrel; 11-parallel external pipe thread; 12-slot.

Figure 5: Complete force test substructure.

Examples of Analysis and Experimental Verification

The mechanical model was used to analyze the forces on an unbalanced injection string in the third phase of the NP-XX well in the Jidong Nanpu Oilfield in different working conditions. The structure of the string and the positions where the experimental instruments were installed are shown in Figure 6; the azimuthal angle and hole drift angle parameters are shown in Figure 7, and the construction parameters of the well are listed in Table 1.

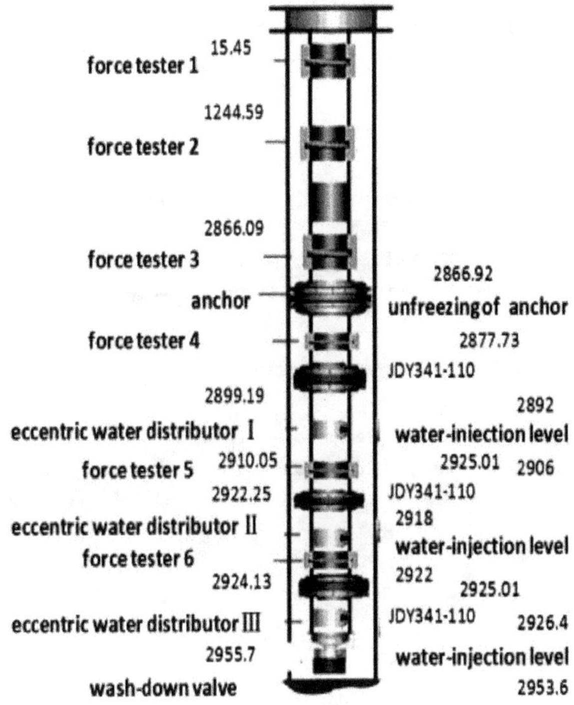

Figure 6: String structure.

Table 1: Construction parameters of NP-XX

Parameter	Value
Inverse well-flushing time	60 min
Injection time	120 min
Injection rate	150 L/min
Inverse well-flushing rate	300 L/min
Surface temperature	22 °C
Fluid	23.5 °C
Injection pressure	10 MPa
Inverse well-flushing pressure	3 MPa
Density of injected liquid	1250 g/cm^3
Setting pressure	15 MPa
JDY341-110 packer 1	2877.73 m
JDY341-110 packer 2	2911.03 m
JDY341-110 packer 3	2925.01 m
Wash-down valve	2955.7 m
Force tester position 1	15.45 m
Force tester position 2	1244.6 m
Force tester position 3	2866.09 m
Force tester position 4	2867.47 m
Force tester position 5	2910.15 m

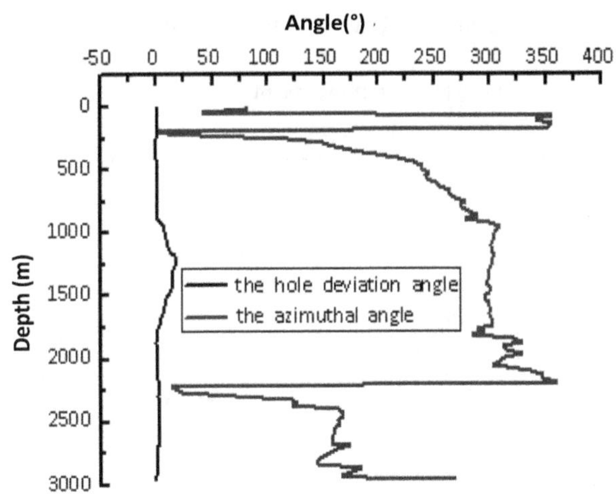

Figure 7: The azimuth angle and hole drift angle parameters.

Figure 8 compares the measurements from the force tester with the computations using the model in different working conditions. It can be seen that the string was subjected to a press by 57 kN force on the wellhead during the process of packer setting, which led to a corresponding 57 kN decrease in tensile force in the wellhead pipe segment during water injection and backwashing.

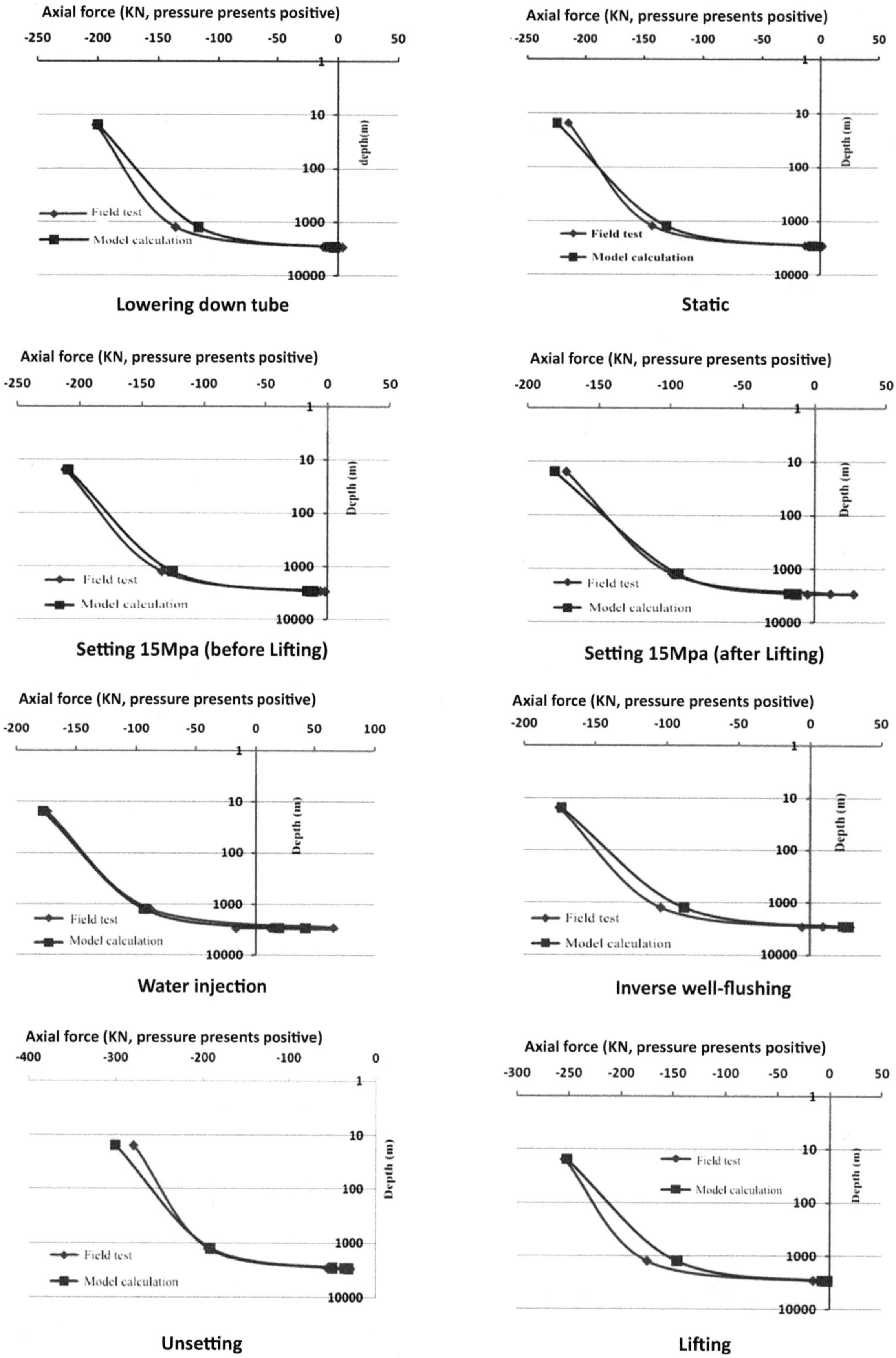

Figure 8: Results in various conditions.

Owing to friction, the effect of pressure on the drill string force decreased with increasing well depth. The error in the calculated results compared with the measurements in the upper part of the tool section was small (<10%), while in the lower part of the tool section, owing to the small force and the limits on measuring conditions, the errors were larger. However, the calculated values of the drill string force were basically consistent with the values measured in the test.

Table 2: Comparison of measured and calculated forces for eight different loading conditions

	Segments	Field Test	Calculation	Error
1	Oil casing stage injection	45	40.8	-9.33%
2	One-stage two-pass	45	46.4	3.11%
3	Two-stage thee-pass	45	42.5	-5.56%
4	Three-stage three-pass	45	42.7	-5.11%
5	Two-stage two-pass	48	44.6	-7.62%
6	Three-stage four-pass	45	47.4	5.33%
7	Two-stage three-pass	42	44.5	5.95%
8	Three -stage four-pass	45	44.0	-2.22%

Table 2 compares the calculated results with the measured values of the drill string force when the strings were loaded together with the associated errors for eight different loading conditions. It is obvious that the errors in the calculated values of the lifting force on the string were within 10%, and the model is basically suitable for an engineering application.

Table 3: Force on packers

Packer number	Depth (m)	Force (kN) (Downward Presents Positive)
1	2877.73	22.092
2	2911.03	-0.434
3	2925.01	-24.415

Table 3 shows the forces on the packers. In the condition of water injection, the packer at the first level was subjected to a downward force of 22.092 kN and the bottom packer to an upward force of 24.415 kN, which means that the pressures on the packers at both ends are greater than that on the intermediate packer. The pressure difference between the top and bottom packers generates a force that pushes the packers upwards and downwards. However, when the middle layer is injected with water, the upper and lower packers form a "self-balancing" system which will be pushed, so the upward and downward movement of the packer group will be prevented. Therefore, this corresponds well to practical engineering experiences.

CONCLUSIONS

(1) A mechanical model has been developed for analyzing the forces on an injection string with multiple packers in a highly deviated well with restrictions imposed by the presence of downhole tools. The model calculates the forces on the downhole tools and the string under various operating conditions, so the model provides a theoretical basis for the design and construction of injection strings.

(2) The combined effects of the temperature field, the pressure field, friction, and the restrictions by the downhole tools have been taken into account in the design of the model. Making a comparison

the design of the model. Making a comparison between the model calculations and field test results indicates
that the computational accuracy of the model can satisfy practical engineering requirements.

(3) The results of the calculations provide a reasonable explanation for the "upward and downward push" phenomenon in which the force on the packers at both ends is greater than that on the intermediate packer. In practice, the ends of the packers should be anchored to prevent the motion of the packer group.

ACKNOWLEDGMENTS

This work has been supported in part by the National Science Foundation through Grant 2011ZX05050.

REFERENCES

1. Yafeng J., Jiuzheng Y., and Gengcheng Y., "The Research on Multilayer Fine Separated Layer Water Injection Technology in Jiyuan Oil Field," *Science Technology and Engineering*, **2012**, *18*, 4504-4506.
2. Haiquan W., Xianming S., and Haitao Gen., "Concentric Layered Water Injection Technology Applied in Nanpu High Angle Wells," *China Petroleum Machinery*, **2011**, *9*, 560-566.
3. Tao X. and Qingyou L., "Analysis of High Angle Well Water Injection Wellbore Pressure Calculation Model," *Petrochemical Industry Application*, **2015**, *6*, 12-15.
4. Hongwei Z., Jiang L., and Jijunet Y., "Mechanical Analysis on Tubular for Inverse Well Flushing in Deep Deviated Well," *China Petroleum Machinery*, **2014**, *10*, 40-45.
5. Quanli J., "Mechanics Analysis of Layered Water Injection Packer String," *Dissertation*, Xian Petroleum University, **2014**.
6. Baoxia Q., Ceji Q., and Congcong S., "Analysis and Design Research on Layered Water Injection String Mechanics," *China Petroleum and Chemical Industry*, **2009**, *1*, 57-59.
7. Jingfu D., "Research on Remaining Oil Description and Redeveloping Measures of PI-4 Oil Layer after Polymer Flooding in Nanwu Area," Dissertation, Northeast Petroleum University, **2013**.
8. Guohua Z., Jiangtao L., and Buhong J., "The Research and Application of New Type of Hollow Point's Injection Process," *Fault Block Oil and Gas Field*, **2007**, *4*, 80-82.
9. Junhui W., Lan H., and Jinqian Li., "Research of Oilfield Water Injection Mode," **2014**, *1*, 84-88.
10. Yanping Z., "The Present Situation of the Oil Field Injection Well Zonal Injection Technology and the Development Tendency," *China Petroleum and Chemical Standard and Quality*, **2014**, *6*, 108.
11. Sheng S. and Jianfu L., "Oilfield Water Flood Technical Inquiry," *China Petroleum and Chemical Industry Standard and Quality*, **2013**, *20*, 177.
12. Wei C., Xiaohong F., and Guangbing J., "Sea Controlled High Angle Well Liquid Separate Injection Technology," *China Petroleum Machinery*, **2009**, *11*, 78-79.
13. Lubinski A., Althouse W. S., and Logan J. L., "Helical Buckling of Tubing Sealed in Packers," *Journal of Petroleum Technology*, **1962**, *14*, 655-670.
14. Lubinski A., "Developments in Petroleum Engineering," Houston, *TX: Gulf Publishing*, **1987.**
15. Hammerlindl D. J., "Packer-to-Tubing Forces for Intermediate Packers," *Journal of Petroleum*

Technology, **1980**, *32*, 515-527.

16. Salies J. B., Azar J. J., and Sorem J. R., "Experimental and Mathematical Modeling of Helical Buckling of Tubules in Directional Wellbores," in *Proc. International Petroleum Conference and Exhibition of Mexico*, Veracruz, Mexico, **1994**, 433-443.
17. Payne M. L. and Abbassian F., "Advanced Torque and Drag Consideration in Extended-Reach Wells," *Society of Petroleum Engineering, SPE Drilling Completion*, **1997**, *12*(3), 55-62.
18. Hill T. H. and Chandler R. B., "Field Curves for Critical Buckling Loads in Curving Well bores," in Proc. IADC/SPE Drilling Conference, 3-6 March, Dallas, TX, 1998, 259-265.
19. Linqiang Z., "Calculating Model of Torque and Drag in Extended Reach Well," *Fault Block Oil and Gas Field*, **2008**, *2*, 88-91.
20. Xiaohuaa Z., Yuan G., and Shaohu L., "Feasibility of Casing Entering Curved Part in Horizontal Well," *Oil Field Equipment*, **2011**, *4*, 6-8.
21. Zuqing H., Xiaorong Z., and Guorong W., "Research for Water Injection String Temperature Field Calculation Model," *Pioneering with Science and Technology Monthly*, **2007**, *20*(6), 178-179.

5

A Study on Structure and Property of Special Sulfonated Petroleum as a Complicated Anionic Surfactant

Fang Yu

Advanced Materials and Analysis Center, National Institute of Clean-and-Low-Carbon Energy, Beijing 102209, China

ABSTRACT

Alkaline/surfactant/polymer (ASP) flooding has been extensively applied to enhancing oil recovery. As an oil displacement agent, pe1troleum sulfonate (PS) is widely used in ASP pilot experiments. It is very important to find a petroleum sulfonate with a high interfacial activity and low adsorption loss. Anionic surfactant petroleum sulfonate is synthesized in a single-tube film sulfonation reactor. PS is separated into fractions with different average molecular weights, and the structure of the fractions is characterized. It has been found out that the interfacial chemistry property of PS is closely related to the molecular structure. The adsorption reaches the maximum value near the critical micellar concentration. Furthermore, there are significantly different interfacial activity and adsorptive property for each fraction. As the lab test result shows, the oil-displacement efficiency for each fraction is changed. The interfacial tension between crude oil and water is declined, and the oil-displacement efficiency of the system is greatly increased as PS or its fraction is used with electrolytes.

Keywords: Petroleum Sulfonate, Interfacial Activity, Molecular Structure, Adsorption, Average Molecular Weight

INTRODUCTION

Oil is a non-renewable energy; after primary exploitation and secondary oil recovery methods are used, 60-70% of the crude oil is still remained in the reservoir. The residual oil needs to be further exploited by means of physical-chemical methods such as enhanced oil recovery (EOR) because of massive requirement for fossil energy. At present, alkaline/surfactant/polymer (ASP) flooding was widely employed for enhancing oil recovery (EOR), which efficiently increased oil recovery by more than 10% compared to water flooding. ASP flooding increased oil displacement efficiency and enlarged the sweep volume simultaneously. The ASP flooding has been developed into the industrial scale and application stage, with a series of important techniques developed in the field tests [1-8].

A low concentration ASP formula is employed to achieve ultra-low interfacial tension with the help of a surfactant. Petroleum sulfonate is a kind of anionic surfactant synthesized by raw

***Corresponding author**

Fang Yu

Email: yufang@nicenergy.com

material of crude oil or oleum distillates, which is widely applied to ASP pilot experiments. As an oil displacement agent, it owns high interfacial activity, effective compatibility with reservoir fluids, and good water solubility. Furthermore, it is simply produced in the real application situation. Malmberg used nine kinds of petroleum sulfonate samples to perform oil-displacement test and found out that this technology was feasible and had particular advantages [9]. Some patents claimed that petroleum sulfonate was used to discover crude oil efficiently [10-11]. The results of ASP flooding field test in Yumen oil field and Daqing oil field in China showed that oil-recovery was greatly increased when petroleum sulfonate was used [12-13]. There is a good prospect that petroleum sulfonate will be used in enhanced oil recovery. However, the adsorption of surfactant to reservoir rock surface is also an important factor because it reduces oil-displacement efficiency and increases cost. Therefore, it is very important to find the petroleum sulfonate with a high interfacial activity and a low adsorption loss.

The typical property of surfactant mainly depends on the chemical structure [14]. Petroleum sulfonate is a complicated compound containing many anionic surfactants with different molecular structures, which are difficult to be separated into single fractions so as to investigate the interfacial chemical properties [15]. The relationship between the molecular structure and the interfacial property of pure sodium alkylbenzensulfonates have been studied [16-18]. However, the composition of petroleum sulfonate and the relationship between composition and interfacial property have not thoroughly been investigated.

In this paper, a kind of petroleum sulfonate (PS) with an appropriate average molecular weight and a high surface activity was synthesized in a single-tube film sulfonation reactor. PS was separated into fractions with different average molecular weights, and the molecular structure of the fractions was characterized by NMR spectrometry. The interfacial tension (IFT) between PS aqueous solutions and Shengli crude oil, and the adsorption loss of PS on Shengli reservoir sand were determined. The oil-displacement property of PS and its component were also investigated in core-flooding experiments, and the effect of the composition of PS on interfacial properties was discussed.

EXPERIMENTAL PROCEDURES

Chemical Reagents and Materials

PS is synthesized by the raw material of refinery vacuum gas oil (VGO) and SO_3 sulfonation agent. The typical properties of VGO are shown in Table 1.

Table 1: Typical properties of VGO.

Kinematic Viscosity (mm²/s)		Density (g/cm³)	Solidification point (°C)	Elemental composition (wt.%)			Chemical composition (wt.%)		
40 °C	100 °C	20 °C	-14	H	C	N	Saturate	Aromatic	Resin
55.40	5.73	0.9383		12.06	87.30	0.24	62.00	37.14	0.86

Sodium hydroxide, ethanol, benzene, isopropanol, chloroform, sec-butyl alcohol, and hydrochloric acid were all analytical reagents. Hyamine 1622, dimidium bromide, and Erioglaucine A were supplied by Fluka Chemical Company.

The crude oil in the experiments was Shengli crude oil obtained from Shengli oil field with an acid number of 1.5 mg KOH/g. Shengli oil field is located in Dongying city, Shandong province in the east of China. The density of the crude oil is 0.9224 g/cm^3 at 30 °C.

The water used in the experiments was the re-injected water of Shengli oil field after filtration. The data of its composition and salinity are tabulated in Table 2.

Table 2: The composition of the re-injected water (mg/L).

K^++Na^+	Ca^{+2}	Mg^{+2}	Cl^-	HCO^{-3}	Total Salinity
4600	106	35	6989	671	12401

The adsorbent in the adsorption experiment is the Shengli reservoir sand from Shengli oil field. The reservoir sand was cleaned with a mixture of methylbenzene/ethanol (weight ratio of 3:1) followed by drying. It was sieved using both 80 and 100 sieve meshes, so its particle diameter was within the range of 0.147-0.175 mm. The composition of reservoir sand was analyzed by X-ray diffraction analysis, it is given in Table 3.

Table 3: The composition of reservoir sand (wt.%).

Quartz	Plagioclase	Potassium feldspar	Clay			
			Kaolinite (K)	Chlorite (Ch)	Illite (I)	Mix-clay (Illite-Montmorillonite)
44	32	22	0.50	0.19	0.32	0.99

In the core flooding experiment, reservoir core (2.5 cm×8 cm) was prepared by using Shengli reservoir sand cemented with epoxide resin and ethylenediamine.

Instruments

PS was synthesized by the single-tube film sulfonation reactor. ^{1}HNMR and ^{13}CNMR spectra were carried out using a Bruker 500-MHz spectrometer. Electrospray ionization mass spectra (ESI-MS, Negative) were obtained using HP1100 MSD. Elemental analyses were performed by an Elementar Vario ELβ. The surface tension was measured by DCAT 21 surface tensionmeter at 20 °C. The IFT between the PS solutions and crude oil was measured at 80 °C by TX500C interfacial tension meter. The core-flooding experiments were performed utilizing a high-temperature high-pressure core-flooding apparatus.

Synthesis and Purification of PS

PS was synthesized in the single-tube film sulfonation reactor using the raw material of refinery vacuum gas oil (VGO) and the SO_3 sulfonation agent. The sulfonation condition was as follows: sulfonation temperature was 35 °C; the ratio of SO_3 to the aromatic amount of substance was 1.25:1; the concentration of SO_3 was 4.7%, and the mass ratio of the thinner to the raw material was 1.4:1.

PS is the mixture of an active component, an inorganic salt, and the unsulfonated oil. It was purified by a solvent extraction method. The inorganic salt and the unsulfonated oil were separated by solvents such as ethanol, petroleum ether, and n-pentane to obtain the NPS active component. The PS was composed of 78.5 wt.% active component, 8.8 wt.% inorganic salt, and 11.3 wt.% unsulfonated oil.

Separation of PS

With reference to the solvent extraction method

determining the equivalent weight distribution of petroleum sulfonate [19], the PS active component was extracted sequentially with different solvents such as benzene, chloroform, and water. A solution of 50 g purified PS (thoroughly deoiled and desalted) in a 750 mL 50 vol.% isopropanol/water mixture was extracted with 700 mL benzene. The benzene layer was removed, and the solvent was evaporated to obtain the benzene extraction part A. The remaining isopropanol/water solution was extracted with 400 mL chloroform to yield chloroform extraction B, and the remaining solution was evaporated to obtain the chloroform residue. The chloroform residue was dissolved in 750 mL sec-butyl alcohol solution saturating with water, and this solution was extracted with 450 mL water to yield water extraction C; the evaporation of the sec-butyl alcohol/water phase yielded the water residue D. The solvents were removed from all the phases, and the residues were thoroughly dried and labeled as fractions A, B, C, and D, which possessed different average molecular weights.

Measurement of Average Molecular Weight of PS and its Fractions

The average molecular weights of the PS and its fractions, namely fractions A, B, C, and D, were measured by the two-phase titration method [20]. The indicator was a mixture solution of dimidium bromide, Erioglaucine A, and H_2SO_4, and the titrating solution was Hyamine1622 solution. The distribution of the average molecular weights of PS and fractions A, B, C, and D were determined by ESI-MS method. Atmospheric pressure ionization electrospray (API-ES) was used as an ionization source, and an 80 vol.% methanol/water solution was regarded as the mobile phase. The flow rate and temperature of the dry gas was 6 L/min and 350 °C respectively, while the nebulizer pressure was 35 psig. The average molecular weights of PS and its fractions were calculated by the following equation:

$$\bar{M} = \sum_{i=a}^{b} M_i \frac{A_i}{\sum_{i=a}^{b} A_i} + 23 \quad (1)$$

where, $\bar{M}$ is the average molecular weight, and M_i is the m/z value; A_i is the abundance.

Characterization of the Molecular Structure of PS and its Fractions

PS and its fractions were reacted with hydrochloric acid to replace Na with H, and petroleum sulfonic acid was obtained. Then, the molecular structure of the petroleum sulfonic acid was respectively characterized by ^{1}HNMR and ^{13}CNMR spectrometry. ^{13}CNMR was obtained with reverse gated decoupling method, and $CDCl_3$ was used as the solvent in NMR measurement

Static Adsorption Measurement of PS and its Fractions

10 grams reservoir sand was mixed with 100 grams PS solution in a ground glass plug Erlenmeyer flask, and it was then shaken for 20 hrs in a water bath at 80 °C. The solution was centrifuged at 3000 rpm for 30 min to separate the reservoir sand. The supernatant was then separated, and the concentration of PS was measured by the two-phase titration. The adsorption amount was calculated according to the concentration change of the PS.

Core-flooding Experiment in Laboratory

The cores were saturated with formation water at a flow rate of 2.0 mL/min at 80 °C to determine pore volume and porosity, and then crude oil was injected until no more water was produced. Then, the formation water was injected until the water content of the produced fluid reached 98%. After water flooding, 1 PV petroleum sulfonate solution was injected to displace the residual oil. At last, the formation water was injected again until the water content of the produced fluid reached 98%. The oil recovery ratio was calculated according to the amount of crude oil displaced.

RESULTS AND DISCUSSION

Molecular Structure of PS and its Fractions

The purified PS was separated into fractions A, B, C, and D which possessed different average molecular weights. The total yield was 99.5%. The average molecular weight of PS and its fractions were measured by two-phase titration method and ESI-MS method simultaneously. The results are given in Table 4. It is clear that the average molecular weight values determined by the two methods are quite close.

The chemical composition and molecular structure of petroleum sulfonate are highly complicated due to the complex composition of the raw materials. Sperling mentioned that it was a compound with a structure changing from $C_{23}H_{38}SO_3H$ to $C_{31}H_{48}SO_3H$, and it was mainly monosulfonate. In this research, the element composition, the ^{1}HNMR, and the ^{13}CNMR spectra of PS and the fractions were measured, and the results are listed in Table 5. Based on the reference method, several important parameters, including average structural parameters of petroleum via NMR spectra, the aromaticity (f_A), and branchiness index (*BI*) were estimated. The average number of hydrogen atoms attached to aromatic carbon (H_{m}) was calculated from the NMR spectra. The f_A value was determined by Brown-Ladner method based on the analysis result of ^{1}HNMR and was double-checked by the number of aromatic carbon and saturated carbon by means of ^{13}CNMR [21]. Then, the range of the average number of carbons(C_T), aromatic carbons(C_A), saturated carbons(C_S), methyls (C_M), and sulfonic groups (N_{SO_3}) were obtained according to the average molecular weight determined by the two-phase titration method.

Table 4: Mass percent and average molecular weight of PS and the fractions.

Sample	Percent (wt.%)	Average molecular weight	
		Two-phase titration method	ESI-MS
PS	100	437	429
A	15.8	666	632
B	35.8	424	412
C	14.3	507	493
D	33.6	402	394

Table 5: The element composition and ^{1}HNMR, ^{13}CNMR data of PS and its fractions.

Sample	H(wt.%)	C(wt.%)	S(wt.%)	n(C)/n(H)	n(C)/n(H) corrected value	H_A	H_α	H_β	H_γ	A_A	A_s
PS	8.22	68.06	8.10	0.69	0.72	0.069	0.167	0.485	0.279	2.00	3.45
A	9.72	75.45	5.15	0.65	0.65	0.050	0.197	0.471	0.282	2.00	5.09
B	8.16	63.73	7.57	0.65	0.66	0.053	0.260	0.433	0.254	2.00	4.80
C	7.56	63.45	9.76	0.70	0.71	0.086	0.295	0.432	0.187	2.00	3.31
D	8.07	65.85	8.09	0.68	0.69	0.064	0.275	0.426	0.235	2.00	3.80

In Table 5, H_A stands for hydrogens attached to aromatic ring, and H_α, H_β, and H_γ respectively represent hydrogens attached to aromatic ring carbons in alkyl substitution α, in alkyl substitution β, and in alkyl substitution γ; they correspond to the peak area with chemical shifts around 6.0-9.0, 2.0-4.0, 1.0-2.0, and 0.5-1.0 ppm respectively in the ^{1}HNMR spectrum. A_A and A_s respectively represent aromatic carbons and saturated carbons, which respectively correspond to the peak area with chemical shifts around 110-160 and 0-50 ppm in ^{13}CNMR spectrum. The structural parameters of PS and its fractions are listed in Table 6. It is shown that f_A values determined by ^{1}HNMR and ^{13}CNMR spectra are very close, which shows good consistency. In this research, f_A value determined by ^{13}CNMR spectra was used, while the other parameters were calculated.

Table 6: Average structure parameters of PS and the fractions.

Sample	f_A		BI	Hm	C_T	C_A	C_S	C_M	N_{SO_3}
	^{1}HNMR	^{13}CNMR							
PS	0.356	0.367	0.285	0.26	23.9-25.7	8.8-9.4	15.1-16.3	3.4-3.6	1.05
A	0.270	0.282	0.282	0.27	40.3-43.3	11.3-12.2	28.9-31.1	6.3-6.8	1.01
B	0.284	0.294	0.244	0.27	22.9-24.7	6.7-7.3	16.2-17.4	3.2-3.4	0.94
C	0.358	0.377	0.171	0.32	28.8-31.0	10.9-11.7	18.0-19.4	2.6-2.8	1.26
D	0.321	0.345	0.224	0.27	21.3-23.0	7.4-7.9	14.0-15.0	2.6-2.8	0.96

$$f_A = \frac{n(C)/n(H) - (H_\alpha + H_\beta + H_\gamma)/2}{n(C)/n(H)} \quad (^1\text{HNMR}) \tag{2}$$

$$f_A = \frac{A_A}{A_A + A_S} \quad (^{13}\text{CNMR}) \tag{3}$$

$$BI = \frac{1/3H_\gamma}{1/2(H_\alpha + H_\beta)} \tag{4}$$

$$H_m = \frac{H_A}{n(C)/n(H) \times f_A} \tag{5}$$

$$C_{T(max)} = (\mathrm{M} - 103)/13 \quad \text{(Assuming all the carbon atoms to be aromatic carbon.)} \tag{6}$$

$$C_{T(min)} = (\mathrm{M} - 103)/14 \quad \text{(Assuming all the carbon atoms to be aliphatic carbon)} \tag{7}$$

$$C_A = C_T \times f_A \quad (8)$$

$$C_S = C_T - C_A \quad (9)$$

$$C_M = C_S \times BI/((1+BI)) \quad (10)$$

$$N_{SO_3} = (M-22) \times (\omega(S)\%)/32 \quad (11)$$

Surface Tension and CMC Value of PS and its Fractions

The surface tension of PS and the fractions in aqueous solutions were measured, and the critical micellar concentration (CMC) was determined from surface tension-concentration curves (Figure 1). The results are shown in Table 7. The surface tension of fraction C is much higher than that of fractions A, B, and D. In addition, the surface tensions of fractions A, B, and D at are very close low concentrations; the surface tension of fraction A is slightly lower than the surface tensions of the others at high concentrations. As found in Table 7, the CMC value of fraction C is higher compared to the other fractions; thus, the surface activity of fraction C is the lowest among the PS fractions.

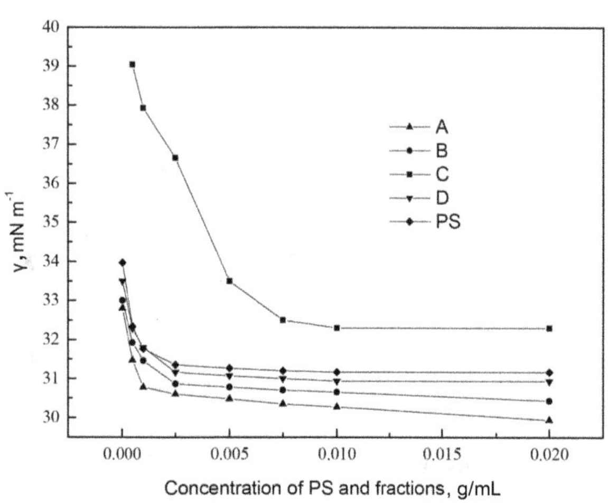

Figure 1: Surface-tension curves of PS and its fractions.

Table 7: CMC and the surface-tension value at the CMC of PS and its fractions.

Sample	CMC (g.mL^{-1})	γ_{CMC} (mN.m^{-1})
PS	0.0025	31.4
A	0.0010	30.8
B	0.0025	30.9
C	0.0075	32.5
D	0.0025	31.2

IFT among PS, the Fraction Solutions, and Crude Oil

Figure 2 shows the IFT among PS; solutions with fractions A, B, C, and D; and crude oil. It was found out that when PS concentration was more than 0.0025 g/mL, the IFT value slightly increased with an increase in concentration. The appearance of the inflection point of IFT curves comes from the formation of a large amount of micelles in surfactant solution, and the concentration at this point is also the critical micellar concentration [14]. Meanwhile, the concentration of petroleum sulfonate monomer reaches the maximum resulting in the lower oil-water IFT. Figure 2 also indicates that the interfacial activity of the fractions shows a significant difference. Among four fractions, the interfacial activity of fraction A is the lowest, while that of fraction C is the highest; the interfacial activities of fractions B and D are quite close. The interfacial activity of the fractions shows the above difference due to their different chemical structures. When surfactant molecules start to form micelles in the solution, and its distribution coefficient between oil and aqueous phase is close to 1, the IFT achieves the minimum value. The interaction between the hydrophobic groups of surfactant molecules induces the formation of micelles. The hydrophobic effect between surfactant molecules is strengthened as the length of hydrophobic chain increases, which enhances the oil-solubility of hydrophobic groups. As a result, surfactant molecules are adsorbed onto oil-water interface more easily, and the CMC of this fraction is reduced. When the number of carbon atoms in hydrocarbon chains of an ionic surfactant is within the range of 8-16, as a homologous compound chain adds one carbon atom, the CMC decreases to half compared with the previous value [14]. Jiang et al. studied the surface properties of sodium polyalkylbenzene sulfonates [17]. They found out that with an increase in the number of carbons in the main chain or the side chains, the degree of CMC reduction is totally different. They reported the influence of the number of carbons of the main chain on the CMC in accordance with the above principles; they also found out that the effect of the side chains in reducing CMC is weaker compared to the main chains. It has been found from Tables 4 and 5 that the more saturated carbon the fraction possesses, the lower the CMC value is, which is satisfied with the above theory. The average number of the saturated carbon of fraction A is much more than that of the other fractions; therefore, the interaction between hydrophobic groups is the strongest, and the oil-solubility is the greatest; the efficiency in reducing IFT is also increased, and CMC value is the lowest. Furthermore, the average number of saturated carbons for fraction A is about fifteen times more than that of fractions B and D, yet the CMC of fraction A is only half of the CMC of fractions B and D, which is perhaps due to the great branchiness index of A. The increase of branched chain enhances the spatial impediment for the petroleum sulfonate molecules to form micelles. Thus, the CMC value of fraction A is slightly lower than that of the other fractions. The average number of the saturated carbons and the branchiness index of fractions B and D are quietly close, and consequently their CMC value is nearly equivalent. The average number of the saturated carbons of fraction C is more than that of fractions B and D, but the CMC value reaches 0.0075 g/mL, which is much higher than the other fractions. Table 4 shows that the number of sulfonic groups of fractions A, B, and D is 1.01, 0.94, and 0.96 respectively, which indicates that they are all monosulfonates. The number of sulfonic groups of fraction C is 1.26, thereby containing a certain percentage of disulfonate. The presence of disulfonate greatly increased the hydrophilicity of petroleum sulfonate molecules and destroyed the original hydrophilic-lipophilic balance. Thus, the average number of the saturated carbon of fraction C is larger compared to the other fractions, while the CMC value is much higher.

The difference in IFT among the PS fractions is also explained by their chemical structures. The structure of the interfacial adsorbed layer, that is the arraying state and the tightness degree of hydrophobic chain present in interfacial adsorbed layer, determines the IFT between surfactant solutions and oil. The difference in the structure of the hydrophobic groups has a significant effect on IFT value, which depends on whether they have an existing branching structure and on the difference in the structure of the end groups. Branching structure can result in a decrease in IFT [14]. Fraction A has a high branchiness index and the greatest average number of methyl. As a result, covering rate (especially CH_3) on oil-water interface is higher, and the array of hydrocarbon chain is much tighter; thus the IFT value is much lower.

Among all the fractions, fraction C has the lowest branchiness index. Furthermore, the presence of disulfonate increases the hydrophilicity of molecules, and the ratio of hydrophilic-lipophilic portion is unbalanced; thus, it is very difficult for the molecules to be adsorbed onto interface. As a result, the IFT of fraction C solution is very high.

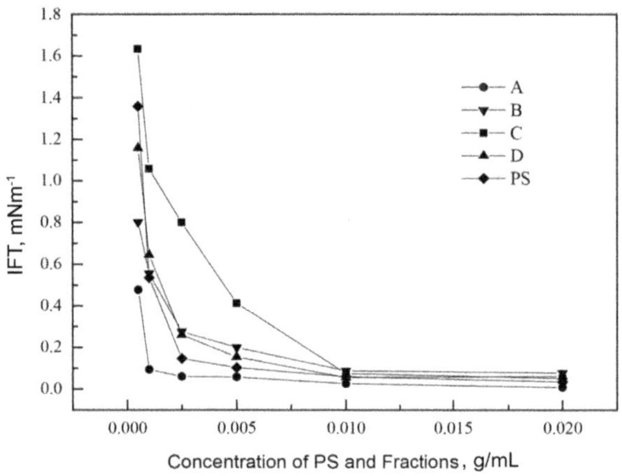

Figure 2: IFT curves of PS and its fractions.

Adsorption Isotherm of PS and the Fractions

Figure 3 shows the adsorption isotherm of PS and the fractions on Shengli reservoir sand at 70 °C. It was found out that within the concentration range of 0.0005-0.0080 g/mL, the adsorption isotherm of fractions A and C is quite similar with that of PS. Their adsorption amount decreased with a further increase in concentration after reaching the maximum value near the concentration of CMC. However, the adsorption amount of fractions B and D decreases after the maximum value, and then rapidly rises. The adsorption isotherm of solid solute was summarized into four types and eighteen varieties [22], and that of fractions B and D belongs to type S, which exhibits the maximum adsorption. The adsorption thermodynamics of anionic surfactant on mineral surfaces have also been studied [23-26]. Liu et al. determined adsorption isotherm of two kinds of sulfonate (ORS-41 and B100) used as oil-displacement agents on oil sand and described their adsorption behavior by the nonuniform surface of two-step adsorption models [23], which showed an adsorption isotherm similar to fractions B and D. Gale separated petroleum sulfonate into three portions with different average molecular weights [27]. They reported that the high average molecular weight component exhibited the strongest interfacial activity and a very big adsorption on clay. It is inferred from Figure 3 that the sequence of the adsorption of the fractions follows the order of water extraction C> water residue D> chloroform extraction B> benzene extraction A. The interfacial activity of fraction A with the highest average molecular weight is the strongest among the four fractions. However, fraction A has the lowest adsorption amount, which is different from the conclusions of Gale. This is due to the difference in composition and structure of sulfonates. Both factors are the intrinsic factors influencing interfacial chemical properties. In the current research, fraction A owns the greatest average molecular weight and the average number of carbons, while Table 4 shows that the average number of aromatic carbons of fraction A is the highest among the fractions, and *H*m value is quite low. *H*m reflects the condensation degree of aromatic rings. The lower the *H*m is, the higher the condensation degree of aromatic rings is. The number of aromatic carbons is great, and the condensation degree of aromatic rings is high. As a result, fraction A contains the most numerous aromatic rings in molecules. Furthermore, its branchiness index is the highest, and the average number of methyls is much greater than that of the others. Thus, there are much more aromatic rings and branched chains in the molecular structure of fraction A. Compared with the other fractions, fraction A contains a greater sectional area, and occupies much larger surface area when it is adsorbed on reservoir sand. Therefore, at the same concentration fraction A has the lowest adsorption amount among the fractions. The number of the aromatic carbon of fraction C is close to that of fraction A, while fraction C has a higher *H*m value, and a lower condensation degree, which enables the number of aromatic carbons of fraction C not to be dominant. Furthermore, fraction C has the lowest branchiness index and the fewest number of methyls. Hence, fraction C does not own the large molecular sectional area. Table 6 shows that disulfonate presents fraction C molecules, in which one hydrocarbon chain is connected to two sulfonic groups, doubling the number of polar groups. When surfactant is adsorbed on a solid surface, the polar groups are adsorbed onto the surface by chemical adsorption or physical adsorption to form the adsorption layer with molecules being in directional arrangement; however, hydrophobic groups stretch into aqueous solution [28]. Thus, the adsorptive capacity of fraction C on a solid surface is stronger, and its adsorption amount is much higher than that of the other fractions.

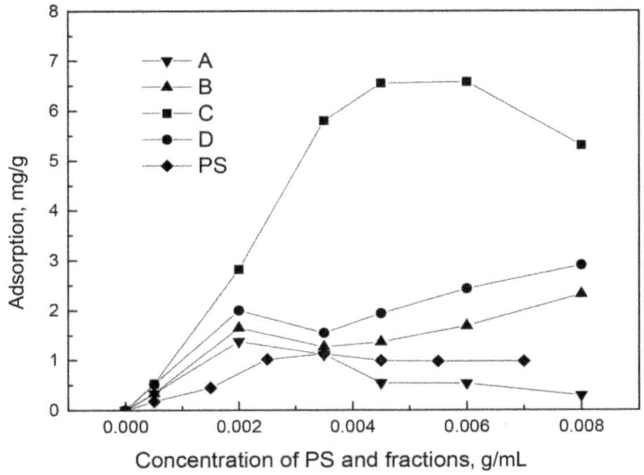

Figure 3: Adsorption isotherm of PS and its fractions.

The Oil Displacement Efficiency of PS and the Fractions in Core Flooding Experiment

The oil-displacement efficiency at a 0.0025 g/mL concentration of PS and the fractions were determined in core-flooding experiments. Furthermore, the different electrolytes such as Na_2CO_3 and Na_2SO_4 were respectively added into PS fraction solutions in order to investigate their effect on the oil-displacement property of PS. The results are shown in Table 8.

Table 8: Oil recovery performance via PS and its fraction with/without electrolytes.

Flooding Solutions	Increasing Oil Recovery (%)		Final Oil Recovery (%)
	Water flooding	Chemical Flooding	
A	38.1	15.8	53.9
B	42.7	10.8	53.5
C	39.2	6.2	45.4
D	39.7	11.2	50.9
PS	38.1	13.9	52.0
A+Na_2CO_3	39.3	31.1	70.4
B+Na_2CO_3	39.6	24.6	64.2
C+Na_2CO_3	38.9	19.9	58.8
D+Na_2CO_3	40.2	25.3	65.5
PS+Na_2CO_3	38.5	28.7	67.2
A+Na_2SO_4	39.7	20.8	60.5
B+Na_2SO_4	39.4	15.2	54.6
C+Na_2SO_4	40.8	10.9	51.7
D+Na_2SO_4	41.1	16.0	57.1
PS+Na_2SO_4	41.5	18.9	60.4

It is obvious from Table 8 that the oil recovery is enhanced to the maximum when fraction A is used, while fraction C shows the poorest performance. The oil-displacement efficiency of petroleum sulfonate mainly depends on the interfacial activity and adsorption property, and a surfactant system with the lower oil-water IFT and adsorption loss owns higher displacement efficiency. From the above research, it is concluded that fraction A has the lowest IFT value and the adsorption amount on the sand among all the fractions studied here; thus, it has the maximum oil recovery value. However, fraction C has the lowest interfacial activity and the highest adsorption amount; thus, it has the lowest oil-displacement efficiency. Furthermore, Table 8 shows that when the electrolytes are added to PS solution, the oil recovery increases, and the effect of alkali electrolyte such as Na_2CO_3 is more significant than the others. The reason for the change is that alkali reacts with organic acids in the oil phase to produce surface-active species, which exhibit synergism with PS; this results in a significant increase in the oil recovery value.

CONCLUSIONS

The main conclusions drawn based on the current study are as follows:

(1) PS was separated into four fractions with different average molecular weights. The fractions have a significant difference in their chemical structure. Benzene extraction A contains much more aromatic rings and branched chains. Water extraction C contains a certain percentage of disulfonate, while chloroform extraction B and water residue D exhibit a similar molecular structure.

(2) The PS fractions show a major difference in interfacial activity and adsorption property. Fraction A has a low IFT and CMC value, and it has the largest interfacial activity but the lowest adsorption amount on reservoir sand. Fraction C exhibits the lowest interfacial activity and the highest adsorption loss. The interfacial activity and adsorption amount of fractions B and D are quite close, and they lie between the values of fractions A and C.

(3) Among all the fractions, fraction A has the highest oil-displacement efficiency, and fraction C shows the lowest performance. The oil recovery is obviously enhanced with the addition of electrolytes such as Na_2CO_3 and Na_2SO_4 to PS solution because these electrolytes accelerate the formation of surface-active species in crude oil and increase the concentration of active species on interface. The alkali electrolyte such as Na_2CO_3 has a very significant effect on increasing oil-displacement efficiency.

NOMENCLATURES

BI	: Branchiness index
C_A	: Average number of aromatic carbons
C_M	: Average number of methyls
C_S	: Average number of saturated carbons
C_T	: Average number of carbons
f_A	: Aromaticity
Hm	: Average number of hydrogen atoms attached to aromatic carbon
N_{SO_3}	: Average number of sulfonic groups

ACKNOWLEDGMENTS

The author would like to thank Professor Weiyu Fan for his grateful guidance and help during the preparation of this paper.

REFERENCES

1. Rai K., Johns R. T., and Lake L. W., "Oil-recovery Predictions for Surfactant Polymer Flooding," SPE-124001, **2009**, 1-12.
2. Wyatt K., Pitts M. J., and Surkalo H., "Mature Waterfloods Renew Oil Production by Alkaline-surfactant-polymer Flooding," SPE-78711, **2002**, 1-7.
3. Vargo J., Turner J., and Vergnani B., "Alkaline-surfactant-polymer Flooding of the Cambridge Minnelusa Field," SPE-68285, **2000**, 1-6.
4. Pitts M. J., Dowling P., and Wyatt K., "Alkaline-surfactant-polymer Flood of the Tanner Field," SPE-100004, **2006**, 1-5.
5. Sharma A., Aziz Y. A., and Clayton B., "The Design and Execution of an Alkaline-surfactant-polymer Pilot Test," SPE-154318, **2012**, 423-431.
6. Elraies K. A. and Tan I. M., "Design and Application of a New Acid-alkali Surfactant Flooding Formulation for Malaysian Reservoirs," SPE- 133005, **2010**, 1-6.
7. Wang C. L., Wang B. Y., and Cao X. L., "Application and Design of Alkaline-surfactant-polymer System to Close Well Spacing Pilot Gudong Oil Field," Presented at *SPE Western Regional Meeting*, U.S.A., **1997**, 605-619.
8. Wang D. M., Cheng J. C., and Wu J. Z., "An Alkaline/surfactant/polymer Field Test in a Reservoir With a Long-term 100% Water Cut," presented at *SPE Annual Technical Conference and Exhibition*, U.S.A., **1998**, 305-318..
9. Earl W. M., Carolyn C. G., and David M. F., "Characterization and Oil Recovery Observations on a Series of Synthetic Petroleum Sulfonates," *Society of Petroleum Engineers Journal*, **1982**, *22*, 226-236.
10. Hsieh W. C., "Process for Enhanced Oil Recovery Employing Petroleum Sulfonate Blends," U.S. Patent, 4446036, **1984**.
11. Nuckels N., Byth N. J., and Thompson J. L. "Surfactant Compositions Useful in Enhanced Oil Recovery Process," U. S. Patent, 4532051, **1985**.
12. Wang Y., Qiao Q., and Dong L., "A Low Cost ASP Flooding System for Industrial Extending EOR Test at East Block of District VII in Karamay," *Oilfield Chemistry*, **1999**, *16*, 40-42.
13. Wang J., Luo P. Y., and Zheng Y., "The Performance Properties of ASP and Polymer Flooding Solutions Prepared with Hydrophobically Associating Amphoteric Polymers for EOR in Daqing," *Oilfield Chemistry*, **2000**, *17*, 168-170.
14. Zhao G. X. and Zhu B. Y., "Principles of Surfactant Action (2nd ed.)," China Light Industry Press, [Printed copy], **2003**, 1-764.
15. Chen Y. M., Jiao L. M., and Li Z. P., "Synergism Effect of Polar Fractions in Petroleum Sulfonates," *Acta Petrolei Sinica.*, **1999**, *20*, 73-77.
16. Wang L., Gong Q. T., and Wang D. X., "Synthesis and Characterization of Sodium Branched-alkylbenzene Sulfonates," *Petrochemical Technology*, **2004**, *33*, 104-108.
17. Jiang X. M., Zhang L., and An J. Y., "Surface Properties of Sodium Alkylbenzenesulfonates with Additional Side Chains," *Acta Physico-Chimical Sinica.*, **2005**, *21*, 1426-1430.
18. Nagashima K. and Blum F. D., "Adsorption and Dynamics of Sodium Alkylbenzenesulfonates on Alumina," *Colloids and Surfaces A: Physicochem. Eng. Aspects.*, **2001**, *176*, 17-24.

19. Sandvik E. I. and Gale W. W., "Characterization of Petroleum Sulfonates," *Soc. Pet. Eng. J.*, **1977**, *17*, 184-192.
20. Brewer P. I., "The Determination of Oil-soluble Sulfonates by Two-phase Titration," *J. Inst. Pet.*, **1972**, *58*, 41-46.
21. Brown J. K. and Ladner W. R., "A Study of Hydrogen Distribution in Coal-like Materials by High-resolution Nuclear Magnetic Resonance Spectroscopy II-A Comparison with Infra-red Measurement and the Conversion to Carbon Structure," *Fuel*, **1960**, *39*, 87-96.
22. Lucassen E. H., "Physical Chemistry of Surfactant Action- Anionic Surfactants (2nd ed.)," [Printed copy], New York and Basel, **1998**, 149-150.
23. Liu K. Y., Zhang W. D., and Zhao Ch. J., "Adsorption Thermodynamics of Surfactants for EOR on Oil Sand," *Journal of Beijing University of Chemical Technology*, **1997**, *24*, 9-13.
24. Ziegler V. M. and Handy L. L., "Effect of Temperature on Surfactant Adsorption in Porous Media," Southern California, *Social Petroleum Engineers Journal*, **1981**, *21*, 218-226.
25. Scamehorn J. F. and Schechter R. S., "Adsorption of Surfactants on Mineral Oxide Surfaces from Aqueous Solutions: I: Isomerically Pure Anionic Surfactants," J. *Colloid. & Inter. Sci.*, **1982**, *85*, 463-478.
26. [Su H. Y., Zhang C. G., and Zhu Z. Q., "Adsorption Thermodynamics of Sodium Dodecylbenzene Sulfonate on Minerals," Presented at *the 2nd ISTCEI.*, U.S.A., **1994**, 360-362..
27. Gale W. W. and Sandvik E. I., "Tertiary Surfactant Flooding: Petroleum Sulfonate Composition-efficacy Studies," *Soc. Pet. Eng. J.*, **1973**, *13*, 191-199.
28. Han D. and Shen P. P., "Principles and Application of Surfactant Oil-displacement (2nd ed.)," **2001**, 80.

6

An Investigation of Asphaltene Deposition Mechanisms During Natural Depletion Process by a Two Phase Modeling Using Genetic Algorithm Technique

Sepideh Kashefi[1], Mohammad Nader Lotfollahi[1], and Abbas Shahrabadi[2*]

[1]Faculty of Chemical, Gas, and Petroleum Engineering, Semnan University, Semnan, Iran
[2]Exploration and Production Division, Research Institute of Petroleum Industry (RIPI), Tehran, Iran

ABSTRACT

In this work, the natural depletion process in sandstone and carbonate cores was modeled to investigate the asphaltene deposition mechanisms. A new permeability reduction correlation was proposed based on the Minssieux model that considers a combination of surface deposition, pore throat plugging, and filtration cake mechanisms. The results showed that the filtration cake is a dominant asphaltene deposition mechanism during natural depletion process in both core samples. Therefore, a modified model was proposed with adding formation of filtration cake mechanism due to pore filling to the Wang and Civan deposition model. The absolute average deviation (AAD (%)) for permeability reduction between the results of the three models (including new correlation, the modified model, and Wang and Civan model) and the experimental data were calculated and reported. These values for the three models were 3.28, 2.67, and 4.83% for sandstone core and 3.01, 2.58, and 4.69% for carbonate core respectively. The results showed that the modified model proposed in this study presented good performance for asphaltene deposition prediction

Keywords: Natural Depletion, Asphaltene Deposition, Modeling, Filtration Cake

INTRODUCTION

Asphaltene is a mixture of heavy components in the oil reservoirs that can be precipitated out of oil by changing pressure, temperature, and oil composition. Natural depletion is one of the usual processes of asphaltene deposition. In this process, asphaltene particles set on the rock surface due to pressure changes. Asphaltene precipitation causes permeability reduction and formation damage in oil reservoirs, and it ultimately affects the oil production [1-3]. To predict the amount of precipitated asphaltene, several thermodynamic models such as solubility model based on the Flory-Huggins polymer theory [4-7], solid model based on the solid-liquid equilibrium [8, 9], and thermodynamic models based on the various cubic and association equation of states [10-13] have been presented. On the other hand, several models have been proposed to simulate the permeability reduction as a result of asphaltene deposition in porous media. Gruesbeck and Collins proposed empirical correlations for calculating fluid velocities in pluggable and non-pluggable pathways.

***Corresponding author**
Abbas Shahrabadi
Email: Shahrabadia@ripi.ir

They found out that the permeability reduction is a function of the volumes of deposited asphaltene. They showed that the entrainment mechanism could decrease the production [14]. Wang and Civan developed a one-dimensional, three-phase asphaltene deposition model, including surface deposition, entrainment, and plugging terms [15]. Soulgani et al. investigated the effects of concentration, velocity, and temperature parameters on the asphaltene deposition and developed the surface deposition term based on these parameters [16]. Jafari Behbahani et al. developed a four-component, four-phase model based on the multilayer adsorption theory to describe formation damage during natural depletion and CO_2 injection processes. They showed that the oil and core types can affect the amount of asphaltene deposition during natural depletion process. They concluded that the live oil flooding caused more asphaltene deposition than dead oil flooding in porous media [17-19]. Khalifeh et al. investigated the asphaltene deposition on the carbonate and sandstone rock surfaces during dynamic flow tests. They simulated the permeability reduction by commercial software. They showed that the surface deposition and entrainment mechanisms were more important for asphaltene deposition process in carbonate rock than sandstone rock surfaces, and pore plugging mechanism had the same effect in both types of rocks [20]. Bagherzadeh et al. studied the permeability reduction mechanisms during simultaneous injection of CO_2 and recombined oil into sandstone and carbonate cores at reservoir conditions. They concluded that the surface deposition and pore plugging mechanisms were simultaneously effective on permeability reduction in initial injected pore volumes for both types of the rock. Thus, they presented a new model consisting of surface deposition and pore plugging mechanisms for permeability reduction predictions [21]. Recently Fallahnejad and Kharrat modeled the asphaltene deposition process, and they considered the asphaltene particles in three conditions (precipitated, flocculated, and deposited). They used asphaltene precipitation model (solid model), kinetic model (first-order chemical), and deposition model consisting of adsorption, pore throat plugging, and re-entrainment terms to predict porosity and permeability reduction during the natural depletion process [22].

In this work, the natural depletion process in a sandstone and carbonate core samples from the Iranian oil reservoirs was modeled and simulated under reservoir conditions to investigate the asphaltene deposition mechanisms. A new correlation was proposed based on the Minssieux model to calculate the weights of different deposition mechanisms and find out the dominant mechanism in asphaltene deposition process. A new model was also proposed by the modification of the Wang and Civan deposition model. The dominant mechanism (internal cake filtration due to pore filling) was added to Wang and Civan model to predict the asphaltene deposition process as a two-phase system. Finally, the permeability reduction predictions were validated by experimental data and compared with that of Wang and Civan model.

THEORETICAL CALCULATIONS

The live oil data from an oilfield in the southeast of Iran was used in this study. The oil compositions and properties of two different core flooding tests are given in Tables 1 and 2 respectively [17].

Table 1: Studied live oil compositions

Component	mol.%	Component	mol. %	Component	mol. %	Component	mol. %
H_2S	0	C_2	8.24	i-C_5	1.38	C_8	3.88
N_2	0.3	C_3	4.14	n-C_5	1.59	C_9	2.49
CO_2	1.83	i-C_4	1.19	C_6	6.95	C_{10}	4.03
C_1	22.7	n-C_4	3.61	C_7	4.1	C_{11}	2.85
ρ_{oil}=767.4 kg/m^3, μ_{oil}=4.35 cP, MW C_{12+} =491 g/gmole, SG C_{+12} =0.9853						C_{12+}	28.74
$P_{reservoir}$= 326 bar, $T_{reservoir}$=96 °C, P_{sat}=97 bar, MW $_{bottom\ hole\ live\ oil}$ = 182 g/gmole							

Table 2: Properties of the core flooding tests.

Core type	Length (cm)	Diameter (cm)	Porosity (%)	Permeability (md)	PV	$P_{backpressure}$ (bar)	Q (cm^3/hr.)
Sandstone	5.96	3.7	13.15	22.8	8.5	180	30
Carbonate	4.5	3.7	26	2.7	13.5	180	10

Dominant Mechanism in Asphaltene Deposition

In this section the dominant mechanism in asphaltene deposition process during natural depletion was found by Minssieux model [23]. This model consists of three possible mechanisms, including formation of filtration cake due to pore filling, surface deposition, and pore throat plugging as respectively given in Equations 1-3:

$$\frac{k}{k_0}=\frac{1}{1+A.(PVI)} \quad (1)$$

$$\sqrt{\frac{k}{k_0}}=1-B.(PVI) \quad (2)$$

$$\frac{k}{k_0}=1-C.(PVI) \quad (3)$$

where, $\frac{k}{k_0}$ is permeability reduction, and A, B, and C are filtration cake, surface deposition, and pore throat plugging mechanism's parameters respectively. PVI is the value for pore volume injection. The genetic algorithm method was used to find the model parameters (A, B, and C) and to investigate the dominant process during natural depletion.

A new permeability reduction correlation, which considers a combination of surface deposition, pore throat plugging, and filtration cake mechanisms, was also proposed to calculate the weights of different asphaltene deposition mechanisms as follows:

$$\frac{k}{k_0}=d_1.\left(\frac{1}{1+A.(PVI)}\right)+d_2.(1-B.(PVI))^2 + d_3.(1-C.(PVI)) \quad (4)$$

$$d_1+d_2+d_3=1 \quad (5)$$

where, d_1, d_2, and d_3 are weight fractions of filtration cake, surface deposition, and pore throat plugging mechanisms respectively. These parameters are between 0 and 1, and they are obtained by optimization using a genetic algorithm technique.

Thermodynamic Model for Asphaltene Precipitation

The solid model was employed to model the precipitation of asphaltene. In this model, the oil was divided to precipitating asphaltene in the solid phase and non-precipitating asphaltene in the liquid phase. The asphaltene component in the liquid and solid phases is in thermodynamic equilibrium. The precipitating asphaltene was modeled by multiphase flash calculation using PR and SRK equations of states. The fugacity of a precipitating component in the solid phase was calculated using the following equation [24]:

$$\ln f_s=\ln f_s^*+\frac{v_s(p-p^*)}{RT} \quad (6)$$

where, p is the pressure, and f_s is the solid fugacity; v_s is the solid molar volume, and R stands for the universal gas constant; T represents the temperature, and * superscript indicates the reference condition.

Dynamic Modeling of Asphaltene Deposition

The oil, gas, and asphaltene mass balance equations, momentum equations, porosity, and permeability models were used as a two-phase system at a constant temperature for dynamic modeling the asphaltene deposition in porous media as follows respectively [15]:

$$\frac{\partial}{\partial t}(\varnothing s_l \rho_l w_{ol}) + \frac{\partial}{\partial x}(\rho_l u_l w_{ol}) = 0 \tag{7}$$

$$\frac{\partial}{\partial t}\left(\varnothing s_v \rho_g + \varnothing s_l \rho_l w_{gl}\right) + \frac{\partial}{\partial x}\left(\rho_l u_l w_{gl} + \rho_l u_g\right) = 0 \tag{8}$$

$$\frac{\partial}{\partial t}(\varnothing s_l \rho_A C_A + \varnothing s_l \rho_l w_{Al}) + \frac{\partial}{\partial x}(\rho_l u_l w_{SAl} + \rho_l u_l w_{Al}) = -\rho_A \frac{\partial E_A}{\partial t} \tag{9}$$

$$u_l = -\frac{k k_{rl}}{\mu}\frac{\partial P}{\partial x} \tag{10}$$

$$u_v = -\frac{k k_{rv}}{\mu}\frac{\partial P}{\partial x} \tag{11}$$

$$\varnothing = \varnothing_0 - E_A \tag{12}$$

$$k = k_0\left(\frac{\varnothing}{\varnothing_0}\right)^3 \tag{13}$$

where, E_A is the volume fraction of deposited asphaltene in the bulk volume of the porous media. W_{Al} and W_{SAl} are the mass fractions of the dissolved and suspended asphaltene in the liquid phase respectively. W_{ol} and W_{gl} represent the mass fractions of the oil and gas in the liquid phase respectively. C_A stands for the volume fraction of the suspended asphaltene precipitated in the liquid phase. ρ_l, ρ_g, and ρ_A represent the oil, gas, and asphaltene densities respectively. $\varnothing$ is the porosity, and U_l and U_g stand for the liquid and gas velocity respectively. S_l and S_v are the liquid and vapor phase saturations respectively. μ is the viscosity, and k is the permeability. K_{rl} and K_{rv} represent the relative permeabilities of the liquid and vapor phases respectively.

The Wang and Civan model was used as the asphaltene deposition model [15]:

$$\frac{\partial E_A}{\partial t} = \alpha \varnothing s_l w_{SAl}\frac{\rho_l}{\rho_A} - \beta E_A(v - v_{cr}) + \gamma s_l u_l w_{SAl}\frac{\rho_l}{\rho_A} \tag{14}$$

$$v = \frac{u_l}{\phi} \tag{15}$$

$$\beta = \beta_i \quad \text{if } \upsilon > \upsilon_{cr}, \quad \text{else } \beta = 0 \tag{16}$$

$$\gamma = \gamma_i(1 + \sigma E_A) \quad \text{if } D_{pt} \le D_{ptcr}, \text{ else } \gamma = 0 \tag{17}$$

v and v_{cr} are the interstitial and critical interstitial velocity of liquid phase respectively, and σ represents the snowball-effect deposition constant. D_{pt} and D_{ptcr} stand for the pore throat diameter and critical pore throat diameter respectively. The first, second, and third terms in the asphaltene deposition model (the right hand side of Equation 14) are the surface deposition rate, entrainment rate, and pore throat plugging rate respectively, the parameters of which are correspondingly α, β, and γ.

An auxiliary equation was required to explain the relation between vapor and liquid phase saturations (Equation 18). Saturation is a measure of the fluid content of the porous rock [25]. The liquid and vapor saturations were calculated by flash calculation at different pressures during natural depletion process.

$$s_V + s_l = 1 \tag{18}$$

The boundary and initial conditions can be considered as follows:

$$\varnothing = \varnothing_0, \; k = k_0, \; E_A = 0 \; : 0 \le x \le L, \; t = 0 \tag{19}$$

$$u_l = \frac{Q}{A} \; : x = 0, \; t > 0 \tag{20}$$

$$p = p_{\text{back pressure}} \; : x = L, \; t > 0 \tag{21}$$

Modification of Wang and Civan Deposition Model

A modified model was proposed with adding internal cake filtration mechanism due to pore filling to the Wang and Civan model for predicting the permeability reduction during natural depletion process (Equation 22). The pore filling, following the pore throat plugging, leads to an internal cake formation. This mechanism was presented by Chang and Civan [26], but it has not been used simultaneously with surface deposition, entrainment, and pore throat plugging mechanisms for predicting asphaltene deposition up to now.

$$\frac{\partial E_A}{\partial t} = \alpha \varnothing s_l w_{SAl}\frac{\rho_l}{\rho_A} - \beta E_A(v - v_{cr}) + \gamma s_l u_l w_{SAl}\frac{\rho_l}{\rho_A} + \lambda \varnothing s_l u_l w_{SAl}\frac{\rho_l}{\rho_A} \tag{22}$$

$$\lambda \neq 0 \quad \text{if } t > t_{cr} \text{ or } b < b_{cr}, \text{ else } \lambda = 0 \quad (23)$$

$$b = \frac{D_{pt}}{D_{asp}} \quad (24)$$

The fourth term in the modified asphaltene deposition model (the right hand side of Equation 22), is the pore filling (or internal filtration cake) rate (λ). t_{cr} represents the critical time when the pore throat is first jammed by asphaltene particles. D_{asp} is an asphaltene particle diameter, and b_{cr} is the critical value below which (b_{cr} = 7) the filtration cake is occurred [27].

Values of asphaltene particle size depend on the method of measurement and also on the solvent type. Leontaritis et al. concluded that the asphaltene size distributions before asphaltene flocculation are about 20 to 350 Å, while after flocculation, asphaltene particles larger than 1000 Å are formed [28]. Leontaritis presented asphaltene particle size distribution versus asphaltene fraction in the oil reservoir at different pressures [29]. Therefore, based on the literature studies [28, 29] and asphaltene weight fraction value of using oil (16.3 wt.%), asphaltene particle diameter is considered 1000 Å.

Numerical Solution

A computer program was written and used to solve the above equations simultaneously by backward implicit finite difference method. Numerical simulation was carried out in a Cartesian system with 200 grid blocks and 50 time steps with a Δt of 100 seconds. The genetic algorithm technique was used as the optimization tool for the determination of the model parameters (α, β, v_{cr}, γ_i, σ, and λ). In the genetic algorithm method, the numbers of generation, population size, and elite count were taken 1000, 100, and 2 respectively. Population size determines the number of individuals in each generation. The selection of a large population size in the calculation causes the genetic algorithm to search the solution region completely in order to find the global minimum. Elite count is the number of individuals with the best fitness values in the current generation insured to survive to the next generation. The rank method was used for fitness scaling, while stochastic uniform was used as a method to specify the next generation. In this study, the objective function is a root mean squared error (RMSE):

$$\text{RMSE} = \sqrt{\frac{\text{SSE}}{m}} \quad (25)$$

$$\text{SSE} = \sum_{i=1}^{m} \left(\left(\frac{k}{k_0}\right)_{exp,i} - \left(\frac{k}{k_0}\right)_{model,i} \right)^2 \quad (26)$$

where, SSE is the sum of squared error. $\left(\frac{k}{k_0}\right)_{model,i}$ and $\left(\frac{k}{k_0}\right)_{exp,i}$ represents the model and the experimental values of permeability reduction respectively, and m is the number of experimental points.

RESULTS AND DISCUSSION

In this study, the permeability reduction for different asphaltene deposition mechanisms was calculated by Minssieux model. Table 3 shows the optimization values of Minssieux model parameters.

Table 3: Results of Minssieux model parameters by genetic algorithm.

Parameter (1/s)	Sand stone core	Fitness value (R^2)	Carbonate core	Fitness value (R^2)
Filtration cake parameter (A)	0.18667	0.992	0.48675	0.996
Surface deposition parameter (B)	0.05642	0.973	0.18751	0.987
Pore throat plugging parameter (C)	0.09708	0.984	0.34375	0.989

Figures 1a and 1b show the predictions of the permeability reduction by each mechanism correlation in the sandstone and carbonate cores respectively.

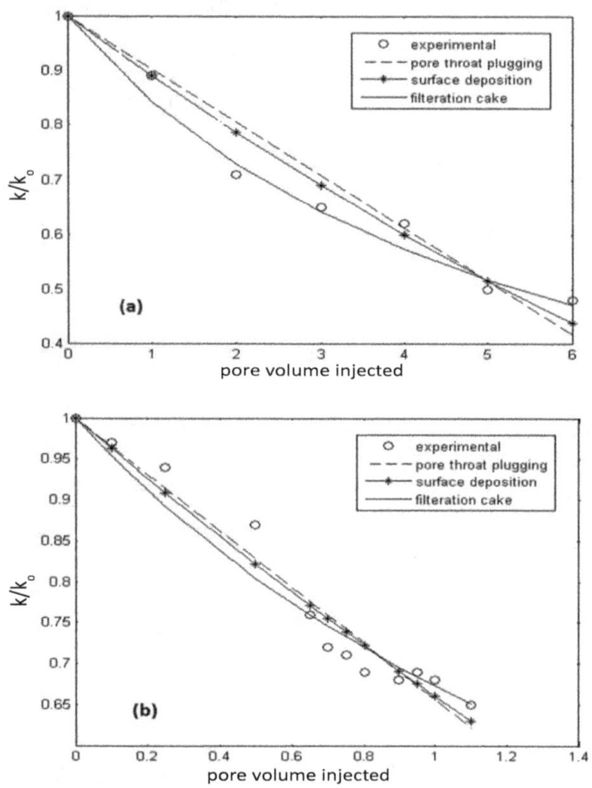

Figure 1: The experimental and calculated permeability reduction by Minssieux model in (a) sandstone core (b) carbonate core.

The results show that the formation of filtration cake due to pore filling is dominant in the deposition of asphaltene with the highest fitting values to the experimental data (R^2>0.99) in both sandstone and carbonate core samples. The weights of asphaltene deposition mechanisms were estimated by the proposed correlation, which considers a combination of surface deposition, pore throat plugging, and filtration cake mechanisms. The proposed correlation parameters (d_1,d_2, and d_3) were optimized by the genetic algorithm technique. Table 4 shows the optimum values of the parameters in the proposed correlation.

Table 4: Results of the proposed correlation parameters by genetic algorithm.

Parameter	Sandstone core	Carbonate core
d_1	0.82	0.59
d_2	0.04	0.19
d_3	0.14	0.22

The results demonstrate that the controlling mechanism during asphaltene deposition followed the order of filtration cake >> pore throat plugging > surface deposition for both sandstone and carbonate core samples. Also, it can be concluded that the surface deposition and pore throat plugging mechanisms have higher effects on asphaltene deposition in carbonate core than in the sandstone core. This may be due to the fact that the carbonate core surface contains polar groups and asphaltene is a polar component. Thus, the π-interaction and hydrogen bonding can form between the asphaltene and core surface polar groups, and they increase the amount of asphaltene deposition on the surface of the carbonate rock samples [30]. The obtained results are in good agreement with literature studies [19-21].

After finding the dominant mechanism in asphaltene deposition process, the precipitation of asphaltene under different pressures was calculated by using solid model and using PR and SRK equations of state. Figure 2 shows the weight fraction of precipitated asphaltene versus pressure. The results show that the same results are obtained by using PR and SRK equations of state. The results of thermodynamic modeling were applied to asphaltene dynamic modeling so as to calculate the precipitated asphaltene.

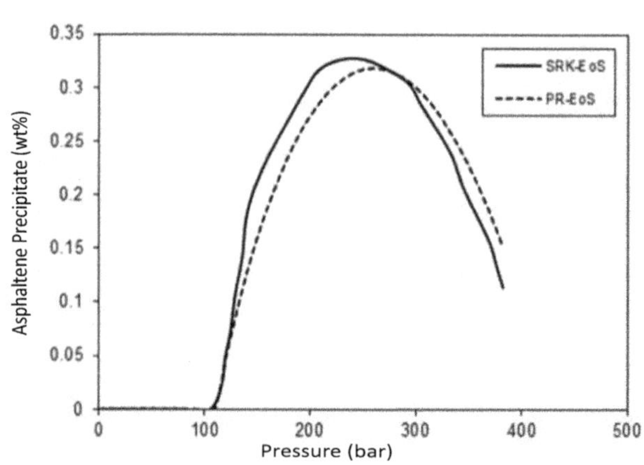

Figure 2: Asphaltene thermodynamic modeling by solid model using PR and SRK equations of state.

For dynamic modeling, a modified model was presented with adding dominant mechanism (internal cake filtration due to pore filling) to Wang and Civan model for predicting asphaltene deposition process. The genetic algorithm technique was used for determining model parameters. Table 5 shows the optimum values of tuning parameters in Wang and Civan and our proposed modified models. Figures 3a and 3b show the predictions of permeability reduction by the proposed correlation, Wang and Civan model, and the proposed modified model during natural depletion process in sandstone and carbonate cores respectively.

Table 5: Asphaltene deposition model parameters obtained by genetic algorithm.

Model para.	Sandstone core		Carbonate core	
	Wang and Civan model	Proposed modified model	Wang and Civan model	Proposed modified model
α (1/s)	0.00085	0.00073	0.00041	0.00049
β (1/cm)	0.04772	0.04485	0.00179	0.03252
U_{cr} ($\frac{cm}{s}$)	0.00194	0.00446	0.00740	0.101494
γ_i (1/cm)	0.00284	0.00217	0.04773	0.06087
σ	0.03994	0.02226	0.01116	0.07060
λ	-	0.08961	-	0.0989

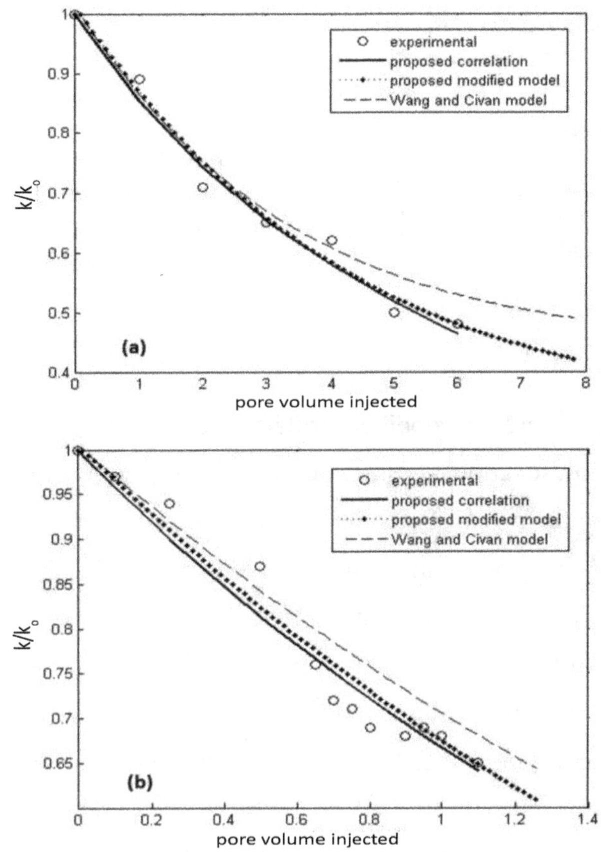

Figure 3: Experimental and calculated permeability reduction by the proposed correlation, Wang and Civan model, and the proposed modified model in (a) sandstone core (b) carbonate core.

Average absolute deviations (AAD (%)) of the predicted permeability reduction from the experimental data using three different models are presented in Table 6. These results confirm that the proposed modified model based on the filtration cake mechanism is capable of correlating the permeability reduction experimental data with an AAD of 2.67% and 2.58% in sandstone and carbonate cores respectively. The proposed correlation based on the combination of three mechanisms (surface deposition, pore throat plugging, and filtration cake) correlated asphaltene deposition experimental data with an AAD of 3.28% and 3.01% in sandstone and carbonate cores respectively. However, the Wang and Civan model has an AAD of 4.83% in sandstone core and an AAD of 4.69% in carbonate core.

Table 6: Average absolute deviation of the correlated permeability reduction from the experimental results by the proposed correlation, proposed modified model, and Wang and Civan model.

Core type	Proposed correlation (AAD (%))	Proposed modified model (AAD (%))	Wang and Civan model (AAD (%))
Sandstone	3.28	2.67	4.83
Carbonate	3.01	2.58	4.69

Therefore, the modified model proposed in this study showed good accuracy for asphaltene deposition prediction in comparison with the other models. This suggests that the filtration cake mechanism has an important role in asphaltene deposition process. Moreover, the obtained results indicate that the proposed modified model has similar effects on asphaltene deposition prediction in carbonate core samples with an AAD of 2.58% in comparison to sandstone core samples with an AAD of 2.67%.

CONCLUSIONS

A new correlation based on the Minssuiex model was proposed to investigate the permeability reduction during natural depletion process. The results showed that the filtration cake formation is a dominant mechanism in formation damage. Furthermore, the surface deposition and pore throat plugging mechanisms indicated more effects on asphaltene deposition process in carbonate core comparing to sandstone core sample. Therefore, the new modified model was proposed with adding filtration cake mechanism, due to pore filling, to Wang and Civan deposition model. It was found that the proposed modified model was more accurate than Wang and Civan model. It can be concluded that the formation of filtration cake mechanism has the highest effect on asphaltene deposition in both sandstone and carbonate core samples.

REFERENCES

1. Danesh A., Krinis D., Henderson G. D., and Peden J. M., "Pore-level Visual Investigation of Miscible and Immiscible Displacements," *Journal of Petroleum Science and Engineering*, **1989**, *2*, 167-177.
2. Burke N. E., Hobbs R. E., and Kashou S. F., "Measurement and Modeling of Asphaltene Precipitation," *Journal of Petroleum Technology*, **1990**, *42*, 1440-1446.
3. De Boer R., Leerlooyer K., Eigner M., and Van Bergen A., "Screening of Crude Oils for Asphalt Precipitation: Theory, Practice, and Selection of Inhibitors," *SPE Production and Facilities*, **1995**, *10*, 55-61.
4. Flory P. J., "Thermodynamics of High Polymer Solutions," *The Journal of Chemical Physics*, **1942**, *10*, 51-61.
5. Mousavi-Dehghani S. A., Mirzayi B., and Vafaie-Sefti M., "Polymer Solution and Lattice Theory Applications for Modeling of Asphaltene Precipitation in Petroleum Mixtures," *Brazilian Journal of Chemical Engineering*, **2008**, *25*, 523-534.
6. Mousavi-Dehghani S. A., Mirzayi B., Mousavi S. M. H., and Fasih M., "An Applied and Efficient Model for Asphaltene Precipitation in Production and Miscible Gas Injection Processes," *Petroleum Science and Technology*, **2010**, *28*, 113-124.
7. Nikookar M., Pazuki G. R., Omidkhah M. R., and Sahranavard L., "Modification of a Thermodynamic Model and an Equation of State for Accurate Calculation of Asphaltene Precipitation Phase Behavior," *Fuel*, **2008**, *87*, 85-91.
8. Thomas F. B., Bennion D. B., Bennion D. W., and Hunter B. E., "Efficient Modeling of Asphaltene Precipitation," *Journal of Canadian Petroleum Technology*, **1992**, *31*, 22.
9. Moradi S., Dabiri M., Dabir B., Rashtchian D., et al., "Investigation of Asphaltene Precipitation in Miscible Gas Injection Processes: Experimental Study and Modeling," *Brazilian Journal of Chemical Engineering*, **2012**, *29*, 665-676.
10. Shirani B., Nikazar M., and Mousavi-Dehghani S.A., "Prediction of Asphaltene Phase Behavior in Live Oil with CPA Equation of State," *Fuel*,

2012, **97**, 89-96.

11. Shirani B., Nikazar M., Naseri A., and Mousavi-Dehghani S. A., "Modeling of Asphaltene Precipitation Utilizing Association Equation of State," *Fuel*, **2012**, *93*, 59-66.
12. Vafaie-Sefti M. and Mousavi-Dehghani S. A., "Application of Association Theory to the Prediction of Asphaltene Deposition: Deposition Due to Natural Depletion and Miscible Gas Injection Processes in Petroleum Reservoirs," *Fluid Phase Equilibria*, **2006**, *247*, 182-189.
13. Jafari Behbahani T., Ghotbi C., Taghikhani V., and Shahrabadi A., "Experimental Investigation and Thermodynamic Modeling of Asphaltene Precipitation," *Scientia Iranica*, **2011**, *18*, 1384-1390.
14. Gruesbeck C. and Collins R. E., "Entrainment and Deposition of Fine Particles in Porous Media," *Society of Petroleum Engineers Journal*, **1982**, *22*, 847-856.
15. Wang S. and Civan F., "Productivity Decline of Vertical and Horizontal Wells by Asphaltene Deposition in Petroleum Reservoirs," SPE international Symposium on Oilfield Chemistry: Society of Petroleum Engineers, Houston, Texas, **2001**.
16. Soulgani B.S., Tohidi B., Jamialahmadi M., and Rashtchian D., "Modeling Formation Damage Due to Asphaltene Deposition in the Porous Media," *Energy & Fuels*, **2011**, *25*, 753-761.
17. Jafari Behbahani T., Ghotbi C., Taghikhani V., and Shahrabadi A., "Asphaltene Deposition under Dynamic Conditions in Porous Media: Theoretical and Experimental Investigation," *Energy & Fuels*, **2013**, *27*, 622-639.
18. Jafari Behbahani T., Ghotbi C., Taghikhani V., and Shahrabadi A., "Investigation on Asphaltene Deposition Mechanisms during CO_2 Flooding Processes in Porous Media: a Novel Experimental Study and a Modified Model Based on Multilayer Theory for Asphaltene Adsorption," *Energy & Fuels*, **2012**, *26*, 5080-5091.
19. Jafari Behbahani T., Ghotbi C., Taghikhani V., and Shahrabadi A., "Experimental Study and Mathematical Modeling of Asphaltene Deposition Mechanism in Core Samples," *Oil & Gas Science and Technology–Revue d'IFP Energies Nouvelles*, **2015**, *70*, 1051-1074.
20. Khalifeh M., Kharrat R., Bagherzadeh H., and Izadi M., "An Experimental and Simulation Study of Asphaltene-induced Permeability Impairment under Natural Depletion Condition," SPE European Formation Damage Conference & Exhibition, *Society of Petroleum Engineers*, **2013**.
21. Bagherzadeh H., Ghazanfari M. H., Kharrat R., and Rashtchian D., "Experimental Investigation and Modeling of Permeability Impairment Mechanisms Due to Asphaltene Precipitation under CO_2 Injection Conditions," *Energy Sources, Part A: Recovery, Utilization, and Environmental Effects*, **2014**, *36*, 591-604.
22. Fallahnejad G. and Kharrat R., "Fully Implicit Compositional Simulator for Modeling of Asphaltene Deposition during Natural Depletion," *Fluid Phase Equilibria*, **2015**, *398*, 15-25.
23. Minssieux L., "Core Damage from Crude Asphaltene Deposition," SPE International Symposium on Oilfield Chemistry, Houston, Texas, **1997.**
24. Nghiem L. X. and Coombe D. A., "Modeling Asphaltene Precipitation during Primary Depletion," *SPE Journal*, **1997**, *2*, 170-176.
25. Civan F., "Reservoir Formation Damage," Gulf Professional Publishing, **2011**.
26. Chang F. F. and Civan F., "Practical Model for Chemically Induced Formation Damage," *Journal of Petroleum Science and Engineering*, **1997**, *17*, 123-137.
27. Pautz J., Crocker M., and Walton C., "Relating Water Quality and Formation Permeability to Loss of Injectivity," SPE Production Operations Symposium, *Society of Petroleum Engineers*, **1989.**
28. Leontaritis K. J., Amaefule J. O., and Charles R. E., "A Systematic Approach for the Prevention and Treatment of Formation Damage Caused by Asphaltene Deposition," *SPE Production & Facilities*, **1994**, *9*, 157-164.
29. Leontaritis K., "Asphaltene Near-well Formation Damage Modeling," paper 39446, Lafayette, LA, SPE Formation Damage Control, **1998**.
30. Hamadou R., Khodja M., Kartout M., and Jada A., "Permeability Reduction by Asphaltenes and Resins Deposition in Porous Media," *Fuel*, **2008**, *87*, 2178-2185.

7

Improved Rheological Model of Oil-Based Drilling Fluid for South-western Iranian Oilfields

Seyed Saied Bahrainian*, **Amin Nabati, and Ebrahim Hajidavalloo**

Department of Mechanical Engineering, Shahid Chamran University of Ahvaz, Ahvaz, Iran

ABSTRACT

In this study, predictive capabilities of apparent viscosity of oil-based drilling fluids which is used in National Iranian South Oilfields Company (NISOC) were evaluated using Newtonian and non-Newtonian models to drive a new suitable equation. The non-Newtonian models include Bingham plastic, Power law, Herschel-Bulkley, Casson, and Robertson-Stiff. To validate the results, the calculated viscosity from rheology models was compared to the fann 35 data of viscometer. The results showed that Robertson-Stiff model has the best prediction of shear stress and viscosity with an absolute average percent error of 3.58. This was followed by Herschel-Bulkley, Casson, Power law, Bingham plastic, and Newtonian with the absolute average percent error of 3.68, 3.77, 9.04, 20.09, and 44.02 respectively. Therefore, the new equation was proposed to predict the shear stress for oil-based drilling fluids which is used in Southwestern Iranian Oilfields. In comparison to the results of the experimental data of this study, it was revealed that the proposed equation has a good agreement with the real shear stress and apparent viscosities.

Keywords: Oil-Based Drilling Fluids, Rheological Models, Fann 35 Viscometer, Shear Stress.

INTRODUCTION

Drilling fluids perform several functions in drilling operations including controlling formation pressures, maintaining hole integrity and stability, cooling and lubricating the drill bit, and the drill string, cleaning the bottom hole, and suspending cuttings in the annulus when circulation is stopped or carrying them to the surface during drilling [1,2]. The rheological behavior of a drilling fluid directly affects all these activities; moreover, its knowledge enables better estimation of flow regimes, frictional pressure losses, equivalent circulating density under down hole conditions, hole-cleaning efficiency, swab/surge pressures; in addition, all of which are of extreme importance for improved drilling efficiency [3]. The determination of drilling fluid rheological properties is important in the calculation of circulating hydraulics and hole cleaning efficiency. Estimation of the rheological properties for the drilling fluids is an essential task for the safety and the economics of drilling a well. Various rheological models have been

***Corresponding author**

Seyed Saied Bahrainian

Email: bahrainian@scu.ac.ir

proposed to describe the rheological behavior of drilling fluid, particularly for drilling applications. The two parameters Bingham plastic model or the power law model is used most often because of its simplicity and the fair agreement of predictions with the rheograms. Although the power law model is useful as a first correction to Newtonian behavior, it may lead to substantial errors if the fluid exhibits yield stress. Other two parameter models like the Casson model have not found wide acceptance [4]. Three constant parameter models have been proposed by Herschel and Bulkley [4,5] and by Robertson and Stiff [4,6].

The proposed model and other models such as Bingham, power law, Herschel-Bulkley, Casson, and Robertson-Stiff have been investigated by Gucuyener [7]; moreover, they were tested with the measured data with a total number of seventy one drilling fluids; in addition, cement slurries of which twenty of them are presented in this study. A continuous regularization for the viscosity function which has been widely used in the numerical simulations of viscoplastic fluid flows, owing to its easy computational implementatio was introduced by Papanastasiou [8]. Shear stress/shear rate data were analyzed by Alderman et al [9]; in addition, the Herschel-Bulkley, Power, and Casson models were considered. The behavior of each rheological parameter in these models with respect to changes in temperature and pressure was investigated. The effectiveness of four different rheological models (Bingham plastic, Power Law, Robertson-Stiff, and Herschel-Bulkley) in prediction of the rheological behavior and the onset of turbulence of various muds flowing in pipes were analyzed by Khataniar et al [10]. They concluded that the Robertson-Stiff model provided the best fit of rheological data. It was shown by Davison et al [11] that Bingham plastic and Herschel-Bulkley models fit the data of low toxicity oil-based mud. Herschel-Bulkley and Casson models accurately fit the mud over a wide range of temperature, pressure, and shear rates. A new rheological model, called the Rational Polynomial model that could represent virtually any time-independent non-Newtonian fluid, was presented by Pilehvari et al [12]. A new unified rheological model was detailed in the papers which have been written by Zamora and Power [13]. The rheological parameters for this model are the plastic viscosity, yield point, and yield stress. The effect of two polymers on the rheological properties of KCl/polymer type drilling fluids was investigated by Versan and Tolga [14]. Moreover, Bingham Plastic, Power Law, Casson, Herschel Bulkley, and Robertson Stiff models for modeling rheology properties have been used by Versan and Tolga [14]. It was found by Kelessidis et al [4] that the Hershel-Bulkley model can effectively correlate the rheological data of several drilling fluids and proposed an improved method for the determination of the Herschel-Bulkley model parameters. In addition, the Golden Section search methodology was used by Kelessidis et al [4] to estimate the best value of the yield stress while the fluid consistency and fluid behavior indices are determined using linear regression on the transformed rheometric data. The effects of temperature on the rheological properties of two types of high-density water-based drilling fluids under high temperature and high pressure with a fann 50 SL rheometer were examined using Wang et al [15]. Four rheological models, the Bingham plastic, Power Law, Casson, and Herschel-Bulkley models were employed to fit the rheological parameters. The cubic splines were used by Bui and Tutuncu [16] to fit the experimental data obtained from field viscometers. The results showed that a very good fit to experimental data in comparison to Bingham, power law, and Herschel-Bulkley models was provided by the proposed model. The effect of black myrobalan rheological properties of flocculated bentonite mud was investigated by Neshat and Shadizadeh [17]. Moreover, Bingham Plastic, Herschel Bulkley, and Robertson Stiff models for modeling rheological properties were used by Neshat and Shadizadeh [17]. The results showed that Herschel- Bulkely and Robertson-Stiff models were more accurate than Bingham plastic in describing bentonite dispersion including black myrobalan. A rigorous predictive model for estimating drilling fluid density at wellbore conditions was suggested by Ahmadi et al [18]. Also, a couple of particle swarm optimization (PSO) and artificial neural network (ANN) were utilized to suggest a high-performance model for prediction of the drilling fluid density.

In this study, the Newtonian model and five major non-Newtonian rheological models were investigated to determine more alternatives for selecting the best model which represents accurately the shear stress-shear rate relationship for drilling fluids. These models are the Bingham, Power law, Herschel-Bulkley, Casson, and Robertson-Stiff. To determine the best rheological model fitting the behavior of the drilling fluids, the graph of the shear stress versus shear rate data of the drilling fluid was plotted. In this study, it was assumed that the model which gives the lowest mean absolute average percent error (EMAP) between the measured and calculated shear stresses is the best model for the drilling fluids. Finally, the new equation was proposed to predict the shear stress for six oil-based drilling fluids used in National Iranian South Oilfields Company.

Theory

The rheological behavior of drilling fluids is very complex. Yield point, apparent viscosity, plastic viscosity, and gel points are the common rheological properties of the drilling fluid. Rheological models intend to provide assistance in characterizing the flow of fluid. Rheological models are mathematical equations used to predict the fluid behavior across a wide range of shear rates [19]. Commonly used model completely describes the rheological characteristics of drilling fluids over its entire shear-rate range. The combination of knowledge of rheological models and practical experience is necessary to fully understand the fluid performance. A plot of shear stress versus shear rate (rheogram) is often used to graphically depict a rheological model.

Newtonian Model

Newtonian fluids follow a simple relationship between shear stress and shear rate (Equation 1). Their viscosities are constant as described by the slope of the line on a linear plot of shear stress versus shear rate (Figure 1). An equation describing a Newtonian fluid is given as:

$$\tau = \mu \dot{\gamma} \tag{1}$$

When the shear stress (τ) of a Newtonian fluid is plotted against the shear rate (γ) in linear coordinates, a straight line through the origin is resulted. The Newtonian viscosity (μ) is the slope of this line [20].

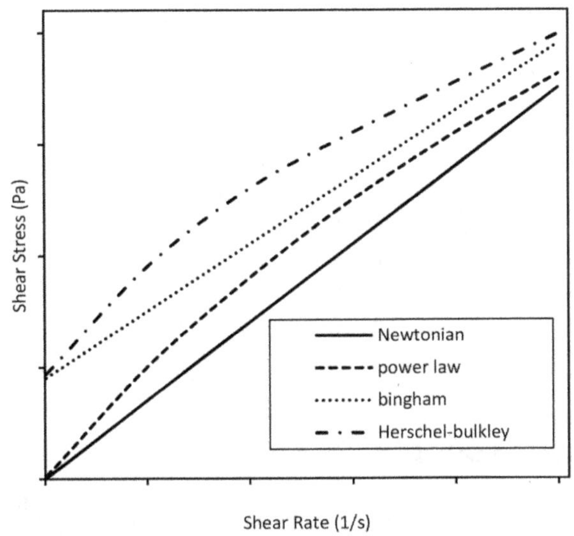

Figure 1: Rheograms of Newtonian, Bingham, Power Law, and Herschel-Bulkley model.

Bingham Plastic Model

This model describes fluids in which the shear stress/shear rate ratio are linear as the specific shear stress has being exceeded. Two parameters: plastic viscosity and yield point are used to describe this model. A rheogram of the Bingham plastic model on rectilinear coordinates is a straight line that intersects the zero shear-rate axes at a shear stress greater than zero (yield point) (Figure 1). The basic Bingham plastic equation is given as:

$$\begin{aligned} \tau &= \tau_B + \mu_B \dot{\gamma} \qquad |\tau| \succ |\tau_B| \\ \dot{\gamma} &= 0 \qquad |\tau| \prec |\tau_B| \end{aligned} \quad (2)$$

where τ_B and μ_B are defined as the yield stress and plastic viscosity respectively [14,21].

Power Law Model

The Power Law is used to describe the flow of shear thinning or pseudoplastic drilling fluids. This model describes fluids in which the rheogram is a straight line when plotted on a log-log graph (Figure 2). Such a line has no intercept, so a true power law fluid does not exhibit a yield stress.

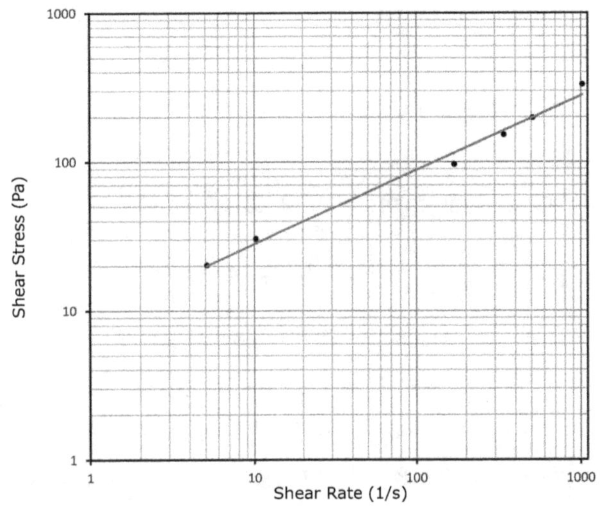

Figure 2: log-log graph of Power Law model.

The mathematical relationship of Power Law model is expressed as:

$$\tau = m\dot{\gamma}^n \quad (3)$$

where m and n are known as the consistency coefficient and flow index respectively [13].

$$log\,\tau = \log m + n \log \dot{\gamma} \quad (4)$$

$$n = \frac{N\sum(\log\dot{\gamma}.\log\tau) - \sum(\log\dot{\gamma}) \times \sum(\log\tau)}{N\sum(\log\dot{\gamma})^2 - (\sum \log\dot{\gamma})^2} \quad (5)$$

$$m = \exp[\frac{\sum(\log\tau) - n\sum(\log\dot{\gamma})}{N}] \quad (6)$$

where N is data number. This procedure is employed for the parameter calculations of the power law model.

m is a measure of the consistency of the fluid, the higher the value of m the more viscous the fluid; moreover, *n* is a measure of the degree of non-Newtonian behavior of the fluid.

Herschel-bulkley Model

This model is used to describe the flow of pseudoplastic drilling fluids which require a yield stress to initiate flow. A rheogram of shear stress minus yield stress versus shear rate is a straight line on log-log coordinates (Figure 3). This model is widely used because it describes the flow behavior of most drilling fluids, and it includes a yield stress value and the Bingham plastic model and power law as special cases [14,22,23].

$$\tau = \tau_H + \mu_H \dot{\gamma}^n \qquad |\tau| \succ |\tau_H|$$
$$\dot{\gamma} = 0 \qquad |\tau| \prec |\tau_H| \tag{7}$$

where τ is the magnitude of the stress tensor, τ_H is the yield stress, $\dot{\gamma}$ is the magnitude of the rate of strain, μ_H is a consistency index, and n is the flow index.

Since this is a three-parameter model, an initial calculation of τ_H is required for other parameter calculations. τ_H is calculated as [14]:

$$\tau_H = \frac{\tau^{*2} - \tau_{min} \times \tau_{max}}{2\tau^* - \tau_{min} - \tau_{max}} \tag{8}$$

where τ* is the shear stress value corresponding to the geometric mean of the shear rate, γ* [14].

$$\gamma^* = \sqrt{\gamma_{min} \times \gamma_{max}} \tag{9}$$

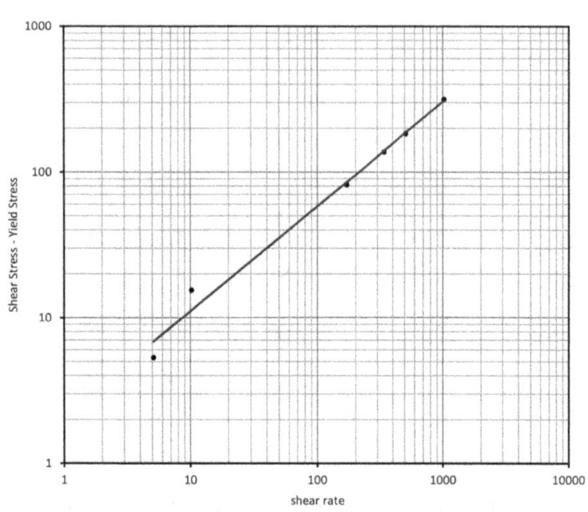

Figure 3: log-log graph of Herschel-Bulkley model.

Casson Model

Casson model is a two-parameter model which is widely used in various industries. This model was represented by Casson to predict the behavior of oil suspensions. The mathematical relationship of Casson model is expressed as [24]:

$$\tau^{1/2} = \tau_C^{1/2} + (\mu_C \dot{\gamma})^{1/2} \qquad |\tau| \succ |\tau_C|$$
$$\dot{\gamma} = 0 \qquad |\tau| \prec |\tau_C| \tag{10}$$

Robertson-Stiff Model

Robertson and Stiff [6] developed a general model to describe the rheological behavior of drilling fluids and cement slurries. The basic equation is given as [6]:

$$\tau = m\left(\dot{\gamma}_y + \dot{\gamma}\right)^n \tag{11}$$

where m, n, and $\dot{\gamma}_y$ are the model parameters. Also, m and n can be considered similar to the parameters m and n of the Power-law model. The third parameter $\dot{\gamma}_y$ is a correction factor for the shear rate. Robertson – Stiff model is a three-parameter model and an initial calculation of $\dot{\gamma}_y$ is required for other parameter calculations. $\dot{\gamma}_y$ is calculated as [6]:

$$\dot{\gamma}_y = \frac{\dot{\gamma}_{min} \times \dot{\gamma}_{max} - \dot{\gamma}^{*2}}{2\dot{\gamma}^* - \dot{\gamma}_{min} - \dot{\gamma}_{max}} \tag{12}$$

where $\dot{\gamma}^*$ is the shear rate value corresponding to the geometric mean of the shear stress, τ*. The geometric mean of the shear stress (τ*) is then calculated from [6]:

$$\tau^* = \sqrt{\tau_{min} \times \tau_{max}} \tag{13}$$

Proposed Equation

The determination of drilling fluid rheological properties is important in the calculation of circulating hydraulics, and hole cleaning efficiency. Therefore, if we are able to determine accurate relationship between shear rate and shear stress; it will be a direct impact over the economics of exploitation projects for wellbore hydraulic management. For the proposed equation, a non-linear regression is used. We attempted to fulfill the following requirements in the development of the proposed equation:

1. The equation should fit closely to the experimental data;
2. The equation should exhibit the yield pseudo plastic character and if possible, should be a general case of the former ones;
3. The equation should contain minimum number of parameters; moreover,
4. The appropriate parameters should be easily determined.

It is evident that most available rheological models are originally developed as empirical correlations. Therefore, we attempted proposed new equation with minimum number of parameters using experimental data. After investigated various equations (such as sinus, exponential, power, liner, logarithmic, polynomial), the follow equation has good results.

$$\tau^n = \tau_p{}^n + \left(\mu_p \dot{\gamma}\right)^n \qquad |\tau| \succ |\tau_p|$$
$$\dot{\gamma} = 0 \qquad |\tau| \prec |\tau_p| \tag{14}$$

where τ_p, μ_p, and n are the rheological coefficient characteristics of the fluid. To minimize the number of parameters and particularly for drilling applications, 20 values (0.05, 0.1, 0.15 . . . 1) investigated for n, and the most appropriate n was selected. It was assumed that the value of 'n' which gives the lowest mean absolute average percent error between the measured and calculated shear stresses is the best value of 'n' for the proposed equation.

EXPERIMENTAL PROCEDURES

Different drilling fluids are available and they are classified according to the base fluid in the mud. The drilling mud can be broadly classified as water based mud (WBM), oil based mud (OBM), synthetic based mud (SBM), emulsions, invert emulsions, air, foam fluids, and etc [25,26]. The type of drilling fluid to be used is selected based on the properties of the formation being drilled, environmental considerations, and cost.

Table 1: Component of oil based for 100 bbl drilling mud.

Material	Fluid A	Fluid B	Fluid C	Fluid D	Fluid E	Fluid F
Density (pcf)	77.5	64	78	70	75	81
Gasoil (bbl)	60	68	60	65	62	58
F.L-C (SX)	20	18	20	18	20	20
Primary Emulsion (DR)	2	1	2	1	2	2
Lime (SX)	20	18	20	18	20	20
Water $CaCl_2$ (bbl)	21	24	21	23	21	20
secondary Emulsion (DR)	0.8	0.8	0.8	0.8	0.8	0.8
Viscosities (SX)	2	1	2	1-2	1-2	2
L.S.P (TN)	4-5	2	4-5	3	4	5

Oil-based drilling fluids have excellent properties such as stability, lubricity, and temperature stability. Though, the excessive use of oil-based drilling fluids harms the environment [17]. Oil base mud is composed primarily of diesel oil or mineral oil and additives [27]. Oil-based drilling fluids (drilling muds) are used to meet demanding applications, including those requiring the highest degree of thermal stability [28]. In the present study, six oil-based muds drilling used in Southwestern Iranian oilfields were investigated. The chemical composition and weight of oil-based mud is given in Table 1.

Furthermore, Fann 35 Viscometer was used for the rheological analysis of drilling fluids (Figure 4). The rotor of this rheometer is driven by an electric motor with six standard rotational speeds (that is, 3, 6, 100, 200, 300, and 600 rpm). Corresponding to the rotor rotation, the dial readings of Fann Viscometer were measured in degrees. The standard dimensions of Fann35 Viscometer are: bob radius (r_1) 0.017245m, rotor radius (r_2) 0.018415m, bob height (h) 0.038 m, and spring constant (k) 3.87×10^5 N/m. For consistent Fann35 Viscometer readings, the experiments were repeated three times daily. Dial reading (that is, bob deflection) measures the torque, which is accurately converted using Equation 15 to the shear stress [1].

The test fluid is contained in the annular space or shear gap between the cylinders. Rotation of the outer cylinder at known velocities is accomplished through precision gearing. The viscous drag exerted by the fluid creates a torque on the inner cylinder or bob. This torque is transmitted to a precision spring where its deflection is measured and then related to the test conditions and instrument constants.

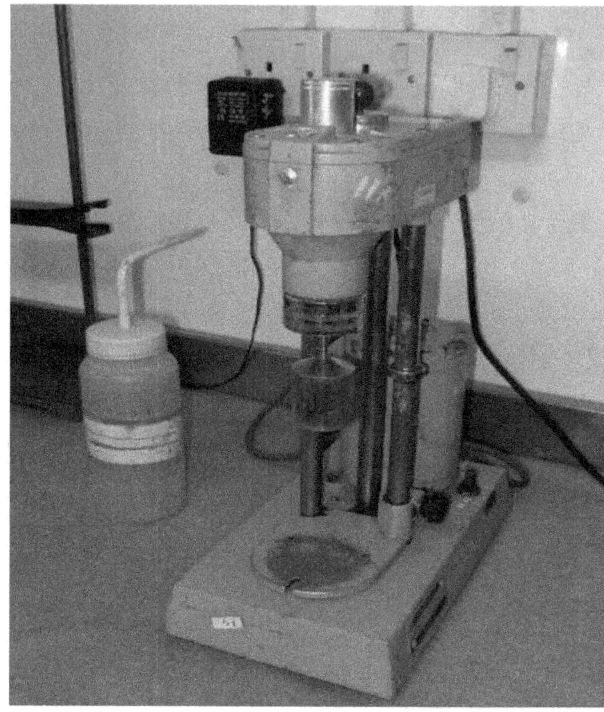

(a)

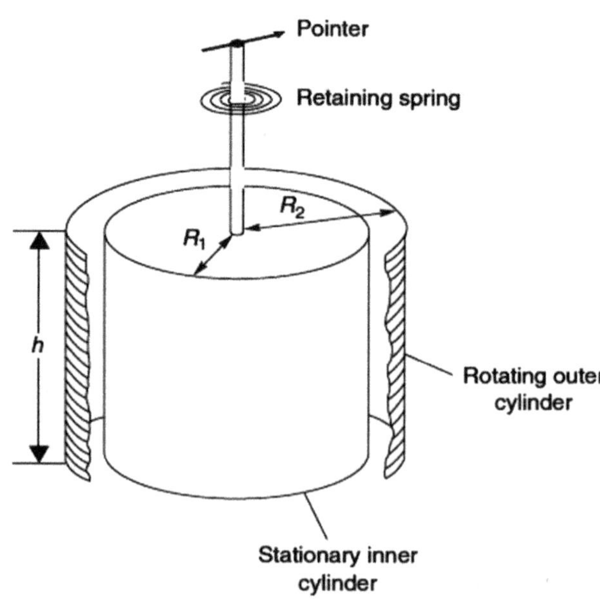

(b)

Figure 4: (a) Fann viscometer Model 35SA, (b) Schematic diagram of viscometer.

To determine the rate of shear between two vertical coaxial cylinders, the outer (cup) one of which is rotating with constant angular velocity and the inner cylinder (bob) is stationary, following assumptions are made: (i) Flow in the annular region between two coaxial cylinders is purely steady, laminar and tangential, isothermal, incompressible. (ii) There is no flow in the radial and axial direction. (iii) There is no pressure gradient in θ-direction. (iv) The fluid is time-independent. (v) There is no slip at the wall. With these assumption the shear stress at the inner cylinder wall is given as [1]:

$$\tau = \frac{k\theta}{2\pi r_1 h} \tag{15}$$

And the shear rate at the inner cylinder wall is given as [1]:

$$\dot{\gamma} = \frac{4\pi N_F}{60\left[1-\left(\frac{r_1}{r_2}\right)^2\right]} \tag{16}$$

where k is spring constant (N/m), θ is dial reading (degree), r_1 is bob radius (m), h is bob height (m), N_F is rotor rotation speed (rpm), and r_2 is rotor radius (m).

To estimate the mean absolute percent error (E_{MAP}), a statistical method has been used [29]. This method is used based on the measured and calculated shear stresses:

$$E_{MAP} = \frac{100}{N}\sum\left|\frac{\tau_{measured} - \tau_{calculated}}{\tau_{measured}}\right| \tag{17}$$

One set of data of drilling fluid has been used to illustrate the accuracy of this approach for selecting the best rheological model, one with the lowest E_{MAP} value.

The standard deviation is a measure which is used to quantify the amount of variation or dispersion of a set of data. For a finite set of numbers, the standard deviation is found by taking the square root of the average of the squared deviations of the values from their average value. The standard deviation is given as [29]:

$$\sigma = \sqrt{\frac{1}{N}\sum_{i=1}^{N}(x_i - \bar{x})^2} \tag{18}$$

where $\bar{x} = \frac{1}{N}\sum_{i=1}^{N} x_i$

RESULTS AND DISCUSSIONS

Six drilling mud samples were used to evaluate the performance of the rheological models. The experimental data obtained from the viscometer were used to determine the values of shear stress and shear rate. The results of the viscometer data are shown in Table 2. The rheological data were obtained from various drilling holes.

Table 2: Rheological data from viscometer.

Drilling Fluids	θ_3	θ_6	θ_{100}	θ_{200}	θ_{300}	θ_{600}	PV* (cP)	YP** ($lb_f/100ft^2$)
Fluid A	4	6	19	30	39	65	26	12
Fluid B	2	3	9	16	21	37	16	6
Fluid C	17	19	40	56	62	91	30	21
Fluid D	5	6	19	28	38	62	19	8
Fluid E	3	4	16	27	34	56	24	12
Fluid F	6	7	23	38	50	84	36	17

*Plastic Viscosity

**Yield Point

Twenty values (0.05, 0.1, 0.15 . . . 1) for 'n' have been Investigated to evaluate value of 'n' for proposed equation (Equation 14). It was assumed that the value of 'n' which gives the lowest mean absolute average percent error between the measured and calculated shear stresses is the best value of 'n' for the proposed equation. The percent errors for different values of 'n', for each drilling fluid are shown in Figure 5. The results show that in the interval 0.35 to 0.5, the smallest amount of 'n' for the drilling fluids which were studied occurs. To summarize the error of 'n' parameter, an attempt is made to calculate the average percent error for each 'n' parameter. Average percent errors for each 'n' parameter are shown in Figure 6. The results show that n=0.4 (0.4=2/5) is the best value for n. Therefore, the proposed equation is as follows:

$$\begin{aligned} &\tau^{2/5} = \tau_p^{\ 2/5} + \left(\mu_p \dot{\gamma}\right)^{2/5} \qquad && |\tau| \prec |\tau_p| \\ &\dot{\gamma} = 0 && |\tau| \prec |\tau_p| \end{aligned} \tag{19}$$

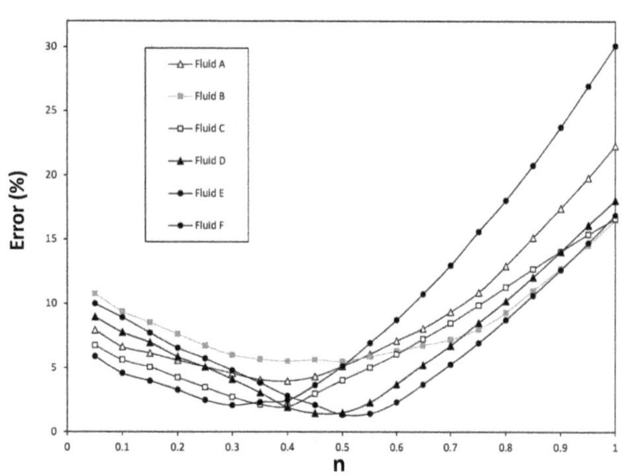

Figure 5: The percent errors for different values of 'n', for each drilling fluid.

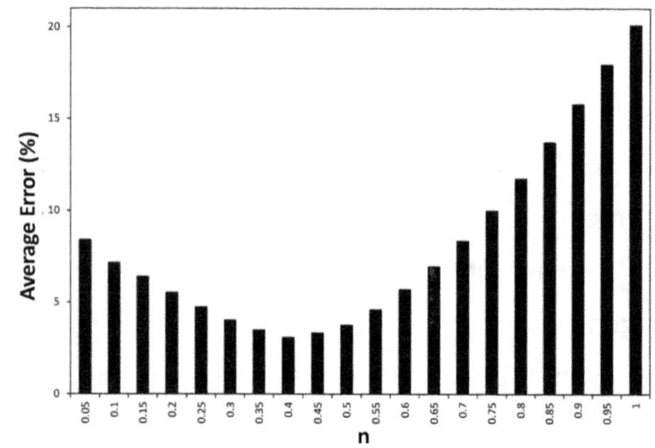

Figure 6: the average percent error for each 'n' parameter.

The viscometer data were used to calculate the parameters of the Newtonian, Bingham, power law, Herschel-Bulkley, Casson, Robertson-Stiff, and proposed models. These seven models have been applied to the reported data set. For each rheogram, the best fit of the data was computed. For the Newtonian model and Bingham plastic fluid, a linear least square method is used, while for the Power Law, Herschel-Bulkley, Casson, Robertson–Stiff models, and proposed equation, a non-linear regression was used. The rheological parameters of each model are given in Table 3. The models were then applied to the original shear rate data to calculate shear stresses, which were compared to the experimental shear stresses.

The plot of experimental data and shear stress predicted by each models versus shear rate for fluid A (oil based mud 77.5 pcf) is shown in Figure 7. These observations are quantified in Table 4 in which the mean absolute percent error and the average standard deviation for each model are listed. However, the proposed equation had the smallest average percent error and the smallest standard deviation, thus indicating the best overall fit. The proposed equation was used to calculate the shear stresses in which an average percent error of 3.95 was obtained; in addition, compared to 5.10% which was obtained in using Casson model and 5.15% using the Herschel-Bulkley model, Newtonian model had the largest average percent error.

Table 3: Constant parameters for rheological models.

Rheological Models	Constant parameters	Fluid A	Fluid B	Fluid C	Fluid D	Fluid E	Fluid F
Newtonian model	μ (pois)	0.35	0.19	0.53	0.33	0.30	0.45
Bingham Plastic	τ_B ($lb_f/100\ ft^2$)	7.306	3.296	24.415	7.663	5.892	8.889
	μ_B (pois)	0.30	0.17	0.36	0.28	0.26	0.39
Power Law	m ($lb_f.s^n/100\ ft^2$)	1.863	0.890	10.161	2.254	1.220	2.521
	n	0.497	0.514	0.305	0.459	0.543	0.486
Herschel-Bulkley	τ_H ($lb_f/100\ ft^2$)	3.122	1.692	15.998	4.442	2.368	5.497
	μ_H ($lb_f.s^n/100\ ft^2$)	0.447	0.165	0.758	0.272	0.261	0.240
	n	0.717	0.775	0.681	0.783	0.783	0.849
Casson	τ_C ($lb_f/100\ ft^2$)	3.076	1.360	15.539	3.580	2.084	3.893
	μ_C (pois)	0.20	0.12	0.17	0.18	0.19	0.26
Robertson-Stiff	m ($lb_f.s^n/100\ ft^2$)	0.574	0.157	3.164	0.510	0.374	0.395
	γ_y (s^{-1})	17.13	27.02	31.68	25.29	15.13	31.45
	n	0.686	0.789	0.489	0.695	0.731	0.778
Proposed equation	τ_P ($lb_f/100\ ft^2$)	2.172	0.952	12.947	2.634	1.387	2.797
	μ_P (pois)	0.15	0.09	0.10	0.13	0.15	0.20

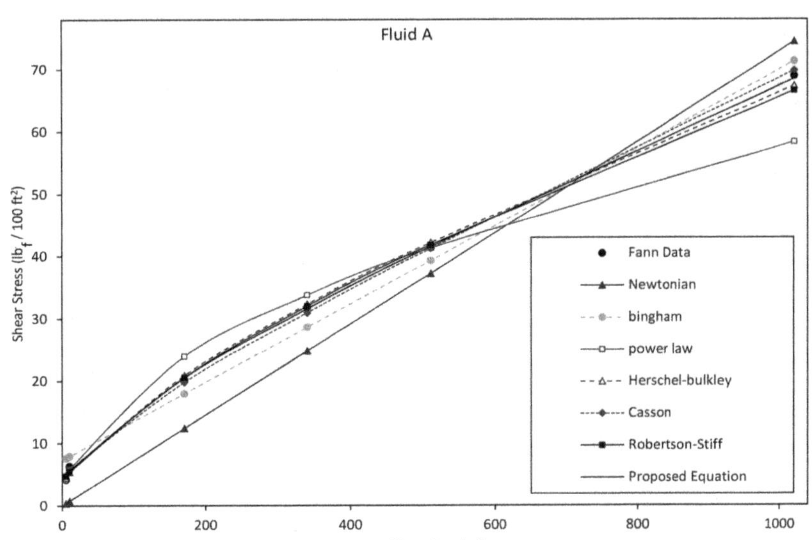

Figure 7: Comparison of rheological models for fluid A.

Table 4: Results of percent error for fluid A.

Shear Rate (s^{-1})	Newtonian model	Bingham Plastic	Power Law	Herschel-Bulkley	Casson	Robertson-Stiff	Proposed equation
5.1069	91.22	79.78	1.25	7.51	16.27	12.27	9.96
10.2138	88.30	24.88	7.09	13.78	8.40	13.75	10.29
170.23	38.41	10.80	18.77	3.67	1.75	1.98	1.63
340.46	21.99	9.98	6.16	1.61	2.63	0.63	0.73
510.69	9.99	4.96	0.11	1.98	0.27	1.11	0.41
1021.38	8.01	3.45	15.42	2.34	1.29	3.49	0.69
Ave. error. (%)	42.99	22.31	8.13	5.15	5.10	5.54	3.95
Standard deviation	34.53	26.62	6.87	4.34	5.64	5.37	4.38

The rheogram for fluid B (oil based mud 64 pcf) is shown in Figure 8. Table 5 shows the average percent error and the average standard deviation for each model. The Proposed equation and Casson model were used to calculate the shear stresses and an average percent error of 5.52 was obtained.

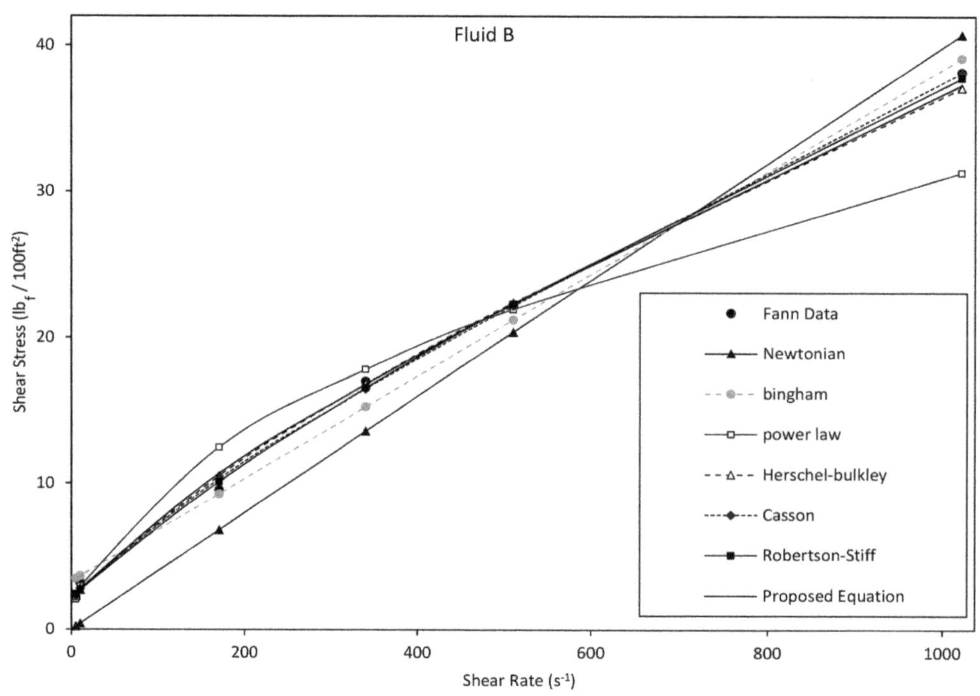

Figure 8: Comparison of rheological models for fluid B.

Table 5: Results of percent error for fluid B.

Shear Rate (s^{-1})	Newtonian model	Bingham Plastic	Power Law	Herschel-Bulkley	Casson	Robertson - Stiff	Proposed equation
5.1069	90.39	63.95	3.07	7.28	9.01	13.84	4.30
10.2138	87.19	14.93	7.73	15.44	12.62	14.74	13.47
170.23	28.85	2.86	30.61	10.25	8.07	5.91	11.89
340.46	19.95	10.15	4.91	0.97	2.91	2.66	1.14
510.69	8.52	4.72	1.54	0.51	0.44	0.14	0.03
1021.38	6.73	2.52	17.98	2.83	0.08	1.07	2.30
Ave. error. (%)	40.27	16.52	10.98	6.21	5.52	6.39	5.52
Standard deviation	35.10	21.66	10.27	5.39	4.68	5.87	5.24

Fluid C (oil based mud 78 pcf) was evaluated using Model 35 Fann viscometer, thus the results are shown in Figure 9 and Table 6. The results show that the Proposed equation provides a very good fit to experimental data. The Proposed equation was used to calculate the shear stresses, in which an average percent error of 1.96 was attained compared to 2.41 and 3.64% which were obtained in using the Robertson-Stiff model and Herschel-Bulkley model respectively.

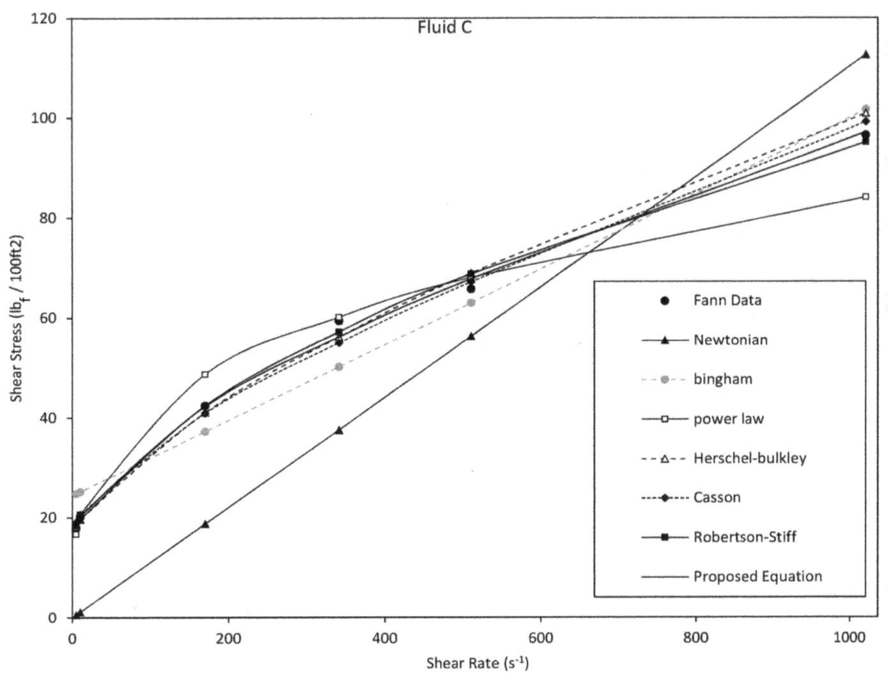

Figure 9: Comparison of rheological models for fluid C.

Table 6: Results of percent error for fluid C.

Shear Rate (s^{-1})	Newtonian model	Bingham Plastic	Power Law	Herschel-Bulkley	Casson	Robertson-Stiff	Proposed equation
5.1069	96.88	37.58	7.32	1.51	5.82	2.32	1.60
10.2138	94.41	25.01	2.45	2.29	2.48	2.44	0.60
170.23	55.74	12.10	14.78	3.20	3.47	0.01	0.36
340.46	36.78	15.54	1.29	5.40	7.38	3.69	5.43
510.69	14.34	4.14	3.53	4.88	2.20	4.58	3.11
1021.38	16.72	5.33	12.86	4.56	2.79	1.44	0.66
Ave. error. (%)	52.48	16.61	7.04	3.64	4.02	2.41	1.96
Standard deviation	33.47	11.65	5.17	1.42	1.92	1.48	1.80

A comparison of the rheological models for fluid D (oil based mud 70 pcf) is presented in Figure 10. The Results of percent error for each model is given in Table 7. This result shows that the Casson model is the best model to represent the rheological properties for this oil-base mud and this is closely followed by the proposed equation, Herschel-Bulkley and Robertson-stiff models, respectively.

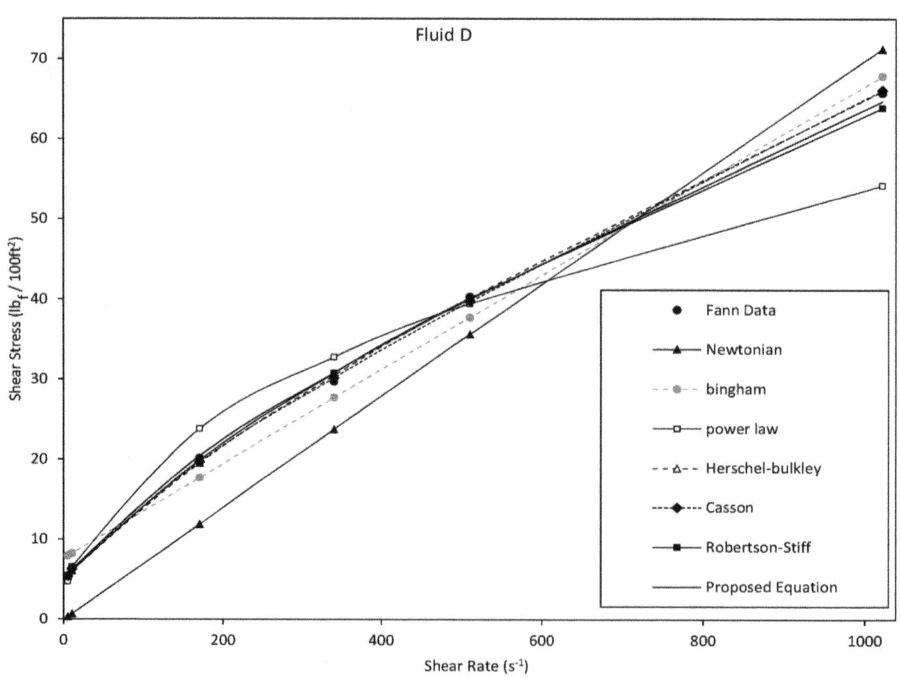

Figure 10: Comparison of rheological models for fluid D.

Table 7: Results of percent error for fluid D.

Shear Rate (s^{-1})	Newtonian model	Bingham Plastic	Power Law	Herschel-Bulkley	Casson	Robertson-Stiff	Proposed equation
5.1069	93.28	50.20	10.16	2.09	2.66	2.92	2.05
10.2138	88.80	29.89	2.91	3.93	0.53	4.46	1.97
170.23	41.06	12.20	18.22	2.89	2.24	1.25	1.35
340.46	20.01	6.66	10.27	2.61	1.61	3.55	3.73
510.69	11.59	6.34	2.13	0.26	1.44	0.49	0.73
1021.38	8.37	3.15	17.54	0.32	0.50	2.89	1.62
Ave. error (%)	43.85	18.07	10.21	2.02	1.5	2.59	1.91
Standard deviation	34.97	16.82	6.28	1.34	0.80	1.34	0.93

The plot of experimental data and shear stress predicted by each models versus the shear rate for fluid E (oil based mud 75 pcf) is shown in Figure 11. Table 8 shows the mean absolute percent error and the average standard deviation for each model. The proposed equation had the smallest average percent error indicating the best overall fit. In calculating the shear stresses using the proposed equation, an average percent error of 2.45 was attained compared to 3.23 and 3.37% observed in using the Robertson-Stiff and Herschel-Bulkley model respectively. Newtonian model and Bingham model have the largest average percent error.

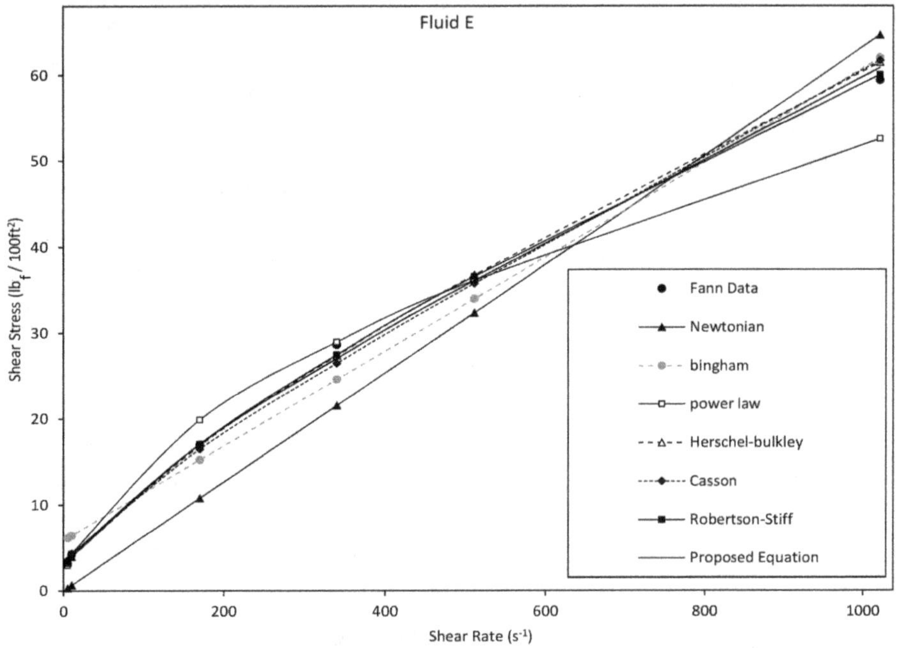

Figure 11: Comparison of rheological models for fluid E.

Table 8: Results of percent error for fluid E.

Shear Rate (s^{-1})	Newtonian model	Bingham Plastic	Power Law	Herschel-Bulkley	Casson	Robertson -Stiff	Proposed equation
5.1069	89.84	94.02	6.98	3.81	13.13	6.21	5.63
10.2138	84.76	52.13	1.64	6.29	2.48	6.10	0.54
170.23	36.51	10.16	17.08	0.39	2.78	0.52	0.43
340.46	24.75	14.10	1.09	4.39	7.52	4.10	5.65
510.69	10.36	5.85	0.05	1.83	0.90	1.37	0.00
1021.38	8.85	4.40	11.50	3.51	3.92	1.07	2.44
Average error (%)	42.51	30.11	6.39	3.37	5.12	3.23	2.45
Standard deviation	33.02	32.85	6.21	1.87	4.12	2.36	2.38

Figure 12 shows the rheological behavior of each model for fluid F (oil based mud 81 pcf). This figure also shows the comparison between the measured data and the calculated data for all models. Table 9 shows the mean absolute percent error and the average standard deviation for each model.

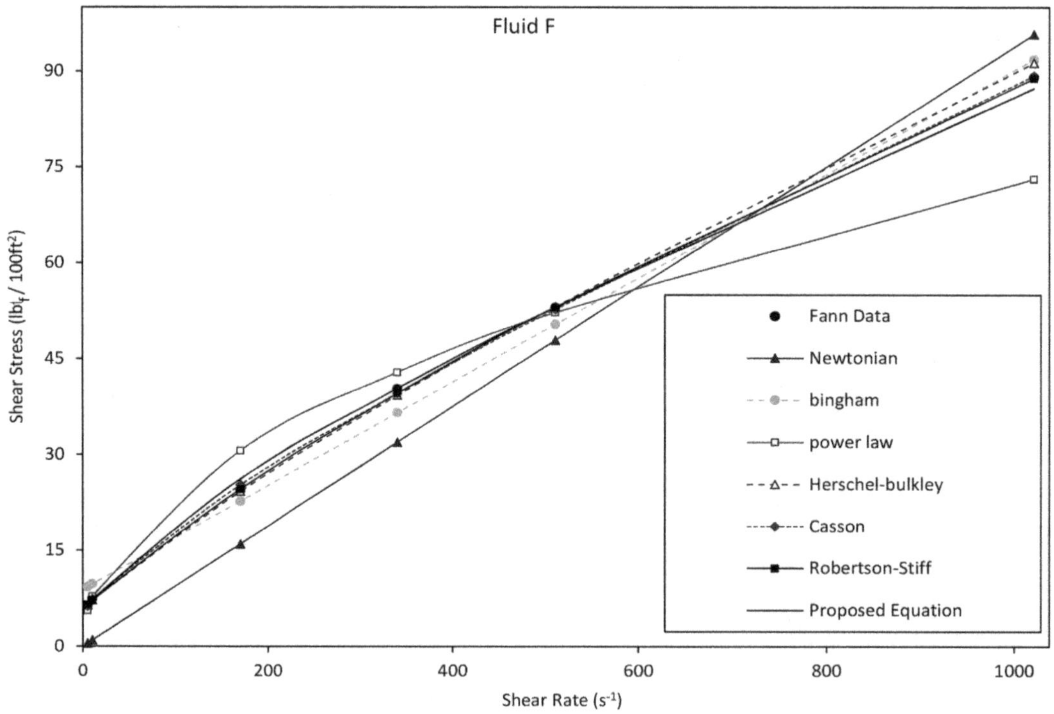

Figure 12: Comparison of rheological models for fluid F.

Table 9: Results of percent error for fluid F.

Shear Rate (s^{-1})	Newtonian model	Bingham Plastic	Power Law	Herschel-Bulkley	Casson	Robertson-Stiff	Proposed equation
5.1069	92.47	46.23	12.48	1.43	1.60	2.26	5.94
10.2138	87.10	30.93	5.06	2.75	0.24	2.96	1.41
170.23	34.55	6.85	25.50	0.57	3.58	0.74	7.30
340.46	20.77	9.30	6.38	2.53	1.79	1.85	0.10
510.69	9.68	4.98	1.54	0.26	0.56	0.01	0.02
1021.38	7.53	3.14	17.91	2.55	0.29	0.23	1.95
Ave. error (%)	42.02	16.90	11.48	1.68	1.34	1.34	2.79
Standard deviation	34.93	16.06	8.21	1.00	1.17	1.09	2.83

The values for the mean absolute average percent error are given for each model in Figure 13. This figure shows that the Newtonian model is not perfect for modeling the behavior of oil-base mud. Also, the results show that the use of Power law and Bingham plastic resulted wrong prediction of shear stress for oil-base mud, while these two models have been used for drilling process in South-western Iranian Oilfields.

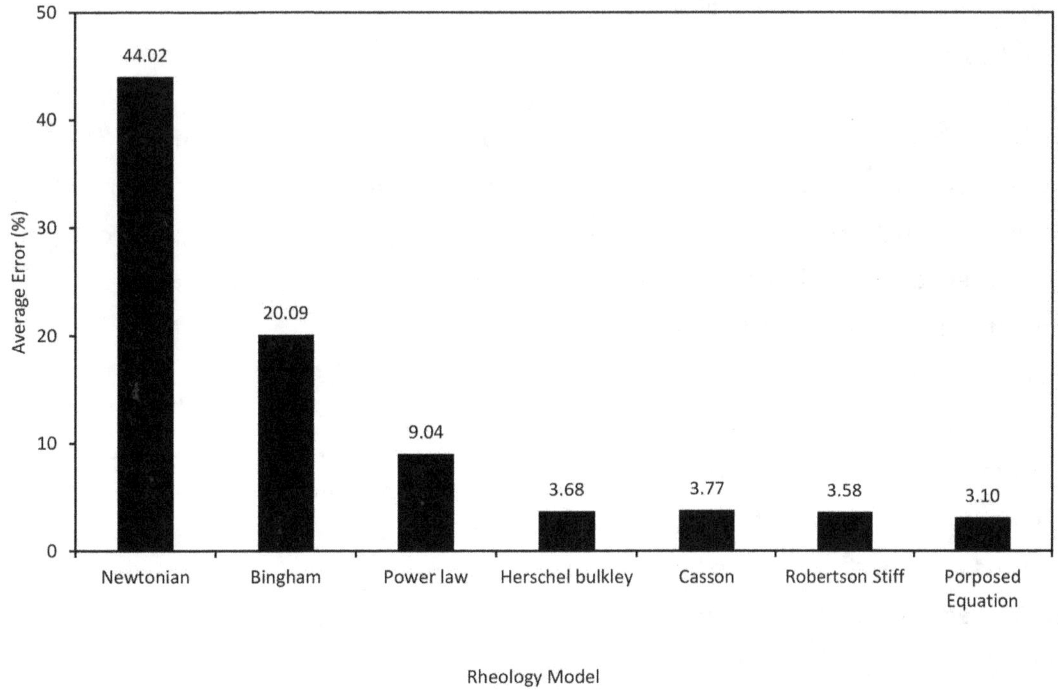

Figure 13: Comparison between mean absolute average percent errors for all models.

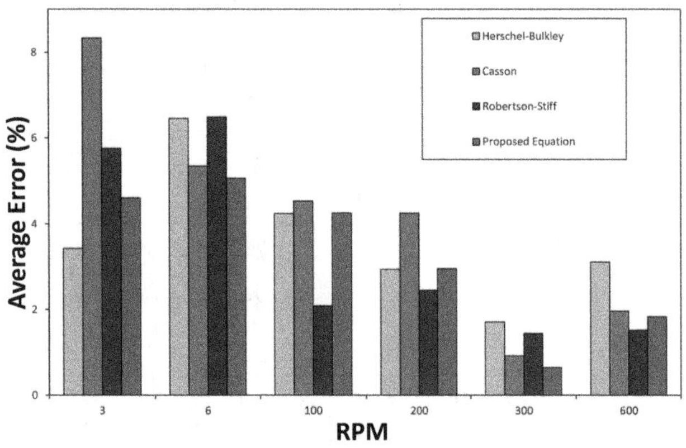

Figure 14: absolute percent errors for Herschel-Bulkley, Casson, Robertson-stiff and proposed equation with different velocity.

Figure 13 shows that the proposed equation is the best model for representation of the rheological properties of oil-base mud, and this is closely followed by Robertson-stiff, Herschel-Bulkley, and Casson models respectively.

Figure 14 shows the results of Herschel-Bulkley, Casson, Robertson-stiff models and the proposed equation with different velocity. It was observed that the percent errors of the rheological models were reduced as the shear rate increases.

CONCLUSIONS

Selection of the best rheological model is of great importance in obtaining correct results for drilling. A simple and direct approach has been presented for selection of the best rheological model for any non-Newtonian fluid according to the lowest EMAP criteria.

The experimental data obtained from the Fann 35 Viscometer were used to determine the values of the shear stress and shear rate. Six oil base mud samples were used to evaluate the performance of the rheological models.

In this study, the rheological models, Newtonian, Bingham plastic, Power law, Herschel-Bulkley, Casson, and Robertson-Stiff were evaluated for accurate representation of the wide range of shear stress/shear rate data. These models were confirmed to describe the rheology of most non-Newtonian fluids, accurately.

The results showed that Robertson-Stiff model had the best prediction of shear stress and viscosity with an absolute average percent error of 3.58. This was followed by Herschel-Bulkley, Casson, Power law, Bingham plastic and Newtonian with the absolute average percent error of 3.68, 3.77, 9.04, 20.09, and 44.02 respectively.

Conclusively, the new equation is proposed to predict the shear stress of six oil-based drilling fluids used in Southwestern Iranian oilfields. By comparing the results obtained with the experimental data, it is revealed that the proposed equation has a good agreement with real shear stress and apparent viscosities with an absolute average percent error of 3.10.

ACKNOWLEDGMENTS

The authors wish to acknowledge the National Iranian South Oilfields Company (NISOC) and the Shahid Chamran University of Ahvaz for providing research funds and Laboratory equipments that enabled the current research.

NOMENCLATURES

ANN	: Artificial Neural Network
EMAP	: Mean Absolute Average Percent Error
NISOC	: National Iranian South Oilfields Company
PSO	: Particle Swarm Optimization

REFERENCES

1. Bourgoyne A. T., Chenevert M. E., Millheim K. K., and Young F. S., "Applied Drilling Engineering," SPE Textbook Series, Richardson, TX, **1991**.

2. Coussot P., Bertrand F., and Herzhaft B., "Rheological Behavior of Drilling Muds, Characterization Using MRI Visualization," *Oil & Gas Science and Technology–Rev. IFP*, **2004**, *59*(1), 23-29.
3. Rooki R., Doulati Ardejani F., Moradzadeh A., Mirzaei H., and et al., "Optimal Determination of Rheological Parameters for Herschel-bulkley Drilling Fluids Using Genetic Algorithms (GAs)," *Korea-Australia Rheology Journal*, **2012**, *24*(3), 163-170.
4. Kelessidis V. C., Maglione R., Tsamantaki C., and Aspirtakis Y., "Optimal Determination of Rheological Parameters for Herschel–Bulkley Drilling Fluids and Impact on Pressure Drop, Velocity Profiles and Penetration Rates during Drilling," *J. Pet. Sci. Eng.*, **2006**, *53*, 203-224.
5. Herschel W. H. and Bulkley R., "Konsistenzmessungen Von Gummi-Benzollosungen," *Kolloid-Z.*, **1926**, *39*, 291-300.
6. Robertson R. E. and Stiff H. A., "An Improved Mathematical Model for Relating Shear Stress to Shear Rate in Drilling Fluid and Cement Slurries," *SPE J.*, **1976**, *16*, 31-36.
7. Gucuyener I. H., "A Rheological Model for Drilling Fluids and Cementing Slurries," paper SPE 11487 Presented at *the Middle East Oil Technical Conference*, Manama-Bahrain, Soc. Pet. Eng., **1983**, 389–386.
8. Papanastasiou T. C., "Flows of Materials with Yield," *J. Rheology.*, **1987**, *31*, 385-404.
9. Alderman N. J., Gavignet A., Guillot D., and Maitland G. C., "High-Temperature, High-Pressure Rheology of Water-Base Mud," SPE 18035, Houston, TX, **1988**, 187-195.
10. Khataniar S., Chukwa G. A., and Xu H., "Evaluation of Rheological Models and Application to Flow Regime Determination," *J. of Petroleum Science and Engineering*, **1994**, *11*, 155-164.
11. Davison J. M., Clary S., Saases A., Allouche M., and et al., "Rheology of Various Drilling Fluid Systems under Deepwater Drilling Conditions and the Importance of Accurate Predictions of Downhole Fluid Hydraulics," SPE 56632, presented at *the SPE Annual Technical Conference and Exhibition held in Houston*, Texas, **1999**, 1-13.
12. Pilehvari A., Serth R., and Largad V., "Generalized Hydraulic Calculation Method Using Rational Polynomial Model," SPE 71403, presentation at *the SPE Annual Technical Conference and Exhibition held in New Orleans*, Louisiana, **2001**, 1-13.
13. Zamora M. and Power D., "Making a Case for AADE Hydraulics and the Unified Rheological Model," Paper AADE-02-DFWM-HO-13 Presented at *the AADE Technical Conference,* Houston, **2002**.
14. Versan M. and Tolga A., "Effect of Polymers on the Rheological properties of KCl/Polymer Type Drilling Fluid," Energy Sources, **2005**, *27,* 405-415.
15. Wang F., Tan X., Wang R., Sun M., and et al., "High Temperature and High Pressure Rheological Properties of High-density," *Pet. Sci.*, **2012**, *9*, 354-362.
16. Bui B. T., and Tutuncu A. N., "A Generalized Rheological Model for Drilling Fluids Using Cubic Splines," SPE-169527-MS, presentation at the SPE Western North American and Rocky Mountain Joint Meeting, Denver, Colorado, 2014, 1-13.
17. Neshat J. and Shadizadeh S. R., "Evaluation of a Naturally-derived Deflocculant (Terminalia

Chebula) in Bentonite Dispersions," *Iranian Journal of Oil & Gas Science and Technology*, **2016**, *5*(2), 21-44.

18. Ahmadi M. A., Shadizadeh S. R., Shah K., and Bahadori A. R., "An Accurate Model to Pedict Drilling Fluid Density at Wellbore Conditions," *Egyptian Journal of Petroleum*, **2018**, *27*, 1-10.
19. Rabia H., "Well Engineering & Construction," Entrac Consulting Limited, **2002**, 242-245.
20. American Petroleum Institute (API), "Recommended Practice on the Rheology and Hydraulics of Oil-Well Drilling Fluids," API RP13D (4th ed.), Washington, USA, **2009.**
21. Feys D., Verhoeven R., and De Schutter G., "Evaluation of Time Independent Rheological Models Applicable to Fresh Self-compacting Concrete," *Appl. Rheol.*, **2007**, *17*(5), 56244-1 –56244-10
22. Chatzimina M., Georgiou G., and Alexandrou A., "Wall Shear Rates in Circular Couette Flow of a Herschel-Bulkley Fluid," *Appl. Rheol.*, **2009,** *19*(3), 34288.
23. Kelessidis V. and Maglione R., "Shear Rate Corrections for Herschel-Bulkley Fluids in Couette Geometry," *Appl. Rheol.*, **2008**, *18*(3), 34482-1–34482-11
24. Power D. and Zamora M., "Drilling Fluid Yield Stress: Measurement Techniques for Improved Understanding of Critical Drilling Fluid Parameters," paper AADE-03-NTCE-35 Presented at *the AADE Technical Conference,* Houston, **2003**.
25. Shah S. N., Narayan P. E., Shanker H., and Chinenye C., "Future Challenges of Drilling Fluids and Their Rheological Measurements," AADE-10-DF-HO-41, *AADE Fluids Conference and Exhibition Held at the Hilton Houston North*, Houston, Texas, **2010**, 1-15.
26. American Society of Mechanical Engineers (ASME), "Shale Shaker Committee," Drilling Fluids Processing Handbook, Elsevier Inc, **2005**.
27. Goodarznia I. and Esmaeilzadeh F., "Treatment of Oil-contaminatment Drill Cuttings of South Parts Gas Field in Iran Using Supercritical Carbon Dioxide," *Iranian Journal of Science & Technology*, Transaction B, Engineering, 30 (B5), **2006**.
28. Sedaghatzadeh M., Fazel Abdulabadi B., and Shahbazi K., "Thermal and Rheological Properties Improvement of Oil-based Drilling Fluids Using Multi-walled Carbon Nanotubes (MWCNT)," *Journal of Petroleum Science and Technology*, **2016**, *6*(2), 37-44.
29. Myttenaere A., Golden B., Grand B., and Rossi F., "Using the Mean Absolute Percentage Error for Regression Models," European Symposium on Artificial Neural Networks, *Computational Intelligence and Machine Learning (ESANN),* Belgium, **2015**.

8

A Novel Bio-based Sulfonic Zwitterionic Surfactant Derived from Transgenic Soybean Oil and its Performance in Surface and Interfacial Activities

Wenjie Xu [1], Pengcheng Bian [1], Hongze Gang[1], Jinfeng Liu[1], Bozhong Mu [1,2], and Shizhong Yang [1]*

[1]State Key Laboratory of Bioreactor Engineering and Institute of Applied Chemistry, East China University of Science and Technology, Shanghai 200237, China

[2]Shanghai Collaborative Innovation Center for Biomanufacturing Technology, Shanghai 200237, China

ABSTRACT

Bio-based surfactants have attracted increasing attention due to their renewable resources and excellent surface properties. In this study, a novel bio-based sulfonic zwitterionic surfactant (BSZS) derived from transgenic soybean oil was prepared using a simple reaction route with two steps of the amidation and quaterisation. The bio-based sulfonic zwitterionic surfactant showed a critical micelle concentration (CMC) as low as 33.34 mg/L with a surface tension of 28.50 mN/m. In addition, it also showed good performance in foamability, emulsibility, and wettability. The excellent performances indicated the great potential applications of the bio-based sulfonic surfactant from vegetable oils in both daily life and industrial fields.

Keywords: Bio-based Surfactant, Zwitterionic Surfactant, Transgenic Soybean Oil

INTRODUCTION

Surfactants have been widely used as detergents in daily life and industrial fields such as daily washing products [1], enhanced oil recovery (EOR) [2-3], and oil spill disposal processing (OSDP) [4-5]. Compared with the traditional surfactants which are generally petroleum-based, bio-based surfactants have received increasing attention because of its renewable resource, outstanding physicochemical property, low costs [6]. Biomass has been used as the raw materials of bio-based surfactants such as proteins [7], carbohydrates [8], alcohols [9-10], vegetable oils, and fatty acids [11-13]. Non-edible vegetable oil is one of the most common biomass and has received a great interest as platform chemicals for surfactants, biodiesel, biopolymers etc. due to their universal availability, inherent biodegradability, and low costs [14-17].

Non-edible vegetable oil has been applied to producing bio-based surfactants such as sodium N-fatty acyl amino acid [16], N-phenyl fatty amidopropyl-N, N dimethyl carboxyl betaine [18], UC18DAB [19], which have shown good surface and interfacial properties. However, the synthetic methods of bio-based surfactant are still defective and could be improvable.

Alkaline hydrolysis, alkylation, and alkaline hydrolysis were commonly used in the synthetic route of bio-based surfactant derived from nonedible vegetable oil, which made the reaction complex with high

***Corresponding author**

Shizhong Yang

Email: meor@ecust.edu.cn

energy consumption and emission of pollutants. For example, after the alkaline hydrolysis, a large amount of acid was consumed to neutralize the alkaline, and a great deal of salt was produced, which increased the burden to the environment. Meanwhile, the complexity of the reaction process greatly increased the energy consumption. On the other hand, SO_2 and HCl would be produced in the acyl-chlorination, because of the use of thionyl chloride. It is neither economic nor environmentally friendly.

In the present study, a novel bio-based sulfonic zwitterionic surfactant was synthesized using renewable transgenic soybean oil as the starting material by a strategy of a facile two-step route, amidation, and quaterisation, which was simpler, was more energy saving, and had lower pollutants. The fatty acid amine obtained from amidation was detected by GC-MS. The novel bio-based sulfonic zwitterionic surfactant (BSZS) was determined by ESI-MS, and its surface behavior; moreover, foamability, emulsibility, and wettability were evaluated in this paper.

EXPERIMENTAL PROCEDURES

Materials

N, N-Dimethyl-1, 3-propane diamine (AR, Aladdin, Shanghai, China), sodium 3-chloro-2-hydroxy propane sulfate (95%, Aladdin, Shanghai, China), KOH (AR, Shanghai Lingfeng Chemical Reagent Co. Ltd., Shanghai, China), acetic ether (AR, Shanghai Lingfeng Chemical Reagent Co. Ltd., Shanghai, China), sulfuric acid (AR, Shanghai Lingfeng Chemical Reagent Co. Ltd., Shanghai, China), methanol (AR, Sinopharm Chemical Reagent Co. Ltd., Shanghai, China), anhydrous sodium sulfate (AR, Sinopharm Chemical Reagent Co. Ltd., Shanghai, China), 1, 4-dioxane (AR, Sinopharm Chemical Reagent Co. Ltd., Shanghai, China), and isopropanol (AR, Shanghai Lingfeng Chemical Reagent Co. Ltd., Shanghai, China) were used without further purification. Transgenic soybean oil was purchased from supermarkets, and its saponification value was determined to be 194.93 mg KOH/g.

Synthesis Methods

BSZS was synthesized from transgenic soybean oil through two processes, namely amidation, and quaterisation.

Amidation

(1) Direct Amidation

A certain amount of transgenic in which the content of main fatty acids were listed in Table 1, N, N-Dimethyl-1, 3-propane diamine, and KOH were added into a flask and were then refluxed at temperatures of 120, 130, and 140°C for several hours.

(2) Transesterification-amidation

8.97 grams (1.03×10^{-2} mol) transgenic soybean oil, 0.18 grams (3.26×10^{-3} mol) KOH, and 12.0 mL methyl alcohol were added into a flask, and refluxed at 65 °C for 2 hours. The resulted mixtures were then remained quiescently for a while until layering. The supernate was a mixture of the methyl alcohol, KOH (potassium hydroxide), and fatty acids methyl ester produced in the reaction and glycerol ester which were not react completely, and the lower layer was glycerol. The supernate was then transferred to another flask and refluxed at 110°C to recycle methyl alcohol for 1 hour; then, the process was followed by the addition of 8.97 grams (8.61×10^{-2} mol) N, N-Dimethyl-1, 3-propane diamine, and 0.19 grams KOH into the flask, and refluxing at 140°C for 8 hours.

After the amidation reaction, the remained N,

N-Dimethyl-1, and 3-propane diamine were recycled by being evaporated under reduced pressure. 5 mL of a 15 (wt.%) NaCl solution was added into the reaction mixture to separate glycerol and KOH. Then 15 mL ethyl acetate was used to extract the organic phase for 3 times. The fatty acid amides were obtained from the organic layer.

Quaterisation

1.00 grams fatty acid amide and 0.68 grams sodium 3-chloro-2-hydroxy propane sulfate were added into a flask along with 0.034 grams NaOH as the catalyst and isopropanol/water mixture as solvent. Under the condition of a reaction time equal to 4 hrs and variable temperature, $V_{isopropanol}/V_{water}$, and amount of solvent, the optimum condition was obtained by the method of orthogonal experiments. After the reaction, the solvent was evaporated under reduced pressure, and then extracted in ethyl alcohol to filtrate and remove the insoluble sodium 3-chloro-2-hydroxy propane sulfate and NaCl. After that, the ethyl alcohol was evaporated under reduced pressure to obtain the crude BSZS, and then the crude BSZS was weighed. Finally, the crude BSZS was extracted in ethyl acetate to remove the soluble fatty acid amine, and obtain the insoluble BSZS. The removed soluble fatty acid amine was washed by water for three times and dried to obtain the unreacted fatty acid amine.

Characterization

Gas chromatography mass spectrometry (GC-MS) was recorded on Agilent 6890N Network GC system and 5978 inter Mass Selective Detector. Electrospray ionization high resolution mass spectrometry (ESI HRMS) was recorded on the Waters LCT Premier XE Mass Spectrometers [17].

Measurements of Surface Tensions

BSZS aqueous solutions of different concentrations ranging from 1.7×10^{-6} mol/L to 5.9×10^{-3} mol/L were prepared by using double distilled water. The surface tensions (SFT) of these solutions were measured by the plate method using a DCAT 21 tensiometer (Dataphysics, Germany) at 25.0 ± 0.1°C [20] for three times, and the average value was obtained. The surface tension between air and double distilled water was 71.9 mN/m at 25 °C. Surface excess at CMC (Γ_{max}) was derived from the Gibbs adsorption isotherm equation[21].

$$\Gamma_{max} = -\frac{1}{2.303\eta RT}\left(\frac{\partial SFT}{\partial \lg C}\right)_T \quad (1)$$

where, C is the concentration of the surfactant in aqueous solution, and R is equal to 8.314 J/mol/K; T was set at 298.15 K; SFT is expressed in mN/m, and η was set to 1 for zwitterionic surfactants [22]. The area occupied per surfactant molecule (A_{min}) at CMC on air/water interface is related to the surface excess Γ_{CMC} as follows [21].

$$A_{min} = (N_A \Gamma_{max})^{-1} \times 10^{16} \quad (2)$$

where, N_A is the Avogadro's constant.

Measurements of Interfacial Tensions

Solutions of BSZS were prepared at different concentrations ranging from 0.005 g/L to 0.500 g/L in Daqing oil field simulated formation water; the concentrations of main ions were shown in Table 2.

Table 1: The content of main fatty acids in transgenic soybean oil.

	Palmitic Acid C16: 0	Stearic Acid C18: 0	Oleic Acid C18: 1	Linoleic Acid C18: 2	Linolenic Acid C18: 3
%.wt	10.68	4.99	25.96	50.72	7.65

Table 2: The concentrations of main ions in formation water of Daqing oil field.

Salinity	Cl^-	SO_4^{-2}	HCO_3^-	CH_3COO^-	Na^+	Ca^{+2}	Mg^{+2}
5318	1068	12	2307	216	1665	41	0.45

Interfacial tensions (IFT) between Daqing crude oil and these solutions were measured by the spinning-drop method at 50 ± 0.1 °C (the average stratum temperature of Daqing oilfield, China) using a SVT 20 tensiometer (Dataphysics, Germany) [23]. The interfacial tension between Daqing crude oil and the simulated formation water was 9.40 mN/m at 50 °C.

Measurement of Contact Angle

The contact angles of 0.500 g/L BSZS solution were measured by sessile drop technique at 25 °C [24] for 5 times for each sample and the average value was figured out. Double distilled water showed an average contact angle of 92.04° on the hydrophobic acrylic substrate at 25 °C [24].

Measurement of Foaming Property

The foaming property of 0.500 g/L BSZS solution was obtained according to Ross-Miles test [25] at 40 °C [7]. The foam heights at the beginning and after 10 min were recorded (hr./mm). The experiments were repeated for 3 times.

Measurement of Emulsifying Property

10 mL BSZS aqueous solution of 0.50 g/L concentration and 10 mL liquid paraffin were added into 20 mL graduated test tube at room temperature. The tube was shaken violently for 2 min, and was then left quietly for 1 min. The time was recorded when the volume of the separated water was 3 mL (15% volume of the emulsion) and 5 mL (25% volume of the emulsion) [17].

Measurement of Krafft Point

5 mL of a 1% (m/m) BSZS aqueous solution was added into the test tube, which was placed in a thermostat water bath at 50 °C. Then, water temperature was reduced slowly until solid appeared in the aqueous solution.

RESULTS AND DISCUSSION

The Structural Characterization and Yields of Amidation

The GC spectrogram of the product from amidation reaction is shown in Figure 1, in which peaks with the retention time of 12.25, 13.86, 13.92, 14.05, and 14.35 min represent the different fatty acid amides, corresponding to the fatty acid amides with hydrophobic carbon chains of C16:0, C18:2, C18:1, C18:3, C18:0 respectively.

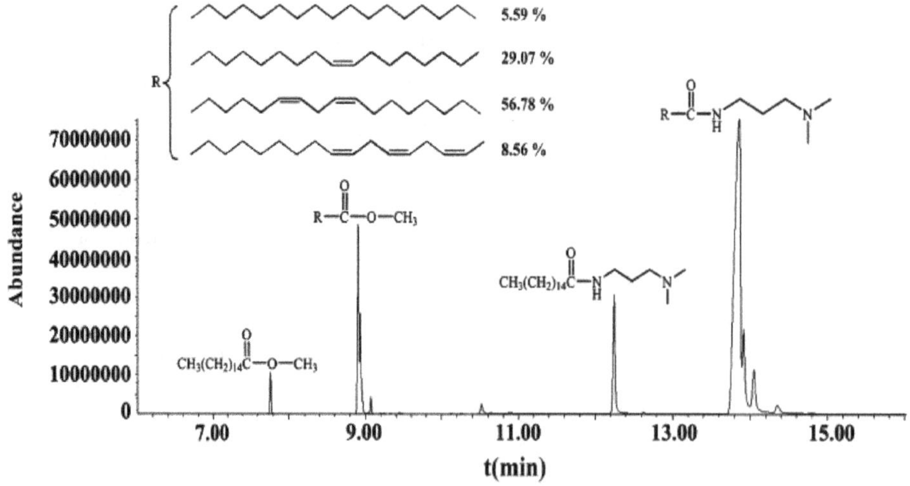

Figure 1: The GC spectrogram of the product from amidation reaction.

Moreover, peaks with retention time of 7.77, 8.90, 8.93, and 9.07 min represented the different fatty acid methyl esters respectively. Furthermore, the molecular weights of each fatty acid amide with hydrophobic carbon chain of C16:0, C18:2, C18:1, C18:3, and C18:0 were detected by the mass spectrum shown in Figure 2.

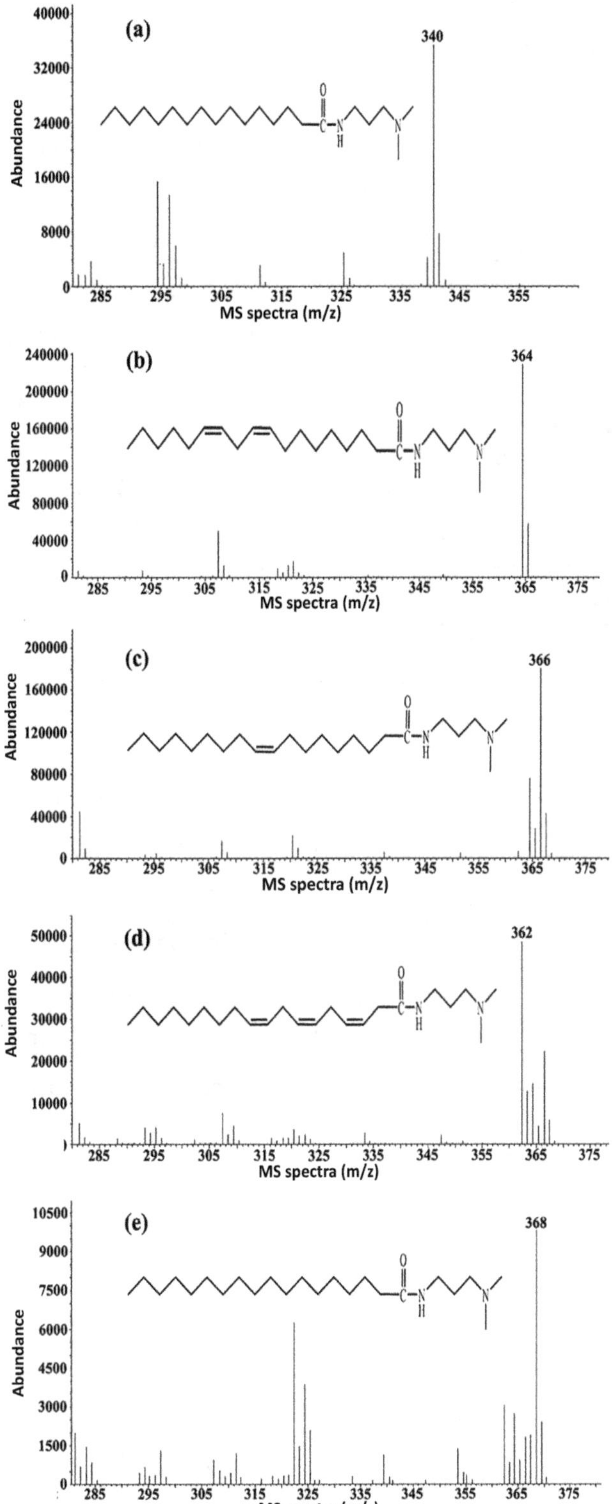

Figure 2: The MS spectrogram of the fatty acid amides.

The condition and yields of amidation orthogonal experiment are shown in Table 3. Under the condition of number of transgenic soybean oil to the number of N, N-Dimethyl-1, 3-propane diamine equal to 1:3, the variables were temperature, reaction time, and the amount of KOH. In addition, the optimum condition was obtained by the method of orthogonal experiment. *A*, *B*, and C in Table 3 represent the temperature, reaction time, the mole ratio of KOH to transgenic soybean oil respectively.

Table 3: The yields of amidation in different conditions.

	A T (°C)	*B* t (hr.)	*C* KOH /transgenic soybean oil (n/n)	*K* Yield (%)
1	120	2	6	1.67
2	120	3	3	9.35
3	120	4	1.5	10.40
4	130	2	3	3.36
5	130	3	1.5	7.66
6	130	4	6	11.72
7	140	2	1.5	10.98
8	140	3	6	3.25
9	140	4	3	26.96
*K*1	7.14	5.34	9.68	-
*K*2	7.58	6.75	13.23	-
*K*3	13.73	16.36	5.55	-
R	6.59	11.02	7.68	-

K corresponds to the yield under different conditions, and R is the range of yield in different conditions. The results show that the higher the temperature is, the higher the yield is. Also, a higher yield was obtained at longer times. Moreover, when the mole ratio of KOH to transgenic soybean oil was 3, the yield was the highest. In addition, $R_B > R_C > R_A$, which means that the reaction time was the most influential factor; moreover, the mole ratio of KOH to transgenic soybean oil took the second place. Additionally, temperature was the least influential factor and had a little effect on the yield.

The yield of amidation was very low, and KOH was not distributed well in the reactants in the absence of solvent. To promote the yield, 3 different solvents at different concentrations were used to optimize the reaction condition listed in Table 4.

Table 4: The yields of amidation using different solvents.

Solvent	Volume	Yield (%)
Dioxane	4 mL (0.0469 mol)	32.21
methanol	4 mL (0.0988 mol)	32.13
N, N-Dimethyl-1, 3-propane diamine	3.8 mL (0.0304 mol)	74.11

It is clear that the yields were obviously increased differently when different solvents were used. When both of the dioxane and methanol were used as the solvents, the yield was very low. KOH could not be dissolved in dioxane in the reaction system, and it may hinder the smooth progress of the reaction. On the other hand, KOH may be dissolved in methanol. However, since the low boiling point of methanol was 64.7 °C, the reaction temperature could not reach 140 °C when the reaction refluxed. As a result, the yield was still low. When N, N-dimethyl-1, and 3-propane diamine was used as the solvent, it acted as a reactant too. Meanwhile, KOH could be dissolved in excess N, N-dimethyl-1, 3-propane diamine, which ensured the smooth progress of the mass transfer. Consequently, the yield at the mole ratio of transgenic soybean oil to N, N-dimethyl-1, and 3-propane diamine to KOH equal to 1:15:3, at a temperature of 140 °C, and for a reflux time of 8 hours reached 74.11%, which was much higher than that obtained in the other conditions.

For the amidation reaction in two steps of transesterification-amidation, the yield was 83.14% which was higher than that of the direct amidation in one step. When methyl alcohol was added into the flask, the transesterification of transgenic soybean oil was first happened through the catalysis of KOH; fatty acid methyl ester was then produced, which reacted with N, N-dimethyl-1, 3-propane diamine through amidation. Compared with transgenic soybean oil, fatty acid methyl ester more easily reacted with N, N-dimethyl-1, 3-propane diamine. However, transesterification-amidation needed complicated operation and more energy consumption; for example, glycerol and methanol needed to be removed respectively. As a result, the direct amidation was more practical and economical. In addition, in comparison with the method reported for the synthesis of nonedible fatty acid amide [16], the direct amidation is simpler, is more environmentally friendly, and has lower pollutant emission since the glycerol produced in the reaction could be used as an industrial raw material; this is significant for its application to industrial production.

The Yields of Quaterisation and Structural Characterization

Table 5 lists yields (*K*) at different temperatures (*A*), volume ratios of isopropanol to water (*B*), and volumes of solvent (*C*) in quaterisation reaction optimized by the orthogonal experiment.

Table 5: The yields of quaterisation in different conditions.

	A **T (°C)**	*B* **Isopropanol /Water (v/v)**	*C* **$V_{solvent}$ (mL)**	*K* **Yield/%**
1	70	2: 1	10	87.14
2	70	1: 1	7	85.05
3	70	1: 2	5	86.57
4	80	2: 1	7	92.59
5	80	1: 1	5	89.25
6	80	1: 2	10	84.37
7	90	2: 1	5	85.92
8	90	1: 1	7	92.93
9	90	1: 2	10	91.38
*K*1	86.25	88.55	87.63	-
*K*2	88.74	89.08	90.19	-
*K*3	90.08	87.44	87.25	-
R	4.55	1.64	2.94	-

It is suggested that the higher the temperature should be, the higher the yield should be. Also, when isopropanol to water ratio (v/v) was 1:1, the yield was the highest. $V_{solvent}$ = 7 mL per 1 gram of fatty acid amine was the best. $R_A > R_C > R_B$ indicated that temperature was the most significant factor, and $V_{solvent}$ took the second place, while isopropanol to water ratio (v/v) was the least influential factor. As a result, the optimum reaction condition for quaterisation was a temperature of 90 °C, an isopropanol to water ratio (v/v) of 1: 1, and a $V_{solvent}$ of 7 mL/g, which lead to a yield of 92.93 %.

The molecular weight of the BSZS was determined by ESI-MS, and the ionization way of ESI-MS was positive ion mode with Na^+, as shown in Figure 3.

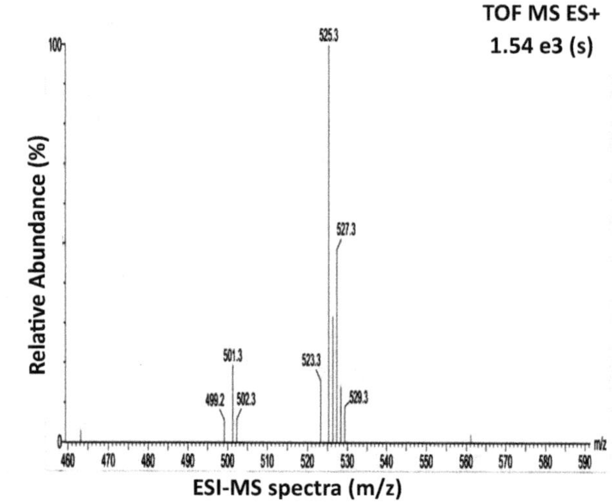

Figure 3: The ESI-MS spectrogram of BSZS.

The mass to charge ratios (*m/z*) of main peaks were 499.2 501.3, 523.3, 525.3, 527.3, and 529.3, which respectively correspond to BSZS with C16:0, C18:3, C18:3, C18:2, C18:1, C18:0. The *m/z* ratios of BSZS with different fatty acid alkyl chains are listed in Table 6.

Table 6: The *m/z* of BSZS with different fatty acid alkyl chains.

Type of fatty acid alkyl chain	Molecular formula of BSZS	M	M+H$^+$	M+Na$^+$	M+K$^+$
C16: 0	$C_{24}H_{48}N_2O_5S$	476.3284	477.3362	499.3182	515.2921
C18: 0	$C_{26}H_{54}N_2O_5S$	506.3753	507.3832	529.3651	545.3391
C18: 1	$C_{26}H_{52}N_2O_5S$	504.3597	505.3675	527.3495	543.3234
C18: 2	$C_{26}H_{50}N_2O_6S$	502.3441	503.3518	525.3339	541.3077
C18: 3	$C_{26}H_{48}N_2O_6S$	500.3285	501.3361	523.3183	539.292

Surface Properties

Equilibrium surface tension as a function of BSZS concentrations at 25.0 °C is illustrated in Figure 4.

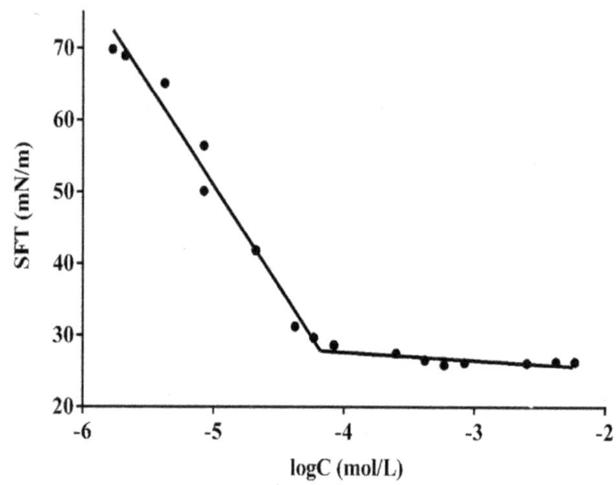

Figure 4: The surface tensions versus the concentration of BSZS at 25.0 °C.

The critical micelle concentration (CMC) and surface tension at CMC (SFT_{CMC}) were estimated from the breakpoints of the plots. The CMC of BSZS was as low as 6.92×10^{-5} mol/L with a SFT_{CMC} of 28.50 mN/m. Moreover, the calculated surface excess (Γ_{max}) and the area occupied per surfactant molecule (A_{min}) at CMC on air/water interface were 4.90 μmol/m^2 and 0.339 nm^2/molecule respectively.

An exceptive phenomenon shown in Figure 4 depicts that the surface tension kept decreasing by raising the BSZS concentration above the CMC point. This may be related to the double bond in the major hydrophobic chain of BSZS (as shown in Scheme 1).

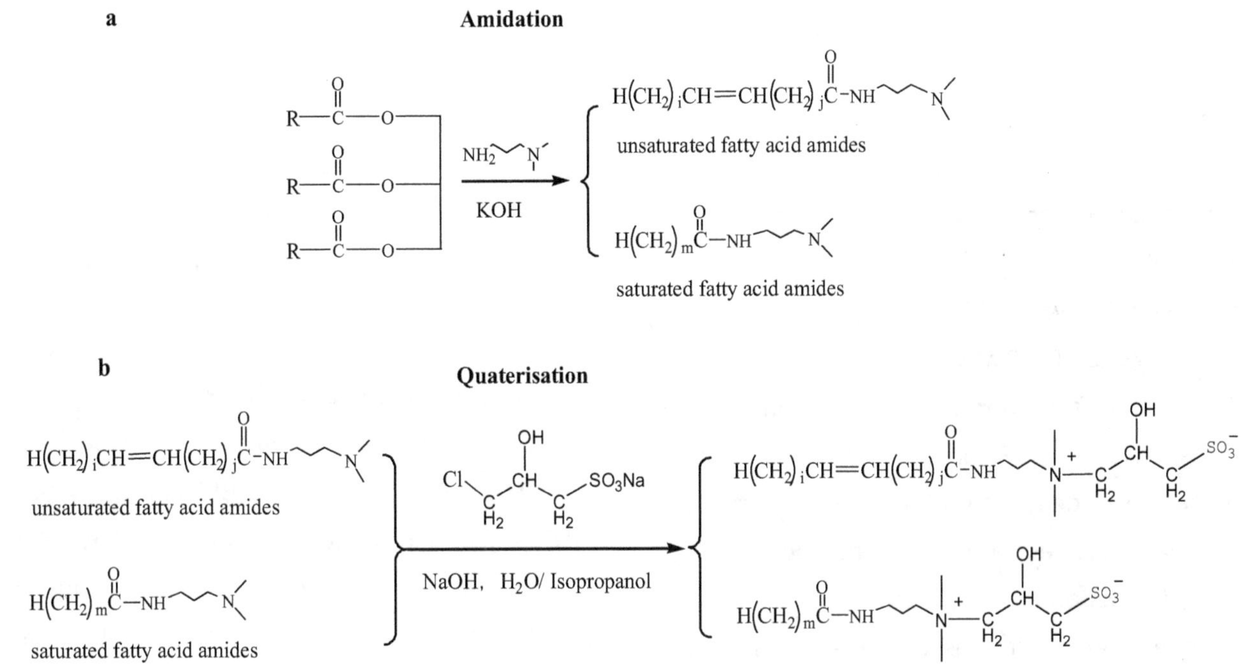

Scheme 1: The synthetic method of BSZS from transgenic soybean oil.

Double bonds are mostly cis-formed; moreover, the hydrophobic chain with a double bond has compressibility; as a result, the arrangement of surfactants molecules is incompact. When reaching the CMC of BSZS, the arrangement of the surfactant molecules on the surface is relatively incompact. When the surfactant concentration was increased, the air–water surface could accommodate more surfactant molecules. Hydrophobic chains with double bonds could also be compressed to decrease space. Surfactant molecule arrangement was more compact, and the surface tension of the solution was lower [18]. A similar phenomenon of anionic surfactants with a double bond hydrophobic chain had been reported elsewhere [26].

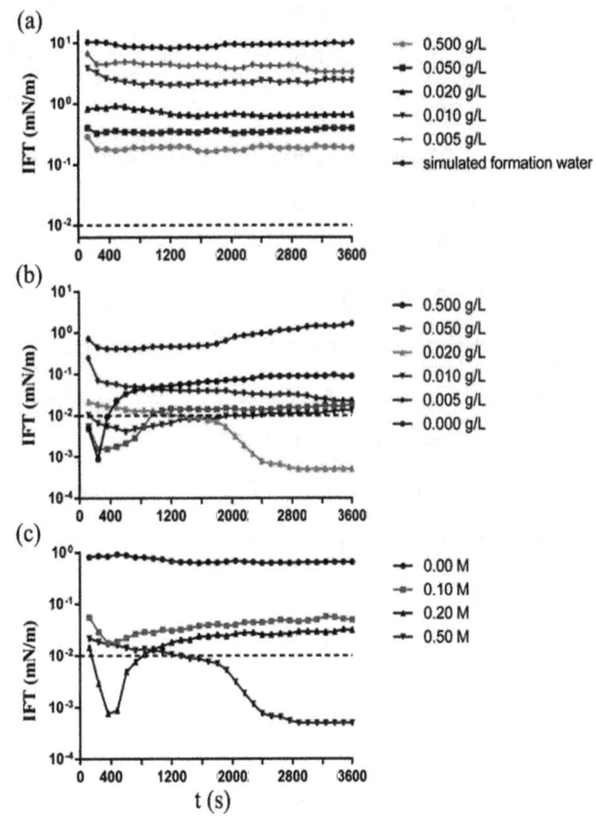

Figure 5: Dynamic interfacial tensions between Daqing crude oil and different concentrations of BSZS in the simulated formation water at 50.0 °C; (a) without Na_2CO_3, (b) with 0.5 M Na_2CO_3 added, (c) dynamic interfacial tensions between Daqing crude oil and 0.02 g/L BSZS in the simulated formation water having different concentrations of Na_2CO_3 at 50.0 °C.

Interfacial Properties

Dynamic interfacial tensions (DIFT) between Daqing crude oil and the simulated formation water at the different concentrations of BSZS solutions and at a temperature of 50.0 °C are displayed in Figure 5(a).

It shows that DIFT is remained unchanged by changing time, but it is decreased by an increase in concentrations. The interfacial tension between crude oil and the simulated formation water solutions decreased to 7.91 mN/m at a BSZS concentration of 0.500 g/L.

When Na_2CO_3 was added into the simulated formation water, DIFT was decreased significantly, and the interfacial properties were improved obviously, as shown in Figure 5(b). A stable ultra-low IFT between crude oil and the simulated formation water was obtained with the system containing 0.010 g/L BSZS and 0.500 M Na_2CO_3 in solutions. Traditional petroleum-based surfactants such as petroleum sulfonates [27] and alkyl benzene sulfonates [28] along alkali (NaOH or Na_2CO_3) are widely used in oil recovery. The acidic components (petroleum acids) in crude oil can react with alkaline in solution to produce in situ surfactant, which can interact with petroleum-based surfactants and reduce the IFT between crude oil and water remarkably [29-32]. Meanwhile, it can be seen in Figure 5(c) that the interfacial properties could be further reduced by an increase in the concentration of Na_2CO_3. The minimum DIFT is in 10^{-3} mN/m orders of magnitude at 0.02 g/L BSZS, which is critically required for the effective displacement of residual crude oil from the pores and capillaries of petroleum reservoirs [33].

Foaming and Emulsification Properties

The Ross–Miles test revealed the initial foam height of 86 mm at a 0.500 g/L BSZS solution; after 5, 10, and 30 min, the foam height was 86, 82, and 80 mm respectively, indicating a little change after 30 minutes. This implied that BSZS inherited very strong foaming ability and stability. The average time for separating 3 mL (15 vol.% of the emulsion) and 5 mL (25 vol.% of the emulsion) water was 283.6 and 504.4 s respectively, which confirmed that BSZS had a strong emulsification property. Moreover, the emulsion produced by BSZS and liquid paraffin was pretty stable because the emulsion layer still existed after 30 min.

Contact Angle and Krafft Point

The average contact angle of air/water/solid was reduced from 92.04 to 44.1° at a BSZS concentration of 0.500 g/L, which showed an excellent wetting property. The Krafft point of 1% surfactant solution was below 0 °C. 1% BSZS in an aqueous solution was clear and transparent at different temperatures greater than 0 °C. The Krafft point is the temperature below which the solubility of surfactant is quite low and insufficient for micellization; moreover, the solubility changes evidently at this point [34]. The Krafft point of $C_{16}H_{33}SO_3Na$ and $C_{18}H_{37}SO_3Na$ were 55.3 and 64.8 °C respectively in a previous research [34]. The Krafft point blew 0 °C indicates that BSZS has better solubility than normal alkyl benzene sulfonates.

CONCLUSIONS

A novel bio-based sulfonic zwitterionic surfactant derived from renewable transgenic soybean oil was successfully synthesized using the strategy of the two-step route of amidation and quaterisation. The novel bio-based sulfonic zwitterionic surfactant showed a CMC as low as 33.34 mg/L with a surface tension of 28.50 mN/m and a higher interfacial activity in aqueous solutions containing Na_2CO_3. The interfacial tension between crude oil and the simulated formation water was greatly reduced to an ultra-low order (≤ 10^{-2} mN/m) at 10 mg/L BSZS in solutions. The novel bio-based sulfonic zwitterionic surfactant also exhibited strong emulsifying (504.4 s for 25 vol.% of the emulsion), foaming ability (a foam height 80 mm after 30 min), and good wetting performance (44.1°), which suggested that the novel bio-based sulfonic zwitterionic surfactant should be a very competitive candidate for surfactant flooding in oil recovery, for oil spill treatment, and for many other industrial applications.

NOMENCLATURES

BSZS	: Bio-based Sulfonic Zwitterionic Surfactant
C	: The Concentration of the Surfactant (mol/L)
CMC	: Critical Micelle Concentration (mol/L)
DIFT	: Dynamic Interfacial Tensions
EOR	: Enhanced Oil Recovery
IFT	: Interfacial Tensions (mN/m)
KOH	: Potassium Hydroxide
OSDP	: Oil Spill Disposal Processing
SFT	: Surface Tensions (mN/m)

ACKNOWLEDGMENTS

This research was supported by National Science Foundation of China (Grant No. 51574125, 21203063), the 863 Program (Grant No. 2013AA064403), and the Fundamental Research Funds for the Central Universities of China (Nos. WK1213003, WJ1214066).

REFERENCES

1. Small L. E., Garrison Jr P. H., Winkler W. M., Seaman S. A., and et al., "Ultra Mild Skin Cleansing Composition," U. S. Patent, **1987**, *4*, 812-253.
2. Shah D. O. and Schchter R. S., "Improved Oil Recovery by Surfactant and Polymer Flooding (1st ed.)," Academic Press of the United Kingdom, **1997**, 1-205.
3. Banat I. M., "Bio-surfactants Production and Possible Uses in Microbial Enhanced Oil Recovery and Oil Pollution Remediation: A Review," *Bioresource Technology*, **1995**, *51*, 1-12.
4. Urum K., Grigson S., Pekdemir T., and McMenamy S., "A Comparison of the Efficiency of Different Surfactants for Removal of Crude Oil from Contaminated Soils," *Chemosphere*, **2006**, *62*, 1403-1410.
5. Owoseni O., Nyankson E., Zhang Y., Adams S. J., and et al., "Release of Surfactant Cargo from Interfacially-active Halloysite Clay Nanotubes for Oil Spill Remediation," *Langmuir*, **2014**, *30*, 13533-13541.
6. Foley P., Beach E. S., and Zimmerman J. B., "Derivation and Synthesis of Renewable Surfactants," *Chemical Society Reviews*, **2012**, *41*, 1499-1518.
7. Sreenu M., Rao B. V., Prasad R. B. N., Sujitha P., and et al., "Synthesis, Surface and Biological Properties of Sodium N-Acyl Isoleucines," *European Journal of Lipid Science and Technology*, **2014**, *116*, 193-206.
8. Rajabi F. and Luque R., "An Efficient Renewable-Derived Surfactant for Aqueous Esterification Reactions," *RSC Advances*, **2014**, *4*, 5152-5155.
9. Abdelkader M. B., Azizi N., Chemli M., Chevalier Y., and et al., "Synthesis and Emulsifier Properties of a New Bio-Sourced Surfactant Based on Isosorbide," *Colloids and Surfaces A: Physicochemical and Engineering Aspects*, **2016**, *492*, 1-11.
10. Fan Z., Zhao Y., Preda F., Clacens J. M., and et al., "Preparation of Bio-Based Surfactants from Glycerol and Dodecanol by Direct Etherification," *Green Chemistry*, **2015**, *17*, 882-892.

11. Huang K., Zhang P., Zhang J., Li S., Li M., and et al., "Preparation of Bio-based Epoxies Using Tung Oil Fatty Acid-Derived C21 Diacid and C22 Triacid and Study of Epoxy Properties," *Green Chemistry*, **2013**, *15*, 2466-2475.
12. Negm N. A., El-Farargy A. F., Halim E. A. A., El-Iboudy S., and et al., "Novel Bio-based Nonionic Surfactants: Synthesis, Surface Activity and Corrosion Inhibition Efficiency against Aluminum Alloy Dissolution in Acidic Media," *Journal of Surfactants and Detergents,* **2014**, *17*, 1203-1211.
13. Hu J., Jin Z., Chen T.-Y., Polley J. D., and et al., "Anionic Polymerizable Surfactants from Bio-based Ω-Hydroxy Fatty Acids," *Macromolecules*, **2013**, *47,* 113-120.
14. Miao S., Wang P., Su Z., and Zhang S., "Vegetable-Oil-Based Polymers as Future Polymeric Biomaterials," *Acta Biomaterialia*, **2014**, *10*, 1692-1704.
15. Biermann U., Bornscheuer U., Meier M. A., Metzger J. O., and et al., "Oils and Fats as Renewable Raw Materials in Chemistry," *Angewandte Chemie International Edition*, **2011**, *50*, 3854-3871.
16. Wu M. H., Wan L. Z., and Zhang Y. Q., "A Novel Sodium N-Fatty Acyl Amino Acid Surfactant Using Silkworm Pupae as Stock Material," *Scientific Reports*, **2014**, *4*, 4428.
17. Zhang Q. Q., Cai B. X., Xu W. J., Gang H. Z., and et al., "The Rebirth of Waste Cooking Oil to Novel Bio-Based Surfactants," *Scientific Reports*, **2015**, 5.
18. Zhang Q. Q., Cai B. X., Xu W. J., Gang H. Z., and et al., "Novel Zwitterionic Surfactant Derived from Castor Oil and Its Performance Evaluation for Oil Recovery," *Colloids and Surfaces A: Physicochemical and Engineering Aspects*, **2015**, *483*, 87-95.
19. Feng D., Zhang Y., Chen Q., Wang J., et al., "Synthesis and Surface Activities of Amidobetaine Surfactants with Ultra-Long Unsaturated Hydrophobic Chains," *Journal of Surfactants and Detergents*, **2012**, *15*, 657-661.
20. Wang H., Gang H., Ye R., and Mu B., "Interaction between Bio-surfactant Surfactin and Cationic Surfactant Cetyl Trimethyl Ammonium Bromide in Mixed Micelle," *Colloid and Polymer Science*, **2014**, *292*, 3169-3176.
21. Lu T., Lan Y., Liu C., Huang J. et al., "Surface Properties, Aggregation Behavior and Micellization Thermodynamics of a Class of Gemini Surfactants with Ethyl Ammonium Headgroups," *Journal of Colloid and Interface Science*, **2012**, *377*, 222-230.
22. Eastoe J., Nave S., Downer A., Paul A. et al., "Adsorption of Ionic Surfactants at the Air-Solution Interface," *Langmuir*, **2000**, *16*, 4511-4518.
23. Zhang Q. Q., Cai B. X., Gang H. Z., Yang S. Z., et al., "A Family of Novel Bio-Based Zwitterionic Surfactants Derived from Oleic Acid," *RSC Advances*, **2014**, *4*, 38393-38396.
24. Nedyalkov M., Alexandrova L., Platikanov D., Levecke B., et al., "Wetting Properties of Aqueous Solutions of Hydrophobically Modified Inulin Polymeric Surfactant," *Colloid and Polymer Science*, **2008**, *286*, 713-719.
25. Sreenu M., Rao B. V. S. K., Prasad R. B. N., Sujitha P. et al., "Synthesis, Surface and Biological Properties of Sodium N-Acyl Isoleucines," *European Journal of Lipid Science*

and Technology, **2014**, *116*, 193-206.

26. Sreenu M., Nayak R. R., Prasad R. B. N., and Sreedhar B., "Synthesis, Surface and Micellar Properties of Sodium N-Oleoyl Amino Acids," *Colloids and Surfaces A: Physicochemical and Engineering Aspects*, **2014**, *449*, 74-81.
27. Ng W., Rana D., Neale G., and Hornof V., "Physicochemical Behavior of Mixed Surfactant Systems: Petroleum Sulfonate and Lignosulfonate," *Journal of Applied Polymer Science*, **2003**, *88*, 860-865.
28. Dai X., Suo J., Duan X., Bai Z., and et al., "A Study of the Technical Route of the Synthesis of Alkylbenzenes for Use as an Oil Displacement Agent," *Journal of Surfactants and Detergents*, **2008**, *11*, 111-115.
29. Chen L., Zhang G., Ge J., Jiang P., and et al., "Research of the Heavy Oil Displacement Mechanism by Using Alkaline/Surfactant Flooding System," *Colloids and Surfaces A: Physicochemical and Engineering Aspects*, **2013**, *434*, 63-71.
30. Tang M., Zhang G., Ge J., Jiang P., and et al., "Investigation into the Mechanisms of Heavy Oil Recovery by Novel Alkaline Flooding," *Colloids and Surfaces A: Physicochemical and Engineering Aspects*, **2013**, *421*, 91-100.
31. Zhao X., Bai Y., Wang Z., Shang X., and et al., "Low Interfacial Tension Behavior between Organic Alkali/Surfactant/Polymer System and Crude Oil," *Journal of Dispersion Science and Technology*, **2013**, *34*, 756-763.
32. Liu Q., Dong M., Yue X., and Hou J., "Synergy of Alkali and Surfactant in Emulsification of Heavy Oil in Brine," *Colloids and Surfaces A: Physicochemical and Engineering Aspects*, **2006**, *273*, 219-228.
33. Qiao W., Li J., Zhu Y., and Cai H., "Interfacial Tension Behavior of Double Long-Chain 1, 3, 5-Triazine Surfactants for Enhanced Oil Recovery," *Fuel*, **2012**, *96*, 220-225.
34. Fekarcha L. and Tazerouti A., "Surface Activities, Foam Properties, HLB, and Krafft Point of Some N-Alkanesulfonates (C14–C18) with Different Isomeric Distributions," *Journal of Surfactants and Detergents*, **2012**, *15*, 419-431.

9

Geochemical Modeling with the Use of Vertical and Horizontal Relative Concentrations of Oil Compounds for the Heavy Oil Fields

Artyom E. Chemodanov[1]*, Robert R. Akhmadullin[2], Vladislav A. Sudakov[1], Sergey A. Usmanov[1], and Rinat K. Khayrtdinov[1]

[1] Institute of Geology and Petroleum Technology, Kazan (Volga Region) Federal University, Russia

[2] PAO "Tatneft", Russia

ABSTRACT

The purpose of this study is the detection of lateral and vertical gradients of relative concentrations of compounds presented in oil. In addition, the detection allows us to assess potential drainage zones in a reservoir during the reservoir production by steam injection. In this paper, a new method for monitoring of steam chamber development in a 2D model was created and tested.

The methodology is used to consider the total hydrocarbon fraction. In addition, the total hydrocarbon fraction has been isolated from core extracts and has been analyzed by GCMS method (TIC) for detection of various compounds and assessment of lateral and vertical gradients of their concentration in lateral.

It has been found out that the ratio of 4- and 1-methyldibenzothiophenes (MDBT) changes in lateral and in vertical directions. These changes are caused by biodegradation of organic matter. Also laboratory research shows that 1-MDBT/4-MDBT ratio in native reservoir rocks is stable under high temperatures and pressures and this can be easily measured by GC-MS. This measurement will allow assessment of location and direction of steam chamber propagation.

In a recent work, the authors have developed a geochemical model which can be used for assessment of oil flow directions during the development of heavy oil fields by SAGD method.

Keywords: SAGD Modeling, Biodegradation of Oil, Beteroatomic Compounds, Isomers, Oil-water Transition Zone

INTRODUCTION

The development prospects of the oil industry have recently been linked to the development of deposits of heavy oil and natural bitumen. In the mid-1980s, Canada developed a Steam Assisted Gravity Drainage (SAGD) technology [1].

A large number of works have been published [2-4] about the modeling of developing super viscous oil deposits using the SAGD method, but many of the parameters characterizing the nature of the reservoir and the basic properties of the oil have been simplified. This may cause significant differences in the model and the real object it describes, which will leads to an inaccurate

***Corresponding author**

Artyom E. Chemodanov

Email: chemodanov41659@mail.ru

prediction of the oil production process.

One promising solution to this problem is the geochemical modeling of the fields, based on the study of the composition zonation and the properties of the oil in the deposit. As we known, any oil contains a unique set of compounds; the ratio between them can both maintain constant values throughout the reservoir as well as show lateral and horizontal gradients. The zoning of the composition and properties of oil in shallow deposits are mainly associated with the biodegradation of organic compounds and is significantly different from the changes caused by physical processes such as washing out and fractionation. For example, the article [5] shows that hopanes, methyladamantanes do not have horizontal gradients by the content, while the concentration of mono-, di-and tri-aromatic hydrocarbons (including heteroatomic) decreases with depth. In [6] authors demonstrated that concentration of 6H-Farnesol in oil may be increased with depth. Specifying dependency of the distribution of any compounds over the whole field will certainly clarify the location and direction of the steam chamber by using GCMS analysis of the extracted oil.

Thus, the establishment of a method for monitoring the development of deposits by natural oil markers is a matter of urgency, which will make it possible to clarify existing patterns of deposits and improve their predictive capacity.

The purpose of this work is to identify the gradients of ratios of compound molecules and show their possible use for the development of super viscous oil deposits in the case of the Nizhne-Karmal'skoe anticline of the Cheremshanskoe oil field (Tatarstan, Russia), where the SAGD technology is currently being used.

EXPERIMENTAL PROCEDURES

The research objects are core samples, selected from the 12 assessment wells of the Nizhne-Karmal'skoe anticline of the Cheremshanskoe oil field (Figure 1). In addition, we analyzed 2 core samples after hydrothermal treatment and 5 samples of oil extracted by the SAGD method.

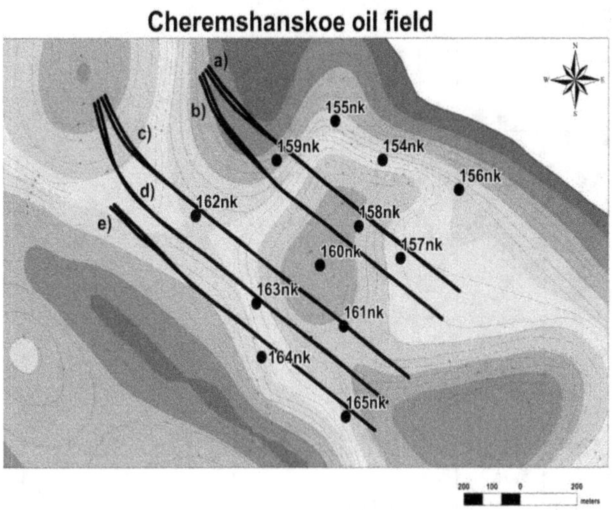

Figure 1: Location of the appraisal and horizontal wells of the Nizhne-Karmal'skoe anticline of the Cheremshanskoe oil field.

The results of the macro description of core show that the oil reservoir being studied has one layer in the entire field, but its depth and thickness vary greatly. The core samples were selected on the whole thickness of the productive reservoir, if possible at the same intervals. As a result, the number of core samples selected from the appraisal wells ranged from 3 to 6 per well, and the thickness of the sampling interval was 4-8 meters, depending on the thickness of the oil reservoir (Table 1). The upper and lower samples were selected as close as possible to the top and the base surface of the reservoir.

Table 1: The number of core samples selected from the assessment wells.

No.	Well	The depth of the oil reservoir, m			Number of samples
		Top	Base surface	Thickness	
1	154	131.5	154.1	22.6	4
2	155	145.4	156.2	10.8	3
3	156	129.6	158.2	28.6	4
4	157	124.4	160.0	35.6	6
5	158	130.2	160.3	30.1	5
6	159	151.8	168.3	16.5	3
7	160	137.1	170.0	32.9	6
8	161	129.8	168.8	39.0	5
9	162	171.0	179.0	8.0	3
10	163	150.0	174.3	24.3	4
11	164	149.8	182.3	32.5	5
12	165	132.6	162.6	30.0	4
Total					**52**

The extraction of bitumen using chloroform was carried out using the Soxhlet extractor. The hydrocarbon fraction, containing saturated and aromatic hydrocarbons, provided using a method of liquid-absorbed chromatography on the siliceous gel (0.25-0.5 mm fraction). Petroleum-ether was used as a movable phase.

The hydrocarbon fraction of bitumen and oil was examined on the gas chromatography/mass spectrometer system. The gas chromatography/mass spectrometer system included the gas chromatograph "Chromatech-Crystal 5000" with a mass-selective ISQ detector. The Xcalibur software has been used for processing the results. The chromatograph is fitted with a capillary column, 30 m long, with a diameter of 0.25 mm. The speed of the flow of carrier gas (helium)-1 mL/min. The injection temperature is 310 °C. The temperature program of the thermostat is a rise from 100 to 150 °C at a speed of 3 °C/min, from 150 to 300 °C at a speed of 12 °C/min followed by its isotherm to the end of the analysis. Electron energy-70 eV, temperature of ion source-250 °C. The compounds have been identified through the electronic library of the NIST spectra and according to the data of the literary sources.

The hydrothermal processing of core samples was performed in the digestion bomb Parr Instruments 4560 (USA), with a volume of 300 cm^3 with mixing under the following conditions: Initial pressure 3 bar, operation pressure 17 bar, operational temperature 200 °C. The duration of the operational process was 6 hours. The core relative water content was 10%.

RESULTS AND DISCUSSION

Search for Compounds with Concentration Gradients

The chromatography-mass-spectrometric analysis of hydrocarbon fractions shows the absence of normal alkanes and the dominance of isoprenoid hydrocarbons in all samples (Table 2).

Table 2: Results of the GCMS (gas chromatography-mass spectrometry) analysis of hydrocarbon fractions of Bitumen.

Compound	m/z	Presence (+/-)	Note
N-alkanes	TIC *, 85	-	Not found even in trace amounts
Isoprenoid alkanes	TIC, 85	+	The contents vary between 60 and 90% rel. (78% on average for all samples)
Monoaromatic hydrocarbons	TIC, 91, 133	+	The content varies between 2 and 20% rel. (10% on average for all samples)
Naphthalene	128	-	-
Monomethylnaphtalene	142	-	-
Dimethylnaphtalene	156	-	-
Trimethylnaphtalene	170	+	-
Tetramethylnaphthalene	184	+	-
Steranes	217	+	The main sterane and hopane geochemical coefficients are the same for all specimens
Hopanes	191	+	

* TIC-Total Ion Current

The absence of n-alkanes in the core sample indicates the degradation of organic matters. There exist two classical scales of oil biodegradation, namely Peters-Moldovan (PM) [7] and Wenger [8] scales, based on which ones can assess the approximate depth of biological destruction of organic matters. Both scales were used for the assessment of the availability of n-alkanes, isoprenoids, steranes, and hopanes. The Wenger scale, unlike the PM scale, also includes the identification of di- and tri-aromatic compounds. Table 2 shows that, apart from the lack of n-alkanes, which we can observe on the total ion current, there are no C_0-C_2-Naphtalenes in the specimens studied, although C_3-and C_4-Naphtalenes are still identifiable. It is also worth noting that the presence of monoaromatic hydrocarbons, hopanes, and will be mentioned. On the basis of the results of the gas chromatography/mass analysis, we concluded that organic matters in the specimens are moderately biodegraded on the PM scale (4-5 units) or heavily biodegraded on the Wenger scale (3 units).

It is necessary to determine which components surviving in the biodegradable environment are characterized by gradients of relative concentrations throughout the deposit. Relative concentrations of isoalkanes and aromatic hydrocarbons do not have any patterns in the distribution of the prospect. Vertical and lateral gradients of hopane and sterane geochemical factors of are also not observed. For example, the value of the coefficient 22S/(22S + 22R) is stable [9], almost identical for all specimens, and is 0.58-0.61 (Figure 2).

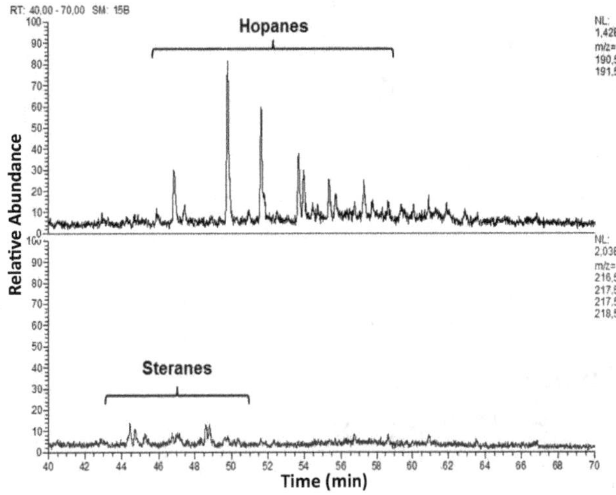

Figure 2: GCMS chromatograms for steranes and hopanes (m/z = 191, 217+218).

Heteroatomic compounds are quite resistant to the processes of degradation [10]. Two of these compounds, which are present in all studied samples, are 1- and 4-methyldibenzothiophenes. In work [11] was shown that 4-methyldibenzothiophene (4-MDBT) was less resistant to degradation than 1-methyldibenzothiophene (1-MDBT) so that the ratio 1-MDBT/4-MDBT is an indirect indicator of the degree of degradation of organic matters. Figure 3 shows mass-chromatograms (by m/z = 198) of two samples of the well No. 163 core, indicating the peaks of 1-MDBT and 4-MDBT.

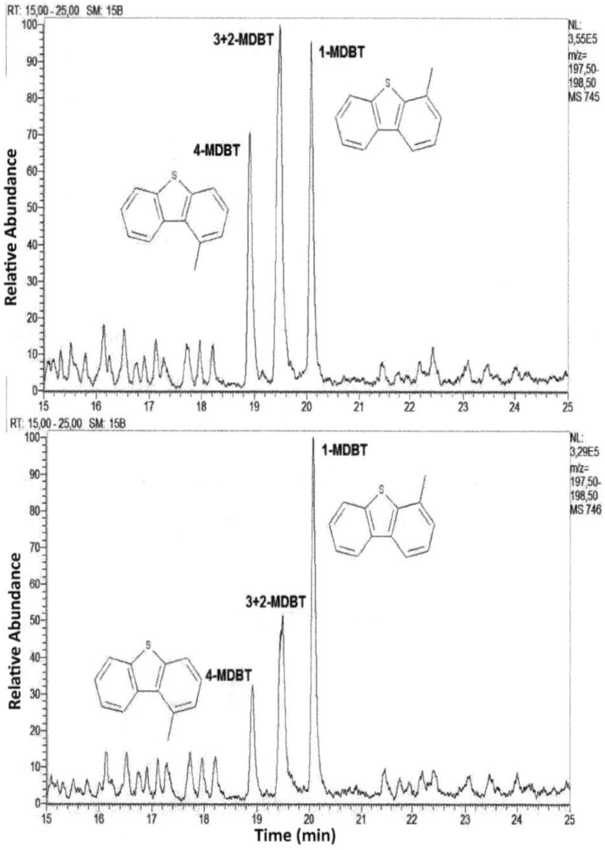

Figure 3: Mass-chromatograms (m/z = 198) of various samples of the well No. 163.

Construction of A Geochemical Model for The Distribution of Ratio 1-MDBT/4-MDBT

For all core samples, a ratio of 1-MDBT/4-MDBT, which varies in the range of 0.4-11.0 (Table 3), has been defined. Average values of 1-MDBT/4-MDBT on wells lie between 0.6 and 9.4.

Table 3: Values of the 1-MDBT/4-MDBT ratio for kern samples from different wells.

Well	The depth of the productive layer, m	1-MDBT/4-MDBT ratio	
		The increase of the sampling depth	Average value of the well
154	131.5 – 154.1	0.9 – 0.9 – 1.9 – 2.1	**1.5**
155	145.4 – 156.2	8.0 – 9.2 – 11.0	**9.4**
156	129.6 – 158.2	1.3 – 1.3 – 2.5 – 0.4	**1.4**
157	124.4 – 160.0	0.5 – 0.7 – 0.6 – 0.7 – 0.7 – 1.6	**0.8**
158	130.2 – 160.3	0.6 – 0.7 – 0.9 – 0.7 –1.8	**0.9**
159	151.8 – 168.3	3.3 – 2.8 – 2.7	**2.9**
160	137.1 – 170.0	0.6 – 0.7 – 0.6 – 0.6 – 0.9 – 2.2	**0.9**
161	129.8 – 168.8	0,6 – 0,5 – 0,8 – 0,9 – 0,4	**0.6**
162	171,0 – 179,0	1,2 – 5,3 – 8,7	**5.1**
163	150,0 – 174,3	1,0 – 1,2 – 1,1 – 2,8	**1.5**
164	149,8 – 182,3	0,9 – 1,1 – 1,1 – 2,6 – 2,2	**1.6**
165	132,6 – 162,6	0,7 – 0,9 – 3,5 – 3,7	**2.2**

According to the lateral distribution of the average ratios 1-MDBT/4-MDBT, the geochemical 2D models of the area of the deposit (Figure 4) were constructed using Schlumberger Petrel 2013 software. Zones with low values of 1-MDBT/4-MDBT are colored blue and dark blue, and the zones with a high ratio are yellow and red. It should be noted that we have a reasonably accurate model using data from six wells. Adding six more wells does not radically change the model which was built, but merely makes it more accurate.

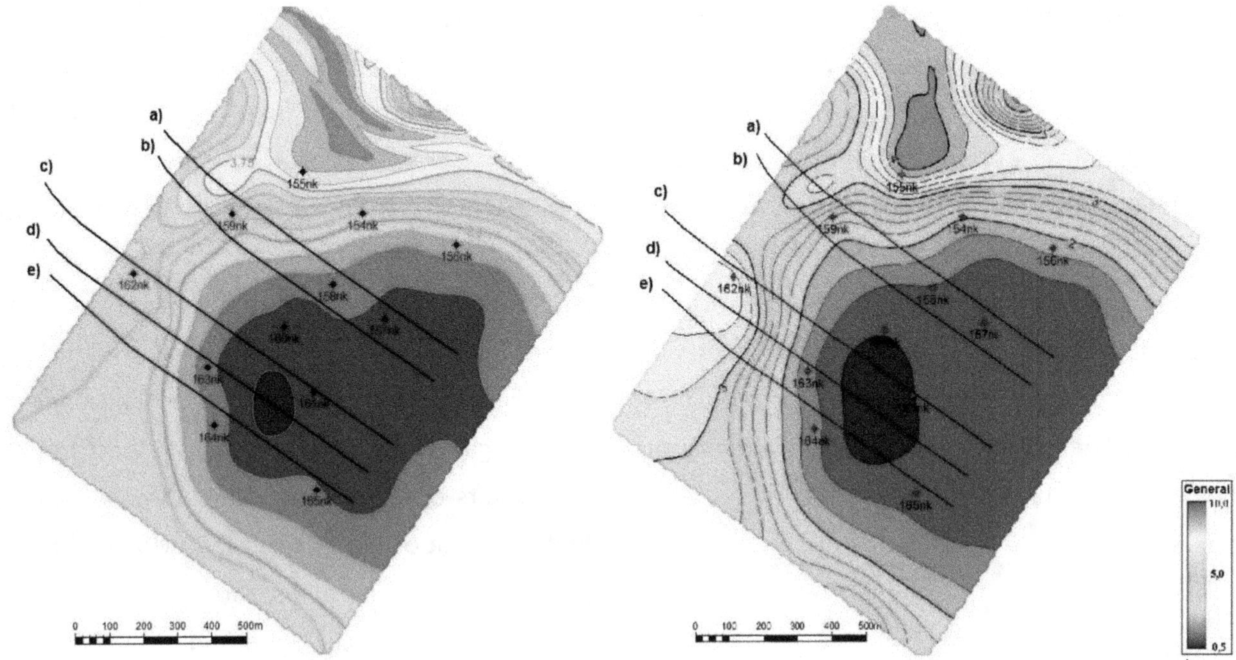

Figure 4: Lateral distribution of the average value of 1-MDBT/4-MDBT by area based on data for 6 wells (left) and 12 wells (right). The black lines indicated by the letters (a) to (e) are the horizontal extracting wells.

According to Table 3, the ratios 1-MDBT/4-MDBT can be quite different even for samples taken from the same well but from different depths. The vertical distribution of the ratios 1-MDBT/4-MDBT for some of the appraisal wells situated close to the production well is presented in Figure 5. In addition these wells have relatively low values of 1-MDBT/4-MDBT at the top of the productive layer. In the lower part of the oil reservoir, there is a sharp increase in the ratio of isomers of methyldibenzothiophenes: at a depth of 150 meters for wells 157, 158 and 160 and at a depth of 160 meters for wells of 160 and 163. It is important to note that this pattern is not characteristic for the well 161, as the entire thickness of its productive layer is low in the values of 1-MDBT/4-MDBT.

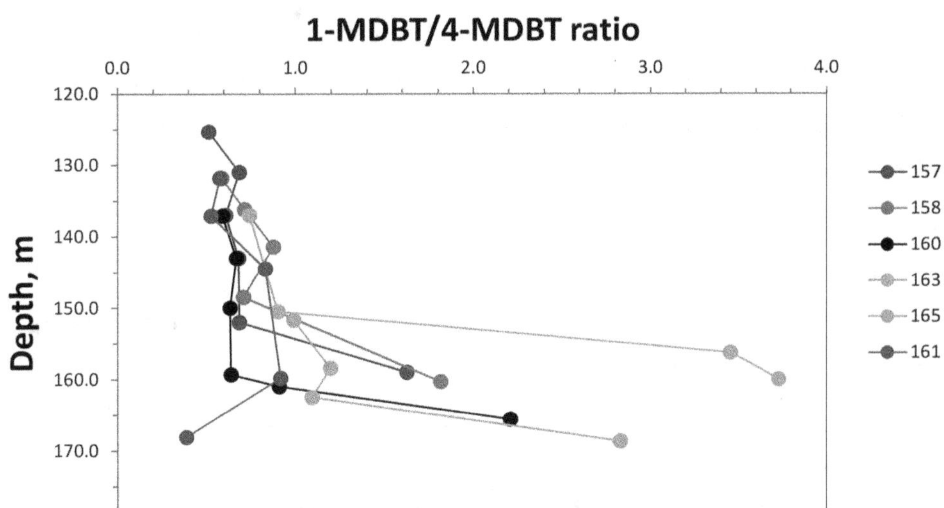

Figure 5: Vertical distribution of 1-MDBT/4-MDBT ratios for some appraisal wells.

The gradients obtained may be related to the presence of oil-water transition zone at a depth of 150-160 meters, where the most intensive biodegradation of organic matters occurs. As the depth of oil decreases, its biological decomposition is also beginning to be suspended. It is also worth noting that the indicator 1-MDBT/4-MDBT correlates with the thickness of the oil reservoir, which is exponential (Figure 6), due to the reduction in biological effects while reducing the thickness of the productive layer.

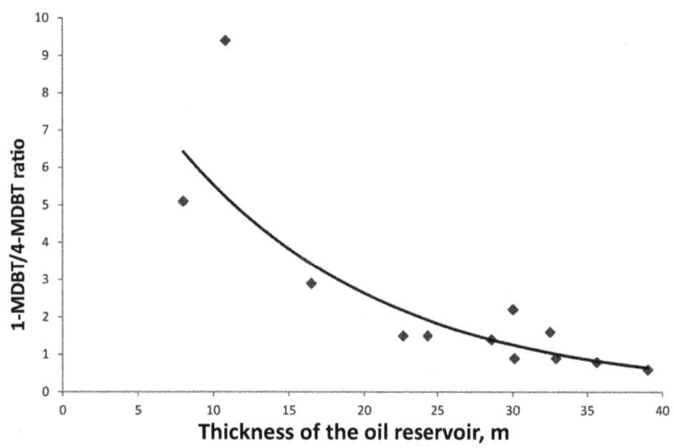

Figure 6: Dependence of average ratio 1-MDBT/4-MDBT of the wells from the thickness of the oil reservoir.

Check of the Reliability of the Received Models

It is clear that the constructed models can assess the likely pathways of super viscous oil inflows by measuring the ratio of 1-MDBT/4-MDBT. However, some heteroatomic compounds may be destroyed in steam-heat processing conditions. For example, in the work [12], it was shown that thiophene is desulphurized even at lower temperatures, with the catalytic effect of the hematite contained in the rock. Afterwards, the impact of hydrothermal treatment on the ratio of isomers of methyldibenzothiophenes in super viscous oil extracted by SAGD technology need to be clarified for the correct application of the models received. A laboratory modeling of reservoir conditions has been chosen as a tool to verify this influence. It is covered by a wide range of publications [13-14]. Two core samples with different values of 1-MDBT/4-MDBT were placed in an autoclave in conditions more stringent than those of the SAGD conditions (they are given in the experimental part of the article). After hydrothermal treatment, a hydrocarbon fraction was identified and analyzed on the gas chromatography/mass spectrometer. The results of the GHMS analysis showed almost identical ratios of 1-MDBT/4-MDBT in the processed specimens compared to the original. On the basis of the experiment, it has been established that hydrothermal processing of the core does not affect the distribution of isomers of methyldibenzothiofenes as part of its organic matter. In addition, a change in the ratio 1-MDBT/4-MDBT in the super viscous oil produced by the SAGD method, seemed to be negligible.

In order to suggest possible ways of inflow of oil to the horizontal wells, samples of the oil from the horizontal wells shown in Figure 1 were marked from (a) to (e) were analyzed. The molecular composition of oil hydrocarbon fractions, according to HMS analysis, showed the same components as found in the core samples (Table 2). However, the 1-MDBT/4-MDBT ratios for samples of extracted oil from the horizontal wells showed the results presented in Table 4.

Table 4: The results of the definition of the ratio 1-MDBT/4-MDBT in the specimens of the oil extracted using the SAGD technology.

Well	1-MDBT/4-MDBT
a)	0.6
b)	0.7
c)	0.5
d)	1.6
e)	0.6

The ratio 1-MDBT/4-MDBT for specimens (a), (b), (c) and (e) is roughly the same, 0.5-0.7, which corresponds to the blue zone on the lateral distribution. In addition, for these horizontal wells, it can excluded that the shift of the steam camera will be down horizontally. For sample (d) the 1-MDBT/4-MDBT ratio is 1.6, and thereby the ratio indicates that there are inflows of super viscous oil to the production well from the lower part of the productive layer (from a depth of 150-160 meters).

CONCLUSIONS

In this paper, the level of biodegradation of organic matter in the Nizhne-Karmal'skoe anticline of the Cheremshanskoe oil field was established. We found the presence of horizontal and lateral gradients of the ratio of methyldibenzothiophenes in the organic matter of the oil reservoir. Moreover, the stability of this ratio in conditions of SAGD process was shown. The presence of these gradients is determined by the depth of the occurrence, the thickness of the oil reservoir, and (by paying attention to) the location of the appraisal wells. Within the developed model, it has been shown that the molecular composition of produced super viscous oil suggests possible ways of influx in the development of the deposits by the SAGD method.

ACKNOWLEDGMENTS

The reported study was performed according to the Russian Government Program of Competitive Growth of Kazan Federal University. The work was supported by the Ministry of Education and Science of the Russian Federation (Project No. 02.G25.31.0170)

NOMENCLATURES

CG-MS	: Gas Chromatography-Mass Spectrometry
COF	: Cheremshanskoe Oil Field
MDBT	: Methyldibenzothiophenes
SAGD	: Steam-Assisted Gravity Drainage
TIC	: Total Ion Current

REFERENCES

1. Butler R. "Horizontal Wells and Steam-assisted Gravity Drainage," *The Canadian Journal of Chemical Engineering*, **1991**, *69*(4), 819-824.
2. Khisamov R., "The Analytical Model for Development of Heavy Oil Deposit by Steam-assisted Gravity Drainage Method," *Neftyanoe Khozyaystvo-Oil Industry*, **2015**, *2*, 62-64.
3. Akin S., "Mathematical Modeling of Steam Assisted Gravity Drainage," *SPE Reservoir Evaluation and Engineering,* **2005**, *8*, 372-376.
4. Alali N., Reza M., and Jabbari H., "A New Semi-analytical Modeling of Steam-assisted Gravity Drainage in Heavy Oil Reservoirs," *Journal of Petroleum Science and Engineering,* **2009**, *69*, 261-270
5. Bennett B. and Adams J. "The Controls on the Composition of Biodegraded Oils in the Deep Subsurface-Part 3," The impact of microorganism distribution on petroleum geochemical gradients in *Biodegraded Petroleum Reservoirs Organic Geochemistry,* **2013**, *56*, 94-105.
6. Chemodanov A. E., Sudakov V. A., Usmanov S. A., and Ahmadullin R. R., "Application of Geochemical Model for Monitoring the Superviscous Oil Deposits Development by the Thermal Steam Methods," *Neftyanoe Khozyaystvo-Oil Industry*, **2017**, *9*, 104-107.
7. Peters K. and Moldowan J., "The Biomarker

Guide: Interpreting Molecular Fossils in Petroleum and Ancient Sediments," Prentice Hall, Englewood Cliffs, NJ, **1993.**

8. Wenger L., Davis C., and Isaksen G., "Multiple Controls on Petroleum Biodegradation and Impact on Oil Quality," *SPE Reservoir Evaluation and Engineering,* **2002**, *5*, 375-383
9. Seifert Wolfgang K., "Application of Biological Marker Chemistry to Petroleum Exploration," *World Oil,* **1980**, *2*, 425-438
10. Peters K., Walters C., and Moldowan J., "The Biomarker Guide," Biomarkers and Isotopes in *Petroleum Exploration and Earth History*, Cambridge University Press, **2005.**
11. Bennett B., Adams J., and Larter S., "Oil Fingerprinting for Production Allocation: Exploiting the Natural Variations in Fluid Properties Encountered in Heavy Oil and Oil Sand Reservoirs," *Frontiers + Innovation CSPG CSEG CWLS*, **2009**, 157-160
12. Khalil M., Lee R., and Liu N., "Hematite Nanoparticles in Aquathermolysis: A Desulfurization Study of Thiophene," *Fuel,* **2015**, *145*, 214-220
13. Sitnov S., Feoktistov D., Kayukova G., Pronin N., and et al., "Catalytic Intensification of In-situ Conversion of High-viscosity Oil in Thermal Steam Extraction Methods," *International Journal of Pharmacy and Technology*, **2016,** *8*(3), 14884-14892
14. Vakhin A., Onishchenko Y., Chemodanov A., Sitdikova L., and et al., "Thermal Transformation of Bitumoid of Domanik Formations of Tatarstan," *Khozyaystvo-Oil Industry,* **2016**, *10,* 32-34

10

Petrographic Characteristics and Fluid Inclusion Study of Carbonate Cements in Permian-Triassic Rock Sequence of Southern Iran: an Implication of Rock-fluid Interactions in Carbonate Reservoir Rocks

Pouran Nazarian Samani*, Maryam Mirshahani, and Navab Khodaei

Petroleum Geology Division, Research Institute of Petroleum Industry, Tehran, Iran

ABSTRACT

The study of the carbonate cements in the Permian-Triassic Dalan-Kangan formations resulted in the identification of five stages of calcite and dolomite cementation, which completely or partially occluded pores. Cement types appear to be early isopachous calcite (i; C1-non-luminescence), equant (ii; C2-dull CL), fairly coarse secondary dolomite rhombs (iii; C3-zoning red CL), coarse sparry calcite (iv; C4-dull CL), and meteoric equant calcite (v; C5-bright orange CL & vi; C6-bright yellow CL). Fluid inclusion analyses invoked three distinct groups of fluids. Group 1 is reflected by fluids with equant calcite (C2) composition, characterized by Th values (126 °C) and salinity of 16 wt.% (NaCl equivalent), and they are interpreted as burial pore-waters. Group 2, which is represented by fluids associated to shallow burial dolomite cementation (C3), has Th values about (127 °C) and salinity of 17 wt.% (NaCl equivalent). The last group, i.e. coarse sparry calcite (C4), is occurred along fractures and vugs and characterized by high Th values (169 °C) and salinity of 17.5 wt.% (NaCl equivalent) which shows that fluids with a higher temperature migrated from deeper parts during deep burial. The hydrocarbon inclusions with yellow fluorescence can be observed in coarse calcite filling fractures in Dalan formation which shows that they are secondary, and hydrocarbon migration predates the precipitation of fracture filling cements.

Key words: Diagenesis, Fluid Inclusion, Kangan, Dalan.

INTRODUCTION

The hydrocarbon reservoirs of the upper Dalan and Kangan ("Khuff equivalent") formations hold some of the most important gas reservoirs in the world [1]. This is demonstrated by the South Pars in Iran and North Dome Gas field in Qatar, which is the largest offshore gas field in the world and is estimated to have over 1100 trillion cubic feet of gas initially-in-place (TCF GIIP) [2]. In particular, the upper Dalan-Kangan reservoirs of onshore and offshore Iran contain some of the largest gas fields in the Middle East region [3]. Recently, these rock sequences, in both outcrop and subsurface, have been more or less well-documented in Iran [4-5] and like their equivalents in Abu Dhabi and Saudi Arabia [6-7].

***Corresponding author**

Pouran Nazarian Samani

Email: samanip@ripi.ir

The studied area is one of the gas fields in the Persian Gulf (Figure 1). More than 60% of the reservoir rock at this field is dolomitized, and anhydritization occurs in close association with the dolomites, reflecting the influence of hypersaline depositional conditions on both calcium sulfate precipitation and dolomitization [8]. This study is conducted to recognize different carbonate cements (calcite and dolomite) occluding the pore spaces, and to document the fluid flow and diagenetic evolution in the Permian-Triassic Dalan-Kangan carbonate reservoirs in the southern Iran. Calcite and dolomite cements formed during burial are found in many carbonate reservoirs [9], associated with chemical compaction commonly occluded porosity to varying degrees in the investigated rock successions. On the contrary, early diagenetic processes such as aragonite dissolution and host rock dolomitization in addition to depositional facies are believed to have played a prominent role in fortifying the porosity and permeability of Dalan-Kangan carbonates reservoir [5, 7, 8, 10, 11, 12, 13, 14, 15]. In this paper, the entire Permian-Triassic reservoir succession is studied in point of petrographic, cathodoluminescence (CL), and fluid inclusion micro-thermometry to reveal: 1) the origin and geochemical evolution of fluids; and 2) the timing of the formation of different cements and their circulation within the studied succession.

EXPERINTAL PROCEDURES

Geological Setting

During the Permian, a shallow restricted evaporite-carbonate platform covered most of the Persian Gulf region [1]. Short-term sea level fluctuations caused recurrent shoaling pulses which culminated in the creation of evaporite sabkha and salinas particularly over highs. The carbonates indicate a shallow-shelf to a coastal-plain depositional environment. In general, a prograding terrigenous to shallow-marine clastic shoreline existed. The sediments of the late Early Permian on the Arabian plate are characterized by regressive sediments such as the Unayzah (A and B) formation. Dolostones appear to constitute most of the Permian rock sequence in the Persian Gulf region. At the end of the Early Permian time, a major marine transgression resulted in the establishment of carbonate platform over much of the region. Following the widespread carbonate deposition during Late Permian [2], arid and semi-arid climate conditions are characterized by the Triassic succession (Figure 2). There was an increase in periodic shallow water carbonate deposition commonly interbedded with anhydrite. After uplift and rifting, shallow-marine conditions became dominant during the Late Triassic time [2].

Sedimentation continued without interruption or facies changes throughout the Permian-Triassic boundary in the Persian Gulf region particularly in the northern regions [16]. The Upper Permian-Triassic carbonate and evaporites of the Persian Gulf basin accumulated on an epeiric shelf.

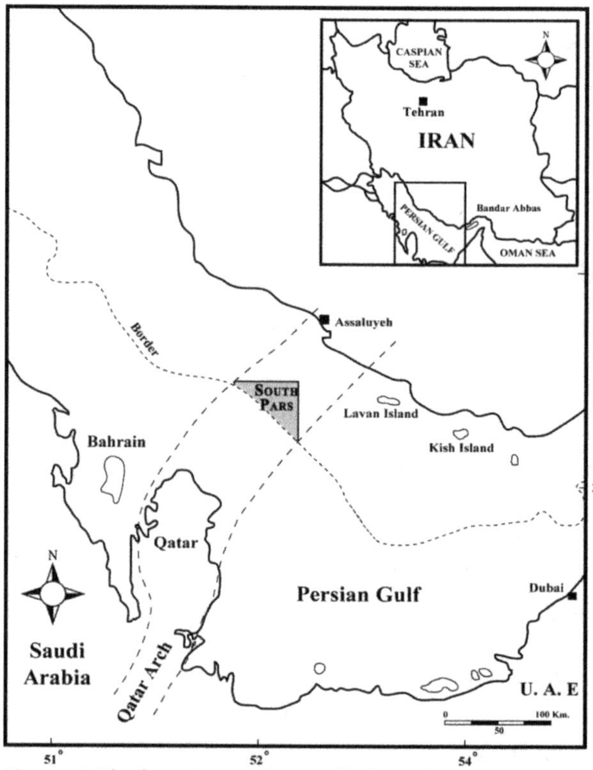

Figure 1: The location of the studied gas field in the Iranian portion of the Persian Gulf is marked by a grey triangle.

interrupted periodically by a series of transgressive and regressive events. In the Zagros basin to the southeast and across the Persian Gulf to northern Oman, thick Permian carbonates pass upward into Triassic carbonates (Figure 2). Hence, the continuous Upper Permian-Lower Triassic carbonates may act as a single reservoir, especially where overlying Triassic Dashtak evaporites served as a caprock [3, 5, 17].

Dalan formation has a total thickness of 600 meters consisting of medium to thick-bedded oolitic to micritic shallow-marine carbonates, locally reefal, with intercalations of evaporites. It extends up into the Lower Triassic Kangan formation in the southern Iran. Moreover, Dalan formation is subdivided into four reservoir units, including K5 (lower Dalan), Nar member, K4, and K3 (upper Dalan), which overlies the Faraghun formation (Figure 2). The transgressive lower Dalan formation was deposited in a more open marine environment as indicated by the presence of the open marine fossils. However, with depositional progradation and less subsidence, the depositional environment became progressively restricted. This eventually led to the deposition of evaporites in the Permian Sea. The main parts of the Dalan carbonates were deposited as a restricted evaporitic, carbonate shelf facies and include subtidal carbonate sand shoals, lagoons, and bars that spread out as sheets across long distances, along with intertidal and supratidal sabkha facies.

Kangan formation represents the first transgression of the Triassic Sea in the Zagros and the northern Persian Gulf region. Subdivided into two reservoir units (K2 and K1), Kangan formation overlays uncomfortably on Dalan formation (Figure 2), and passes upward into the Aghar shale member of Dashtak formation. A more or less fossil-barren zone is visible at the base of Kangan, which is followed by microbial mats and thrombolytic beds across the Zagros basin and Arabian Peninsula [1]. This marker bed specifies the Lower Kangan formation. Kangan formation correlates with Khuff-A member and the lower part of Khuff-B in the southern Persian Gulf [18]. The carbonate sequences and the evaporite facies were deposited in an arid homoclinal ramp platform, with the environmental conditions changing from sabkha to open marine settings.

Methodology

About 900 thin-sections were prepared from the core samples of the upper Dalan and Kangan formations and were used to describe petrographically. The thin-sections were treated with Alizarin-red-s following Dickson [19] methodology. Standard petrographic analyses on the stained thin sections were carried out to aid the recognition of calcite and dolomite cementing phases. Cathodoluminescence was applied to determine different generations of cements using a Cold System CL Microscope (model 8200 MK4). For this study, 30 samples were selected to prepare the polished thin section. The study was conducted by means of CL apparatus in the RIPI laboratory in 15 kv and 400 μA conditions.

Fluid inclusions were analyzed petrographically using the methods of Burruss [20]. Micro thermometric data were collected by standard techniques [21, 22], and the equation of Bodnar [23] was used to convert final ice melting temperatures (Tm) to salinity. For this study, 30 samples were selected for preparing the doubly-polished thick sections, and micro thermometry was carried out on 11 samples using a Linkam THM600 heating-cooling stage and the controller attached to an Olympus petrographic microscope in the RIPI laboratory; the precision was 0.05 °C. The phase transition temperatures measured are calibrated using synthetic inclusions and pure chemicals obtained commercially which are accurate to better than 1.2 °C between 0 °C and 300 °C, and better than 0.3

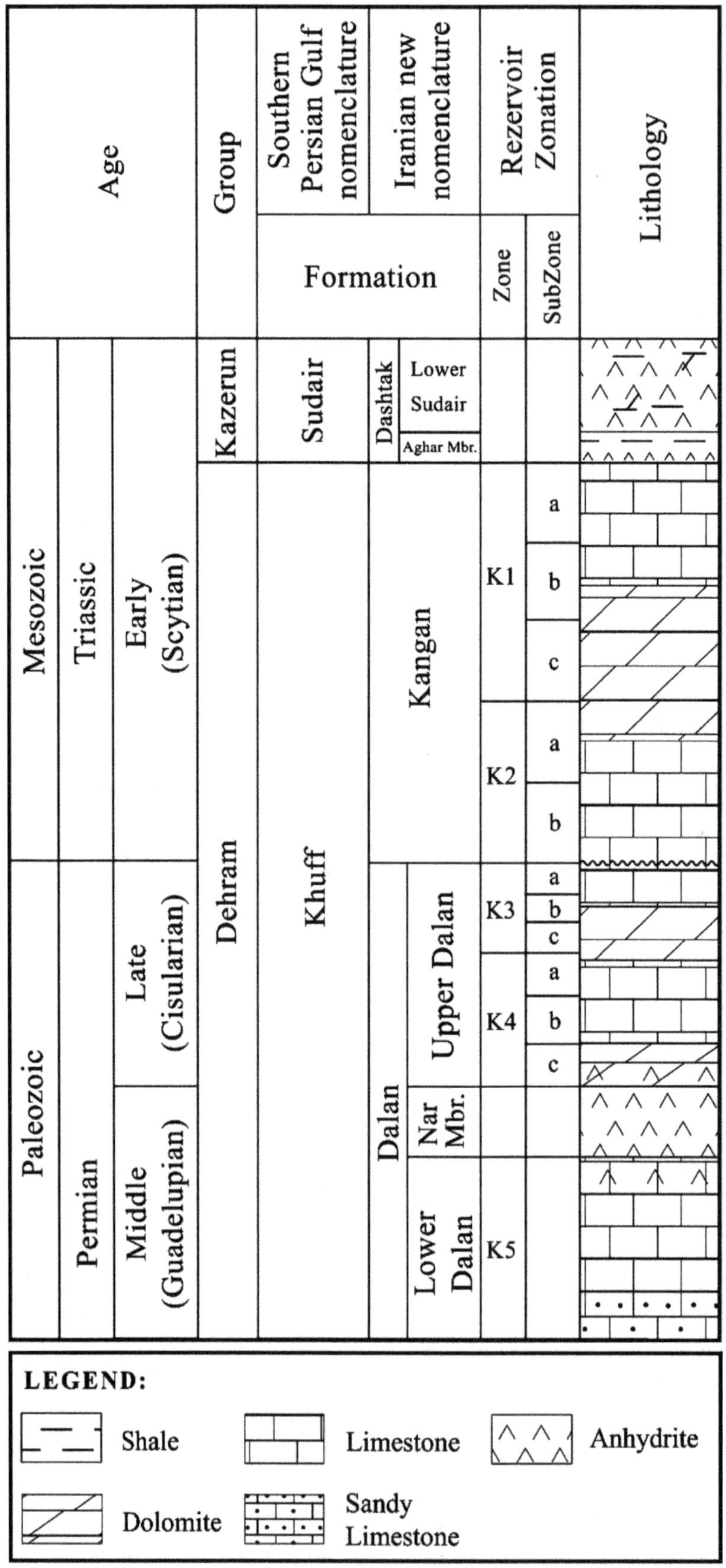

Figure 2: Stratigraphic nomenclature for Permian-Triassic (and underlying) rocks in Iran (modified after [5]).

°C between -150 °C and 0 °C. To measure these temperatures, small inclusion-bearing portions were cut from the wafers and, for each piece, homogenization temperatures were measured in an increasing order before any freezing runs. This minimizes the risk of stretching or bursting the inclusions before the data have been obtained. Petroleum and aqueous inclusions were identified under a combination of incident UV and transmitted plane polarized light on a standard petrographic microscope. API gravities of fluorescent petroleum inclusions were estimated from a correlation of their visible range spectra with spectra from a calibration set of 67 reservoir oil and condensates, spanning the API range of 10-60°. The spectra were obtained using an Olympus UMSP50 spectrometer with a barrier filter at 395 nm, scanning a diffraction grating at 4 nm intervals. Variable sized aperture masks allow spectra to be obtained from individual inclusions.

RESULTS AND DISCUSSION

Petrography and Cathodoluminescence

Kangan and Dalan carbonates are extremely susceptible to mineralogical and textural change, cementation, and dissolution. These alterations can occur at any time from initial deposition to deep burial [24-25]. Moreover, the diagenesis of carbonates can take place in many settings: the marine environment during deposition of the sediment, near the sediment surface where fresh waters penetrate the sediments, or in deeper subsurface. In addition, diagenetic changes may affect porosity, so they must be considered for better exploration in carbonates. Cementation is a major diagenetic process affecting carbonate sediments and rocks in Kangan and Dalan formations in a studied well. Cementation comprises processes leading to the precipitation of minerals in primary or secondary pores and requires the super saturation of pore fluids with respect to the minerals [26]. Cementation is the main result of the chemical precipitation of a binding agent or the chemical welding of adjacent detritus grains. In this case, a chemical precipitate, usually calcium carbonate, fills the pores [27]. Sometimes dolomite precipitates between pore spaces as an initial mineral and acts as a binding agent between the grains. The bulk and mineral compositions change slowly during diagenesis in response to changes in pore water chemistry; thus, they can provide perception in the rock-fluid interaction evolution. Subsequently, various types of cements at different times can be distinguished in the Dalan and Kangan formations. These cements include calcite, anhydrite, and dolomite cements which partially or completely have filled pore spaces. The common type of carbonate cements which has affected the pores are isopachous, granular equant, and coarse blocky dolomite cement.

Isopachous Cement

Isopachous rim cement (C1) forms as a relatively thin film around skeletal grains. This type is not mostly disseminated and appear in many cases leached, indicating a primary metastable mineralogy [28-29]. The nature of this calcite phase indicates precipitation in the marine-phreatic environment, where all pore spaces are filled with marine water [30]. It mostly appears on the outer and inner surfaces of micritized bioclasts, forming an isopachous coating of a variable size (Figures 3A, 3C, and 3H). Under CL, the isopachous cement is non-luminescent.

Granular Calcite

The granular equant calcite cement occurred as pore filling material occluding different types of pore spaces, including fractures (Figure 3E), interparticle pore spaces (Figure 4F), and intraparticle pore spaces

(Figure 3C). Under a cross-polarized light, clear calcite crystals show a straight extinction. Most aragonitic ooids, foraminiferas, and molluscs are represented by molds occupied with granular calcite. The precipitation of this cement type occurred after a period of dissolution, where gastropods, some aragonitic ooids, and foraminiferas have been dissolved. Cement crystals in pores generally increase in size from pore boundary toward pore center (Figure 3G). Granular calcite cements overlay the cloudy isopachous calcite cements (Figure 3H). The granular equant calcite cement have variable luminescence patterns, from dull red, C2, (Figures 4C and 4G), to bright orange, C5, (Figures 4A and 4C), and to brightly yellow, C6, (Figure 4E) luminescent zones, indicating that it was formed under different fluid compositions. By an optical conventional microscope view, equant calcite cements (types C5 and C6), the bright light orange, and yellow luminescent appear to have been formed after the precipitation of dolomite (C3) cements (Figures 3D and 3G and Figure 4C). They appear to be the last cement generations and have petrographic characteristics [31-32] and cathodoluminescent characteristics similar to those of meteoric-phreatic origin.

Coarse Blocky

Coarse blocky cement (C4) occurs as a pore filling material occluding all types of pore spaces including fractures, interparticle pore spaces, and vuggy porosities (Figures 3F and 3H, Figures 4E and 4F). Late stage, deep burial cement crystals are commonly large crystals filling more than one interparticle pore space, facture, and molds. They have dull luminescent (Figure 4E).

Dolomite Cement

By secondary dolomitization, dolomite cement occurred in some molds and vugs (Figures 3B, 3D, 3F, and 3G) in Kangan and Dalan formations, which reveals red zoning luminescent pattern (Figure 3G and Figure 4C). According to the witness coming through microscopic analyses, dolomite crystals postdate the equant C2 and predate coarse blocky calcite C4 (Figure 3F) and equants C5, C6 (Figures 3G and 4C).

Fluid Inclusion Study

Calcite cements, including types C2 and C4 with dolomite cement C3 from the investigated Kangan and Dalan rocks were selected for fluid inclusion analyses. Fluid inclusion assemblages (FIAs) were taken in defining and representing the groups of fluid inclusions that presumably trapped at about the same time [21-22]. The second step involves verifying the fluid inclusions composition (aqueous or hydrocarbon inclusion). Table 1 provides a listing of the analyzed samples within the studied formations. Under UV, most hydrocarbon filled inclusions fluoresce, revealing a particular color depending on the maturity of the hydrocarbon. Aqueous inclusions do not fluoresce [20-32].

Petrographic observation showed that all of fluid inclusions within calcite cements (pore-filling and fracture-filling) were two-phase aqueous inclusions, while in dolomite cements, two types of inclusions were identified: two-phase aqueous inclusions and hydrocarbon inclusions. The homogenization temperature (Th) and final ice-melting temperature (Tm_{ice}) of aqueous inclusions were measured on the some cement samples. The micro thermometric measurements of fluid inclusions for some of these samples were impossible because of their size. Based on petrographic and fluid inclusion analyses, fluids responsible for these cement precipitations in Kangan and Dalan formations can be subdivided into three groups. Group 1 is the fluid responsible for the precipitation of equant II calcite (Th values: 107.2 °C-136.8 °C, mean=126 °C; mean salinity= 16 wt.% (NaCl equivalent)). Group 2 represents a dolomitizing fluid that has a Th value in the range of 107.2°C to 139.3 °C (with a mean at 127 °C) and mean salinity of 17 wt.% (NaCl equivalent). The last fluid inclusion analysis occurring along fractures is characterized by high Th values (155.4°C-181.2°C, mean=169 °C) and mean salinity of 17.5 wt.% (NaCl equivalent). These identified groups are characterized by their specific Th and salinity values of primary inclusions (Figure 5).

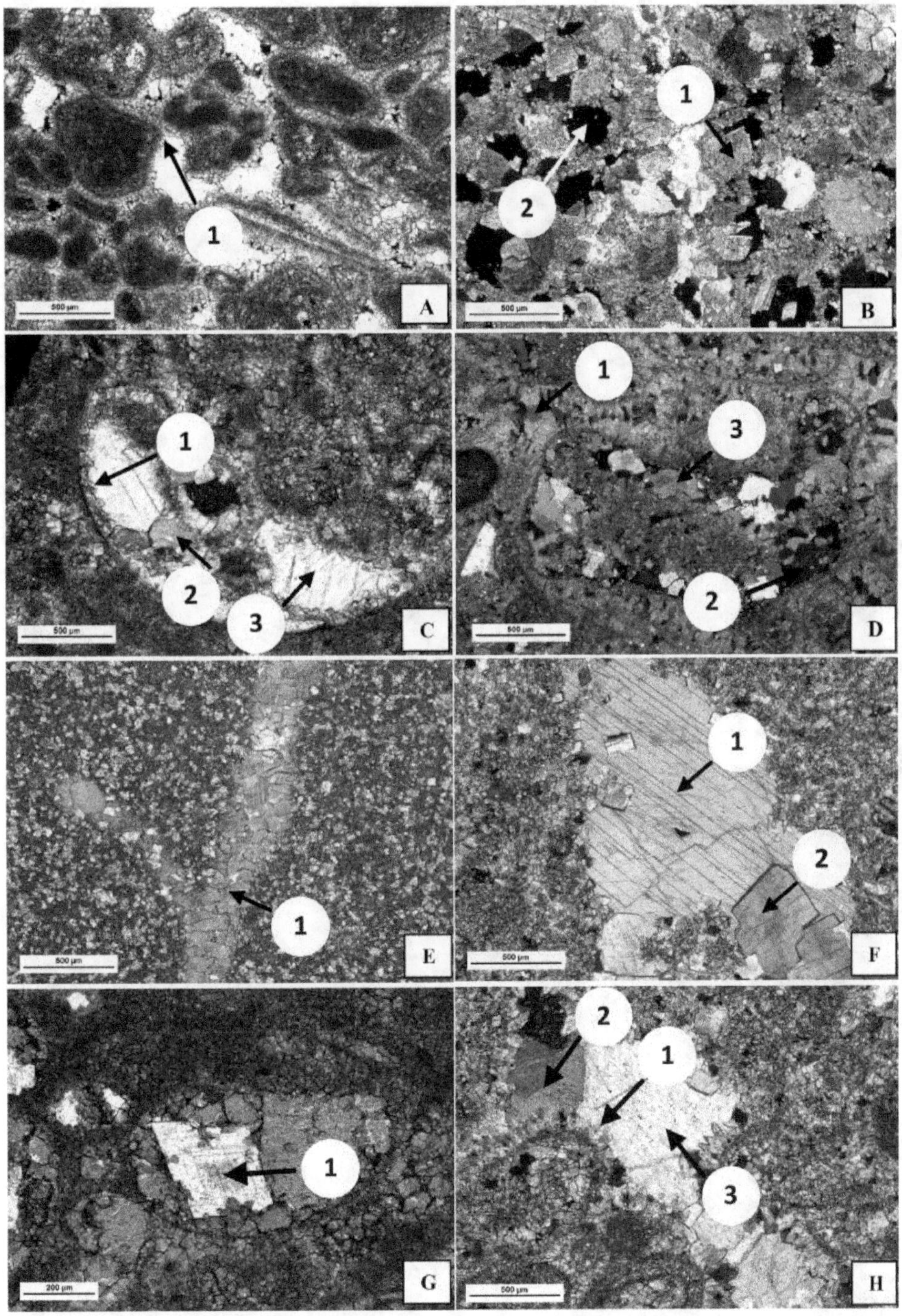

Figure 3: A- Early isopachous rime calcite cement (arrow no. 1), PPL. B-Coarse dolomite cement (arrow no. 1), moldic porosity resulting of dissolution (arrow no. 2) XPL. C-Isopachous rim (arrow no. 1), equant (arrow no. 2), coarse burial (arrow no. 3) cements, XPL. D-equant (arrow no. 1), coarse calcite (arrow no. 2), coarse dolomite cement (arrow no. 3), XPL. E- equant filling fracture (arrow no. 1), PPL. F-coarse burial calcite (arrow no. 1), coarse dolomite (arrow no. 2), XPL. G-granular calcite cement increased in size toward the center, coarse zoned dolomite cement (arrow no. 1) was predated and affected by equant calcite, PPL. H-Isopachous filling (arrow no. 1), equant (arrow no. 2) and coarse burial cement (arrow no. 3), XPL.

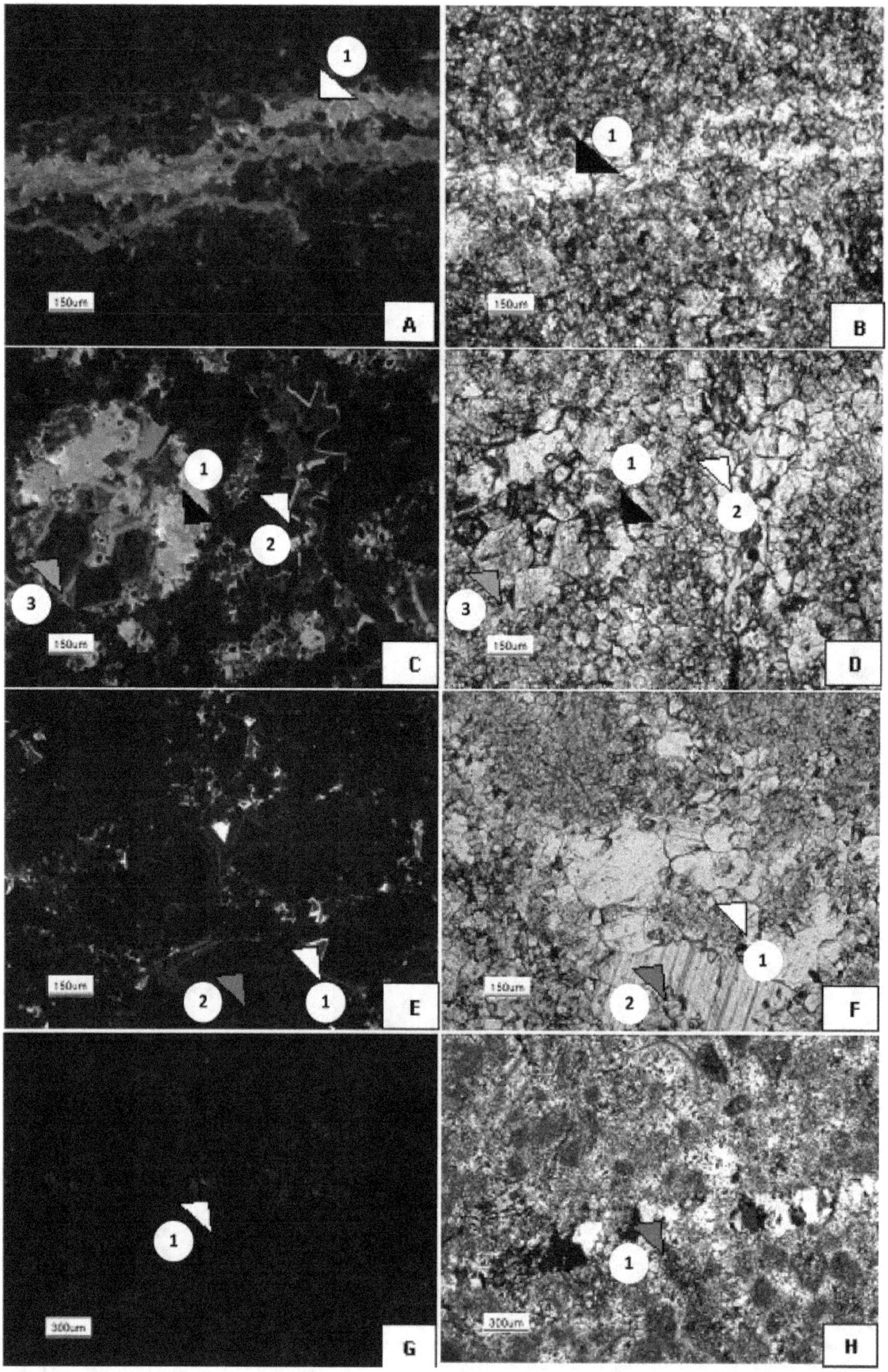

Figure 4: A and B- Granular equant II calcite filling fracture, bright orange CL (arrows no. 1) PPL and. C and D- equant II (arrow no. 1), equant (arrow no. 2) dull red, coarse dolomite (arrow no. 3) red luminescence, corroded dolomite crystals by equant calcite II and III, XPL. E and F- equant III (arrow no. 1) yellow luminescence, coarse burial calcite (arrow no. 2) dull CL, PPL. G and H- fracture filling equant I (arrow no. 1) dull red luminescence and red arrow in XPL view.

In conformity with salinity and temperature results, Fontana has reported similar data from dolomite and coarse sparry calcite cements in the Upper Permian-Lower Triassic Khuff formation [7]. Hydrocarbon inclusions were observed in three samples. All of these samples belong to Dalan formation, and they show yellow fluorescence. Some of photomicrographs of aqueous and hydrocarbon inclusions are shown in Figure 6.

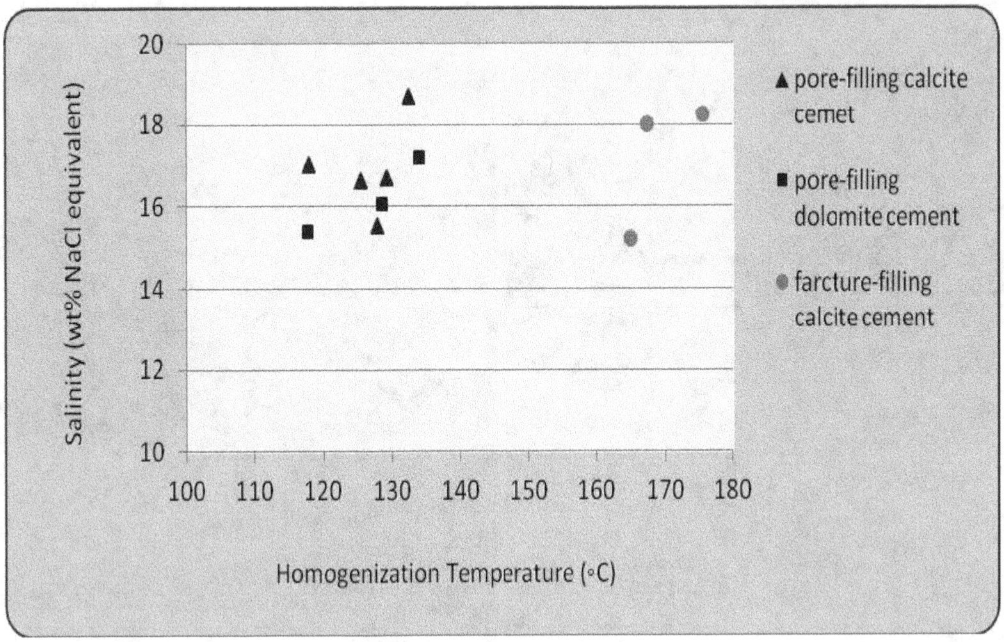

Figure 5: A cross plot of homogenization temperature (Th) and salinity values (wt.% NaCl equivalent) of fluid inclusions measured from coarse-crystalline calcite and dolomite cements from Kangan and Dalan Formations. Each point shows a fluid inclusion assemblages (FIAs), representing of the groups of fluid inclusions that presumably trapped at about the same time.

Impacts on Reservoir Characteristics

Previous studies have investigated the impact of diagenesis on porosity evolution in the Permian-Triassic carbonates of the Arabian Plate (see for example references [34-35]), and this has been well-documented in offshore fields [8, 12, 36, 37]. Core and thin-sections investigations indicate that meteoric dissolution, dolomitization, cementation (calcite cementation and dominantly anhydrite cement), compaction, and fracturing are the most prominent and effective diagenetic processes [5, 16, 12, 37]. The extent and propagate of lithology, sedimentary environment, cement, and porosity of Kangan and Dalan formations in different rock units of the studied well are presented in a sedimentological and petrophysical log in Figure 7. Moreover, their comparative cross plots are shown in Figure 8. The Figures show that each unit represents special characterization.

The K1 Unit

The K1 unit with an apparent thickness of 100 m and average porosity of 6.6% is dominated by dolomite, with intercalations of anhydrite and mud to grain dominated limestones (Figure 7). Oomoldic porosity is the main pore type in limestone parts, while interparticle and intercrystalline porosities are attached to moldic pores in dolomitic parts and increased the reservoir quality. However, the formation of equant and coarse sparry calcite cement in limy parts and dolomite cement in dolomitic parts has reduced porosity and permeability of those related intervals (Figure 7); anhydrite cement is more frequent than carbonate cement, and it has extensively annihilated the reservoir quality.

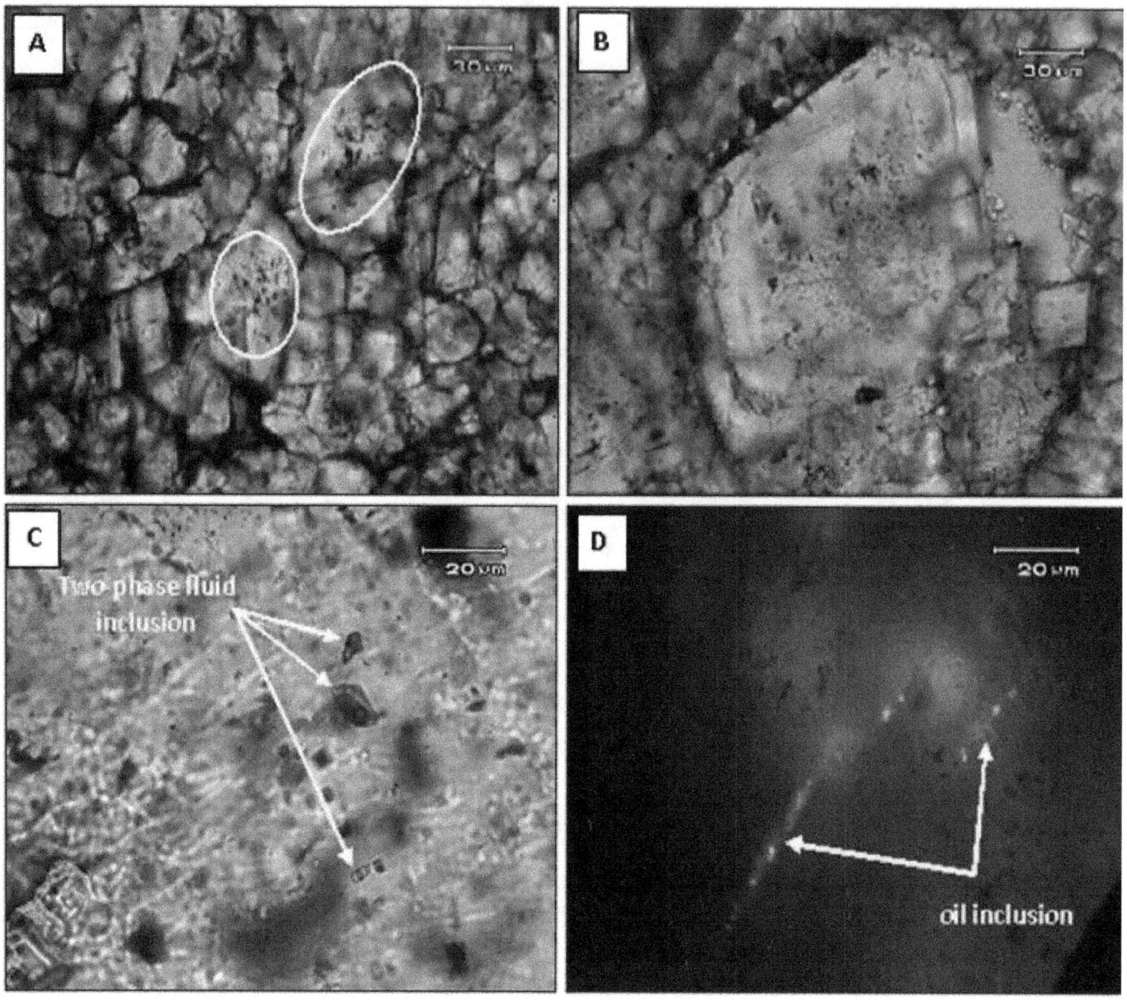

Figure 6: Photomicrographs of fluid and hydrocarbon inclusions within coarse crystalline cements in Kangan and Dalan Formations: A) indicates size in µm inclusions in pore-filling calcite cement, ppl. B) Small fluid inclusions in pore-filling dolomite cement, ppl. C) Medium to large two-phase fluid inclusions in fracture-filling calcite cement, ppl. D) Secondary hydrocarbon inclusions which formed along of micro cracks, UV light.

The K2 Unit

The K2 unit with a thickness of 45 m commonly consists of thrombolite boundstones at the base, overlaid by thick shoal setting ooid grainstones, and thin intercalations of lagoon-peritidal mud dominated facies. The reservoir quality of this rock unit is enhanced toward the upper part of K2 significantly in ooidal grainstones (with average porosity of 9.3%). The abundance of equant and coarse sparry cements in the lower part of K2 has occluded the pore spaces, and has reduced the reservoir properties (Figure 7).

The K3 Unit

The K3 unit is separated by thick anhydrite from the underlying K4. Rahimpour-Bonab [22] identified this anhydritic zone as a barrier to vertical fluid flow in this field. The prominent lithology is dolomite. Limestone is present in the lower and upper parts of K3. This unit mainly consists of lagoonal, shoal, slightly peritidal, and open marine facies (Figure 7). The frequency of calcitic and dolomitic cements is decreased, while the amount of anhydrite cement is increased and has notably reduced the reservoir quality (Figure 7).

The K4 Unit

The K4 unit with a thickness of 135 m and average porosity of 5.5% is characterized by dolomitic intervals with anhydrite bed in the upper part (Figure 7). Dolo ooid grainstones in shoal setting with good reservoir qualities occur in the middle part of the unit. Intercrystalline, interparticle, and moldic porosities are common pore types. The K4 consists entirely of stacked and oomoldic grainstone/packstones [4, 11, 12]. Except for pore occluding anhydrite, carbonate cement is observed scarcely in this unit.

Comparative cross plots (Figures 8A-8D) represent the frequency and average of lithology, cement type, sedimentary environment, and porosity in defined rock units of Kangan and Dalan formations. The K2 unit is not as much as other units under dolomitization and has remained more as a limy part (Figures 7 and 8A). Calcite cements like isopachous, granular equant, and coarse blocky are developed commonly in calcareous part (unit K2), while dolomite cement and associated cement, i.e. anhydrite, has extended in the dolomitized units of K1, K3, and K4 (Figures 7 and 8B). The dominant lithology of the rock can have an effect on the binding agent in that rock. The rock-fluid interaction causes the ions transport and the precipitation of calcite or dolomite cement in the calcitic or dolomitic prevailing lithology respectively; however, the depositional facies and environment are also affected by cementation. As can be seen in Figure 8C, lagoonal facies have higher frequency in units K1, K3, and K4, and, due to their nature of muddy facies with high Mg calcite (HMC), they have been replaced by dolomite.

Oomoldic porosity has frequently been developed in ooidal grainstone as a result of meteoric diagenesis and dissolution (Figure 3B). Also, dolomitization (intercrystalline porosity) may cause pore spaces (moldic and intercrystalline), and consequently porosity and permeability are focused on these intervals (units K1, K2, and K4) as shown in Figure 8D. The plot shows that the average porosity values in K2, K1, K4, and K3 decrease respectively. Meteoric dissolution (largely in ooid grainstones) is the most important diagenetic process in the generation of secondary porosity in the studied interval.

The history of different cementing phases and related dissolution in the Dalan and Kangan carbonates, following deposition, is shown in Figure 9.

Interpretation

The interpretation of the petrographic and cathodoluminescence data in conjunction with data from fluid inclusion studies indicates a complex diagenetic history for the Kangan-Dalan deposits. It has been assumed that cathodoluminescence in carbonate rocks is mainly derived from the incorporation of manganese into the calcite lattice, with the ion commonly acting as a cathodoluminescence quencher [38]. The isopachous cement (C1) on skeletal and non-skeletal particles is generally interpreted as a typical of a marine phreatic environment, where all the pore spaces are filled with marine water [39]. The isopachous cement and their non-luminescent features, together with the fact that they occur as the first stage cement before compaction, would indicate a marine phreatic origin [40-41]. It is important to point out that luminescence is not only controlled by the variations in the Mn^{++} and the Fe^{++} fluid concentrations, but also by crystals growth rates [42]. The bright orange and yellow luminescence of equant C5 and C6 cements suggest that both should have had the same initial probably high magnesium calcite composition.

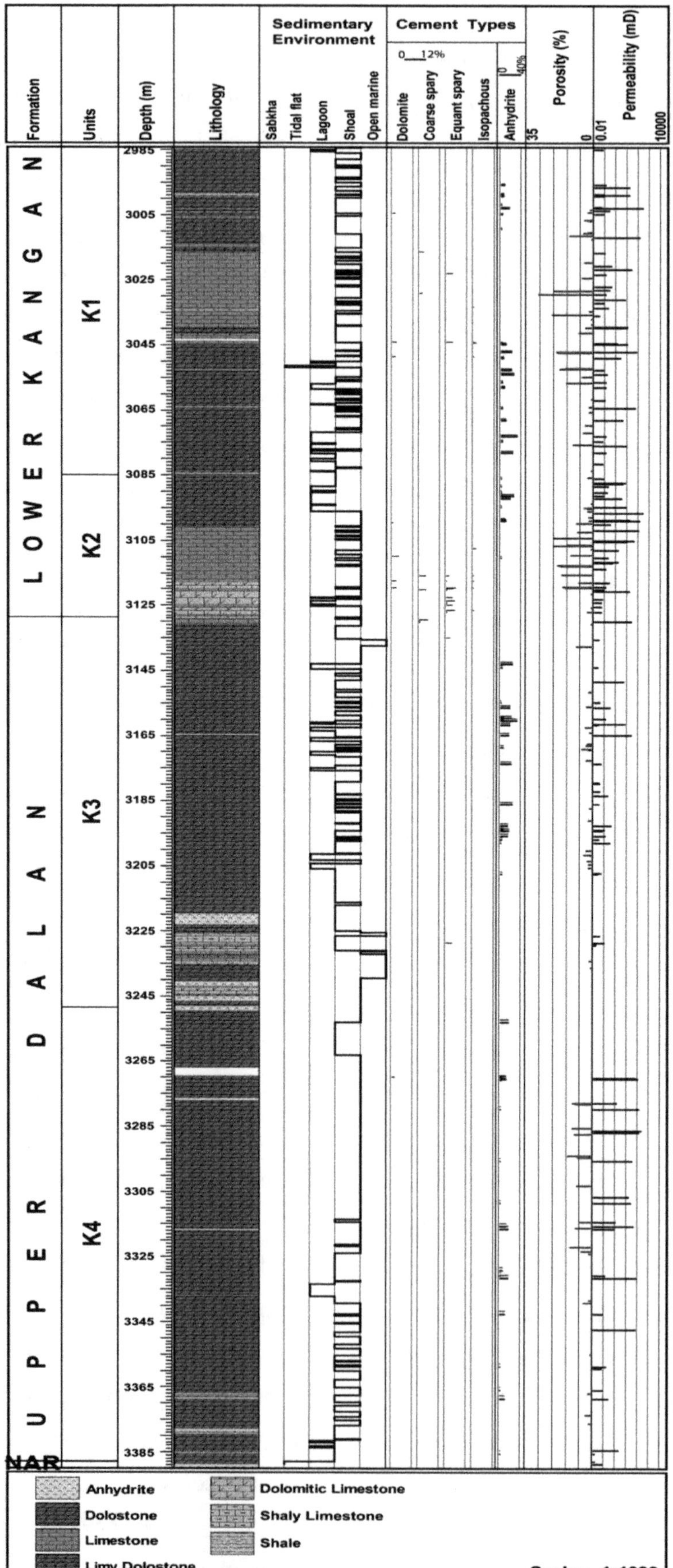

Figure 7: A Sedimentological and petrophysical log of Kangan and Dalan Formations.

Alternatively, the luminescence might be interpreted as having formed in the same way and at the same time on the basis of their similarity in the luminescence pattern; this might be produced by diagenetic alteration, as well as the luminescent features of the microspar calcite cement. The granular calcite cement has variable luminescent according to distinct color, which represents phases of crystal growth. In fact, each color represents the precipitation of calcite from pore waters with different chemical composition [43-44] from dull red (C2) luminescent (DL), to bright orange (C5) luminescent area (BOL), and to brightly yellow (C6) luminescent (BYL) zone. These variations in the luminescence characteristics are probably caused by the periodic incorporation of Mn^{++}. Irregular boundaries between alternating zones suggest rapid changes during the temporal interruption of crystal growth. Moreover, coarse blocky calcite (C4) in fractures and vugs with dull to non-luminescent together with distinct cleavage in large crystals could originate from the deep burial environment as the fluid inclusion reveals that this cement has precipitated from saline brines over a range of temperatures. It means that the fluids with higher temperature have migrated from deeper parts of the basin and have filled fractures during deep burial. It seems that hydrocarbon yellow inclusions are secondary; furthermore, they are distributed along micro cracks that can show the migration path, and it can be interpreted that hydrocarbon migration predates the precipitation of cements.

The dull/bright/yellow CL transition is generally interpreted as being caused by an increase in redox potential (Eh) under conditions of progressive marine burial meteoric-phreatic diagenetic environment [41]. This zonation in luminescence suggests geochemical variations within pore fluids system. The dull red CL cement (C2), under an anoxic condition and based on the first group of fluid inclusion data, is a fluid with calcite composition and is interpreted to have precipitated during burial from pore water. However, the bright orange and yellow luminescence of granular calcite cements (C5 and C6) were probably precipitated under suboxic conditions [41], and it is understood that this type of cements occurred as a consequence of the influence of meteoric water. Leaching and neomorphism (microspar) occurred when meteoric waters were flowed through Kangan Dalan sediments during shallow burial. Cathodoluminescence reveals fine scale oscillatory zonation within granular calcite cement, which is more common in the Kangan Dalan succession. Dolomite cements, under both optical and cathodoluminescence microscopes, demonstrate zoning patterns, which, through the data achieved from vacuoles fluid with dolomite composition, was precipitated possibly during shallow burial to burial. Finally, the sediments originated as shallow marine mainly in a phreatic realm during the first stage of diagenesis. The second stage of diagenesis is interpreted to have occurred in a burial environment by the formation of equant (C2), dolomite (C3), and coarse sparry calcite (C4). The next phase of diagenesis is expected to have happened during meteoric-phreatic area by the precipitation of C5 and C6. The entry of waters will first dissolve aragonitic grains (bioclasts and ooids), and later the saturated waters induce the adjacent precipitation of a stable mineral phase as mosaic intergranular calcite (meteoric equant cement). However, the comparison of Figures 8B and 8D shows that increasing cements are associated to occluding depositional-dissolution related porosities of pertinent unit. In general, as Kangan and Dalan carbonates are buried more deeply, they are commonly in equilibrium with the adjacent subsurface waters; nevertheless, significant diagenetic changes can occur. Since different phases of cementation (Figure 9) with various amounts have occurred, the associated phases of pore space filling in each rock unit have taken place. Therefore, the creation and preservation of porosity as a function of fluid saturation in different units (Figure 8D) of the formation generally will be arisen.

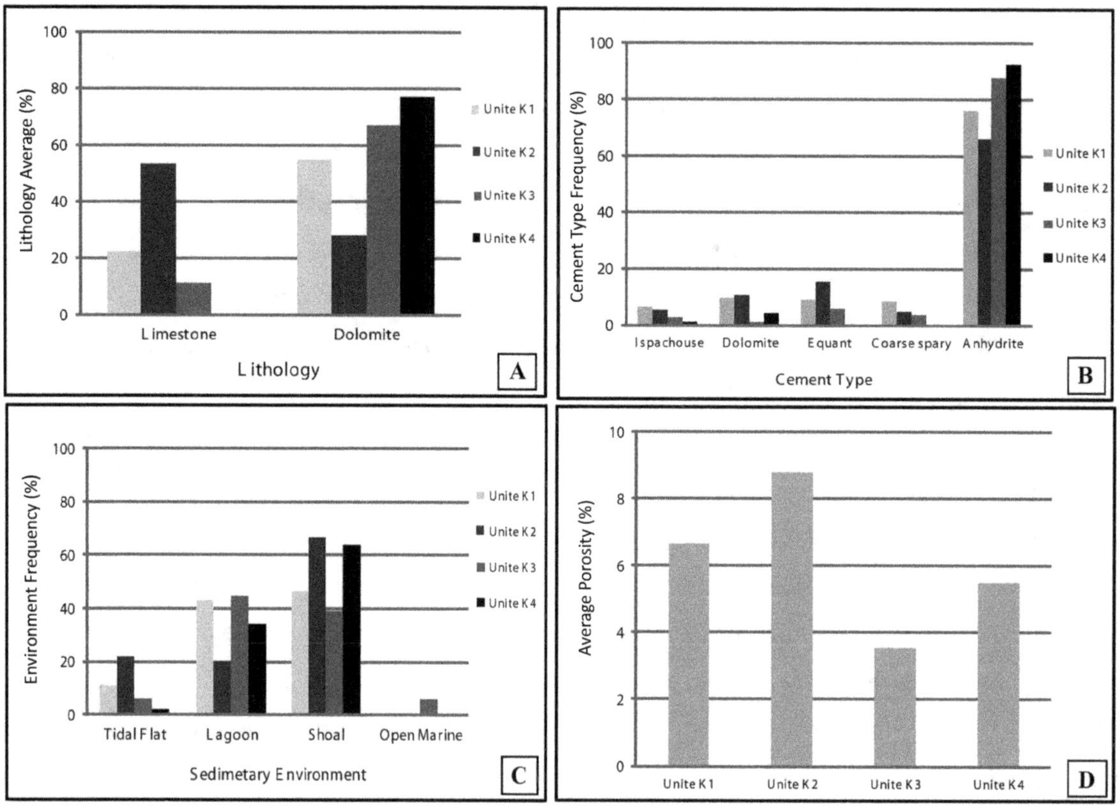

Figure 8: Cross plots of lithology, cement, sedimentary environment and porosity in different units of Kangan and Dalan Formations.

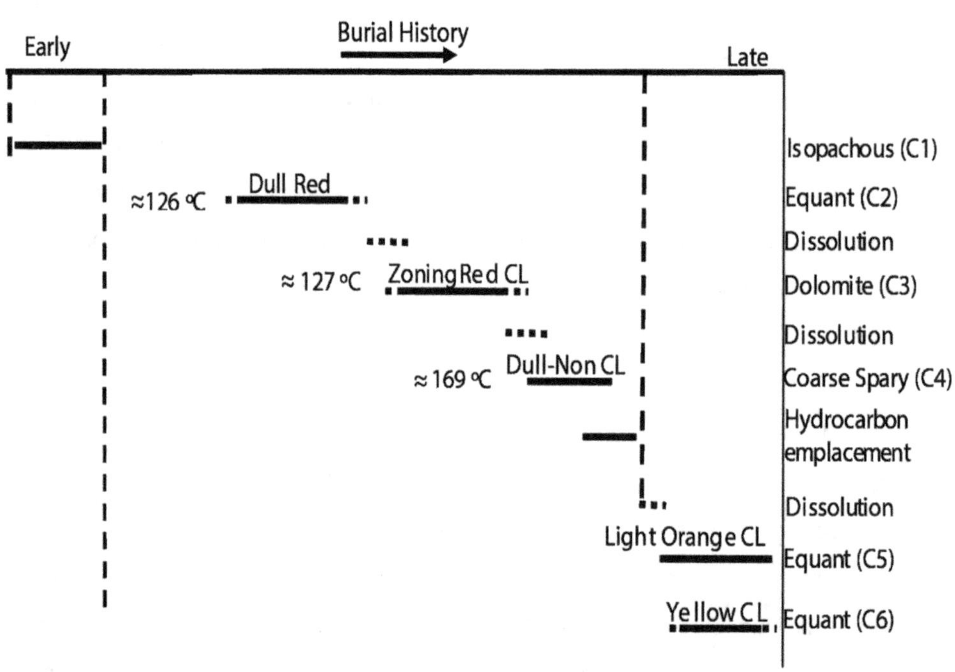

Figure 9: Diagenetic paragenesis of carbonate cement in Kangan and Dalan Formation.

CONCLUSIONS

The Upper Permian–Triassic carbonate platform succession has undergone a complex diagenetic evolution, where several diagenetic phases have recorded in the initial depositional, shallow-deep, and uplift conditions.

Petrography and cathodoluminescence observations revealed that cement types appear to be (i) early granular isopachous calcite (C1) with nonluminescent CL precipitated as early alterations on the sea floor which could be syn-depositional and are interpreted as early marine cements.

The next diagenetic phases of the Kangan-Dalan formations can be summarized as (ii) burial equant calcite (C2-dull red CL), (iii) coarse secondary dolomite (100 < size micron) rhombs (C3-red zoning CL), (iv) burial coarse sparry calcite (C4-dull to non-luminescence). Finally, uplift can be identified by two types of meteoric cements including (v) equant with C5-bright orange CL and (vi) equant with C6-yellow CL. All these processes are indicative of changes from marine phreatic-freshwater to burial to uplift diagenetic conditions.

Fluid inclusion micro thermometric data on calcite and dolomite cements show an alternative trend of increasing and decreasing temperature and salinity through depth and time from shallow to late-burial diagenetic conditions. The high Th value of coarse sparry calcite (155.4 °C to 181.2 °C) is interpreted as fluid migration from deeper parts of the basin and filling fractures during deep burial. Dolomitizing fluid was precipitated possibly during shallow burial and through calcitizing fluids, and equant cements (C5, C6) have precipitated from pore fluids influenced by meteoric water.

Hydrocarbon inclusions were observed only in Dalan formation, and it seems that they are secondary; hydrocarbon migration predates the precipitation of coarse calcite cements. Thus, this study shows that there is some post migration diagenesis which is caused by an unsteady state hydrocarbon migration. Reservoir properties result from the original sedimentary texture and diagenesis. Average porosity values differ from 3.5% in the K3 to 10% in the K2 units, made up of fine to coarse ooidal packstone to grainstone. The early to late diagenetic dissolution of the ooids and cement can presumably account for a significant amount of porosity formation of the studied Permian-Triassic Dalan-Kangan reservoirs.

ACKNOWLEDGMENTS

The authors would like to thank the Research Institute of Petroleum Industry (RIPI) for providing us with cathodoluminescence and fluid inclusion laboratory equipment for the examination. We extend our appreciation to the Petropars Oil Company for permission to publish the result of this paper. This research was financially supported by RIPI. The authors are also grateful to Mr. Motiei, Dr. Nader (Institute de France du Petrol), and Mr. Kasaei for improving the manuscript with their critical reviews and suggestions.

REFERENCES

1. Aleali M., Rahimpour-Bonab H., Moussavi-Harami R., and Jahani D., "Environmental and Sequence Stratigraphic Implications of Anhydrite Textures: A Case from the Lower Triassic of the Central Persian Gulf," *Journal of Asian Earth Sciences*, **2013**, *75*, 110-125.
2. Ghazban F., "Petroleum Geology of the Persian Gulf," *Tehran University and National Iranian oil Company Publications*, **2007**, 707.
3. Kashfi M. S., "Greater Persian Gulf Permian-Triassic Stratigraphic Nomenclature Requires Study," *Oil and Gas Journal*, **2000**, *6*, 36-44.
4. Esrafili-Dizaji B. and Rahimpour-Bonab H., "A

Review of Permian-Triassic Reservoir Rocks in the Zagros Rrea, Southwest of Iran: Influence of the Qatar-Fars arch," *Journal of Petroleum Geology*, **2013**, *36*(3), 257-279.

5. Moradpour M., Zamani Z. and Moallemi S. A., "Controls on Reservoir Quality in the Lower Triassic Kangan Formation, Southern Persian Gulf," *Journal of Petroleum Geology*, **2008**, *31*(4), 367-385.
6. Fontana S., Nader, F.H., Morad S., Ceriani A., et al., "Fluid-rock Interactions Associated with Regional Tectonics and Basin Evolution," *Sedimentology*, **2014**, *61*(3), 660-690.
7. Fontana S., Nader F. H., Morad S., Ceriani A., et al., "Diagenesis of the Khuff Formation (Permian-Triassic), Northern United Arab Emirates," *Arab J. Geosci.*, **2010**, *3*, 351-368.
8. Ehrenberg S. N., "Porosity Destruction in Carbonate Platforms," *Journal of Petroleum Geology*, **2006**, *29*, 41-52.
9. Neilson J. E. and Oxtoby N. H., "The Relationship between Petroleum, Exotic Cements and Reservoir Quality in Carbonates, A review," *Marine and Petroleum Geology*, **2008**, *25*, 778–790.
10. Rahimpour-Bonab, H., "A Procedure for Appraisal of a Hydrocarbon Reservoir Continuity and Quantification of its Heterogeneity," *Journal of Petroleum Science and Engineering*, **2007**, *58*, 1-12.
11. Esrafili-Dizaji B. and Rahimpour-Bonab H., "Effects of Depositional and Dagenetic Characteristics on Carbonate Reservoir Quality: a Case Study from the South Pars Gas Field in the Persian Gulf," *Petroleum Geoscience*, **2009**, *15*, 325-344.
12. Rahimpour-Bonab H., Esrafili-Dizaji B., and Tavakoli V., "Dolomitization and Anhydrite Precipitation in the Permian-Triassic Carbonates at the South Pars Gas Field, Offshore Iran: Controls on Reservoir Quality," *Journal of Petroleum Geology*, **2010**, *33*, 43-66.
13. Ameen M. S., Buhidma I. M., and Rahim Z., "The Function of Factures and in-situ Stresses in the Khuff Reservoir Performance, Onshore Fields, Saudi Arabia," *AAPG Bulletin*, **2010**, *94*, 27-60.
14. Koehrer B., Zeller M., Aigner T., Poeppelreiter M., et al., "Facies and Stratigraphic Framework of a Khuff Outcrop Equivalent: Saiq and Mahil Formations, Al Jabal al-Akhdar, Sultanate of Oman," *GeoArabia*, **2010**, *15*, 91-156.
15. Koehrer B., Aigner T. and Pöppelreiter M., "Field-scale Geometries of Upper Khuff Reservoir Geobodies in an Outcrop Analogue (Oman Mountains, Sultanate of Oman)," *Petroleum Geoscience*, **2011**, *17*, 3-16.
16. Insalaco E., Virgone A., Courme B., Gaillot J., et al., "Upper Dalan member and Kangan Formation between the Zagros Mountains and offshore Fars, Iran: Depositional System, Biostratigraphy and Stratigraphic Architecture," *GeoArabia*, **2006**, *11*(2), 75-176.
17. Edgell H. S., "Salt Tectonism in the Persian Gulf Basin, In: Asop, (Eds.)," *Salt Tectonics, Geological Society of London*, Special Publication, **1996**, *100*, 51-129.
18. Szabo F. and Kheradpir A., "Permian and Triassic stratigraphy of the Zagros Basin, Southwest Iran," *Journal of Petroleum Geology*, **1978**, *1*, 57-82.
19. Dickson J. A. D., "Carbonate Identification and Genesis as Revealed by Staining," *Journal of Sedimentary Petrology*, **1966**, *36*(2), 491-505.
20. Burruss R. C., "Practical Aspects of Fluorescence Microscopy of Petroleum Fluid Inclusions," *Luminescence Microscopy and Spectroscopy*, Society of Economic Paleontologists and Mineralogists Short Course, **1991**, *25*, 1-7.
21. Goldstein R. H., "Fluid Inclusions in Sedimentary

and Diagenetic Systems," *Lithos*, **2001**, *55*, 159-193.

22. Goldestein R. H. and Reynolds T. J., "Systematics of Fluid Inclusions in Diagenetic Minerals: Society of Economic Paleontologists and Mineralogists," *Short Course*, **1994**, *31*, 199 .
23. Bodnar R. J., "Revised Equation and Table for Freezing Point Depressions of H_2O-salt Fluid Inclusions (Abstract): PACROFI," *4th Biennial Pan-American Conference on Research on Fluid Inclusions*, Program and Abstracts, Lake Arrowhead, **1992**, *14*, 15.
24. Ali S. A., Clark W. J., and Moore W. R., "Diagenesis and Reservoir Quality, Oilfield Review," *Schlummberger*, **2010**, *22*(2).
25. Bjorlkkey K., "Petroleum Geoscience: from Sedimentary Environments to Rock Physics," *Springer*, **2010**, 508.
26. Mackenzie F. T., "Sediments, Diagenesis, and Sedimentary Rocks," *Elsevier*, **2005**, 448.
27. Moore C. H., "Carbonate Diagenesis and Porosity," *Elsevier*, **1989**, 338.
28. Palma R. M., Bressan G. S., and Kietzmann D. A., "Diagenesis of a Bioclastic Oyster Deposit from the Lower Cretaceous (Chachao Formation), Neuquen Basin, Mendoza Province, Argentina," *Carbonates and Evaporites*, **2008**, *23*(1), 39-49.
29. Tucker M. E. and Wright V. P., "Carbonate Sedimentology," Blackwell Scientific Publications, Oxford, **1990**, 497.
30. Flugel E., "Microfacies of Carbonate Rocks, Analysis, Interpretation and Application," 2nd edition, *Springer*, **2010**, 984.
31. Longman M. W., "Carbonate Diagenetic Textures from Near Surface Diagenetic Environments," *AAPG Bulletin*, **1980**, *64*, 461-487.
32. Scoffin T. P., "Introduction to Carbonate Sediments and Rocks," Chapman and Hall, New York, **1987**, 274.
33. McLimans R. K., "The Application of Fluid Inclusions to Migration of Oil and Diagenesis in Petroleum Reservoirs," *Applied Geochemistry*, **1987**, *2*(5-6), 585-603.
34. Dasgupta S. N., Hong M., and Al-Jallal I. A., "Accurate Reservoir Characterization to Reduce Drilling Risk in Khuff-C Carbonate, Ghawar field, Saudi Arabia," *GeoArabia*, **2002**, *7*, 81-100.
35. Alsharhan A. S., "Sedimentological Character and Hydrocarbon Parameters of the Middle Permian to Early Triassic Khuff Formation, United Arab Emirates," *Geo Arabia*, **2006**, *11*, 121–158.
36. Tavakoli V., Rahimpour-Bonab H. and Esrafili-Dizaji B., "Diagenetic Controlled Reservoir Quality of South Pars Gas Field, an Integrated Approach, *Comptes Rendus Geoscience*, **2010**, *343*, 55-71.
37. Peyravi M., Kamali M. R., and Kalani M., "Depositional Environments and Sequence Stratigraphy of the Early Triassic Kangan Formation in the Northern Part of the Persian Gulf: Implications for Reservoir Characteristics," *Journal of Petroleum Geology*, **2010**, *33*, 371-386.
38. Meyers W. J., "Carbonate Cement Stratigraphy of the Lake Valley Formation (Mississippian) Sacramento Mountains, New Mexico," *Journal of Sedimentary Petrology*, **1974**, *44*, 837-861.
39. Prezbindowski D. R., "Burial Cementation, Is It Important? A Case Study, Stuart City Trend, South Central Texas," *Society of Economic Paleontologists and Mineralogists*, Special Publication, **1985**, *36*, 241-264.
40. Frank J. R., Carpenter A. B., and Oglesby W., "Cathodoluminescence and Composition of Calcite Cement in the Taum Sauk Limestone (Upper Cambrian), Southeast Missouri," *Journal of Sedimentary Petrology*, **1982**, *52*, 631-638.
41. Grover G. and Read J. F., "Paleoaquifer and Deep Burial Related Cements Defined by

Regional Cathodoluminescence Patterns, Middle Ordovician Carbonates, Virginia," *AAPG Bulletin*, **1983**, *67*, 1275-1303.

42. Reeder R. J. and Grams J. C., "Sector Zoning in Calcite Cement Crystals: Implications for Trace Elements Distributions in Carbonates," *Geochimica et Cosmochimica Acta*, **1987**, *51*, 187-194.
43. Meyers W. J., "Calcite Cement Stratigraphy: an Overview," *SEPM Short Course*, **1991**, *25*, 133-148.
44. Machel H. G., "Application of Cathodoluminescence to Carbonate Diagenesis" Springer-Verlag, Heidelberg, **2000**, 271-301.

11

Evaluation of Barium Sulfate (Barite) Solubility Using Different Chelating Agents at a High Temperature

Badr Salem Bageri*, Mohamed A. Mahmoud, Reyad A. Shawabkeh, and Abdulazeez Abdulraheem

King Fahd University of Petroleum & Minerals, Dhahran, 31261, Kingdom of Saudi Arabia

ABSTRACT

Barium sulfate (barite) is one of the widely used weighting materials in the preparation of drilling fluid for deep oil and gas wells. Barite is not soluble in the regular solvents; such as, hydrochloric acid (HCl) and other acids. Therefore, in this study, we focused on evaluating the dissolution of the industrial barite particles in different chelating agents. Chelating agents; such as, diethylene triamine penta acetic (DTPA), ethylene diamine tetra acetic (EDTA), and hydroxyethyl ethylene diamine tri acetic (HEDTA) acids with high and low pH values were used in this study. The effect of the base of chelating agents, namely alkali and alkali earth hydroxides, on the dissolution of barite was also investigated. For the first time, the optimum pH, concentration, and base (sodium or potassium) of chelating agents which yielded the maximum dissolution were investigated in this study at a high temperature. Previous studies did not consider the wellbore constraints during their experiments and they used chelating agent volume to barite weight ratio, which could not be implemented in the real wells. In this study and for the first time, we considered the wellbore volume (chelating agent volume that can be used) and the barite weight (filter cake) during the dissolution experiments. Based on the results obtained from this study, potassium base DTPA-K_5 and EDTA-K_4 of a concentration of 20 wt.% were found to be the most effective chelating agents to dissolve barite. The solubility of barite was found to be 26.8 g/L in a solution containing 20 wt.% of DTPA-K_5 and 25.6 g/L in a solution containing 20 wt.% of EDTA-K_4 during a soaking time of 24 hrs and a pH value above 11 at 200°F.

Keywords: Barium Sulfate (Barite) Solubility, Barite Solubility, Chelating Agents, DTPA, EDTA

INTRODUCTION

Barium sulfate (barite) is one of the common weighting materials used in drilling fluid for deep oil and gas wells. It is mostly preferred over other weighting materials; such as, iron oxides, calcium carbonate, and others due to its high density which provides adequate weight at a low cost [1]. Consequently, the main source of formation damage during drilling is the weighting material used in drilling fluids [2]. Due to the wide use of barite in field applications, there is a need for deep understanding of the barite dissolvers in field conditions to formulate a treatment which can dissolve the barite solids formed during drilling or production operations.

***Corresponding author**

Badr Salem Bageri

Email: g200804580@kfupm.edu.sa

The low solubility of barium sulfate in water (2.3 mg/L) leads to the formation of barium sulfate scale in oil and gas wells [3]. On the other hand, contacting the high sulfate seawater with barium in the underground formations is another source of barite scale precipitation [4,5].

The solubility of barite in chelating agents has been investigated, but the optimum conditions for dissolution have not completely been addressed. Lakatos et al. [6,7] compared seven different barite dissolvers. Their work provided with the determination of the dissolution capacity of each solvent. The following chelating agents were used in their study; Diethylene triamine penta acetic acid (DTPA), ethylene diamine tetra acetic acid (EDTA), hydroxyethyl ethylene diamine tri acetic acid (HEDTA), dioxaocta methylene dinitrilo tetra acetic acid (DOCTA), nitrile tri acetic (NTA), (DCTA), and tri ethylene tetra amine hexa acetic acid (TTHA). They conducted the experiments at 25°C which does not simulate the downhole conditions in oil and gas wells. Their results showed that the barite particles solubility in the DTPA was about (8 g/L) in a solution containing a DTPA concentration of 0.1 mol/L. The chelating agents were prepared by the sodium base (NaOH) in Lakatos' study [7]. Another study was conducted by Putnis et al. [8,9] to investigate the dissolution rate of barite in DTPA at different concentrations and temperatures. The DTPA in their study was prepared by a potassium base (KOH). The solubility test in their study was carried out using barite with a particle size ranged from 104 to 150 μm and the DTPA solution with a concentration ranged from 0.001 M to 0.5 M. The particle size of the barite they used is completely different than the one used in oil and gas well drilling fluid. In this study, we used industrial grade barite that is typically used in drilling fluids for deep oil and gas wells.

The results obtained by Putnis et al. [8] indicated that the lower concentration of DTPA resulted in a good dissolution rate at a low temperature, where the solubility of barite was found to be 1.17 g/L (650 ppm of Ba) in a 0.05 M DTPA solution during 19 hours at 22°C. Increasing the temperature to 80°C showed that a 0.5 M solution of DTPA gave better results, where 5280 ppm of barite dissolved in 0.5 M DTPA, while 4936 ppm dissolved in a 0.05 M solution.

Two barite dissolvers were evaluated using DTPA or ETDA as the main solvents to dissolve the barite scale by Nasr-El-Din et al. [10]. The solubility test was conducted to test the two formulations and the results showed that the EDTA and DTPA yielded a barite solubility of 23.7 g/L Although the results of the two formulations were effective, the composition was not disclosed.

Several attempts were conducted to increase the solubility of barite in chelating agents, but there is still a need to examine different chelating agents in the presence of alkali and alkali earth hydroxides. In this study, different chelating agents were used to enhance the knowledge of barite dissolution in different chelating agents prepared with different bases (KOH and NaOH) at a high temperature (200 °F). The current study objectives are to (1) evaluate the barite dissolution rate using different chelating agents with different bases (sodium base (NaOH) or potassium base (KOH), (2) present the right approach toward preparing the chelating agent to get the maximum dissolution rate of barite, (3) evaluate the optimum dissolution time, (4) study the effect of the pH on barite dissolution, and (5) understand the mechanism behind the barite dissolution by chelating agents using Fourier Transformation Infra-Red (FTIR) and scanning electron microscopy (SEM). The results of this work will help formulate a treating formulation for cleaning the barite scale from surface and subsurface facilities and cleaning the wellbore after drilling operations of deep oil and gas wells drilled using barite-based mud.

EXPERIMENTAL PROSEDURES

Materials

Industrial grade barite (barium sulfate) was used to carry out the experimental work of this study. The barite particles were sieved using a 75 micron mesh size to insure that the same particle size range of barite (less than 75 micron) is used in all the experimental studies. Eight chelating agents were used in this study; Diethylene triamine penta acetic acid (DTPA) potassium salt and sodium salt base, ethylene diamine tetra acetic acid (EDTA) potassium salt and sodium salt base, hydroxyethyl ethylene diamine triacetic acid (HEDTA) sodium salt and glutamic diacetic acid (GLDA) sodium salt (GLDA). Table 1 lists the details of the chelating agents used in this study. The schematic diagrams of the chelating agent molecular structures of DTPA, EDTA, HEDTA, and GLDA are shown in Figure 1. The molecular structure shows that each chelating agent has a certain number of carboxylate functional groups (ligands) which bind to metal center having different numbers of nitrogen atoms [8,11]. For example, DTPA has five ligands linked with three nitrogen atoms in the center. At a high pH, the ligand forms strong 1:1 chelates in solution [11]. Based on that, the ligand can be either Na or K based on the used base to prepare the solution; in our study, NaOH and KOH were used to prepare the solutions.

Table 1 : Chelating agents used in this study.

Chelating Agents		Appearance/ Form/ Assay	Chemical Formulation
1	DTPA	99% Powder	$C_{14}H_{23}N_3O_{10}$
2	DTPA-Penta sodium (DTPA-Na_5)	40 % aqueous solution	$C_{14}H_{18}N_3\ Na_5O_{10}.5H_2O$
3	DTP-Penta potassium (DTPA-K_5)	40 % aqueous solution	$C_{14}H_{18}N_3\ K_5O_{10}.5H_2O$
4	EDTA	98-100% Powder	$C_{10}H_{16}N_2O_8$
5	EDTA-Di Sodium (EDTA-Na_2)	98-100% Powder	$C_{10}H_{14}N_2Na_2O_8.2H_2O$
6	EDTA-Di Potassium (EDTA-K_2)	99% Powder	$C_{10}H_{14}N_2K_2O_8.2H_2O$
7	EDTA-Tetra Sodium (EDTA-Na_4)	39 % aqueous solution	$C_{10}H_{12}N_2Na_4O_8.4H_2O$
8	EDTA-Tetra Potassium (EDTA-K_4)	39 % aqueous solution	$C_{10}H_{12}N_2K_4O_8.4H_2O$
9	HEDTA-Tri Sodium (HEDTA-Na_3)	40 % aqueous solution	$C_{10}H_{15}N_2Na_3O_7.3H_2O$
10	GLDA-Tetra Sodium (GLDA-Na_4)	38 % aqueous solution	$C_9H_9NNa_4O_8.4H_2O$

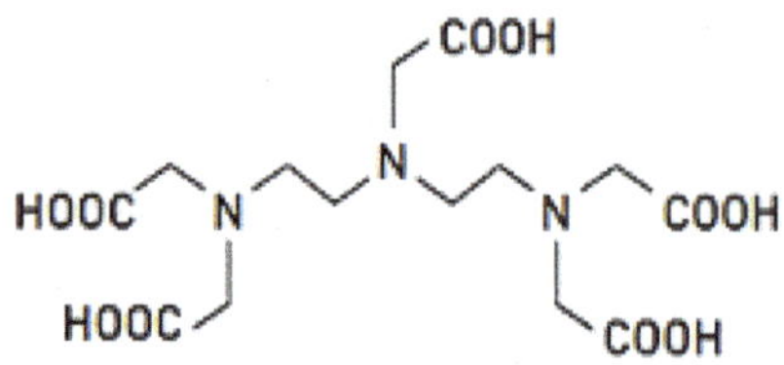

Diethylene triamine penta acetic acid
DTPA-H_5

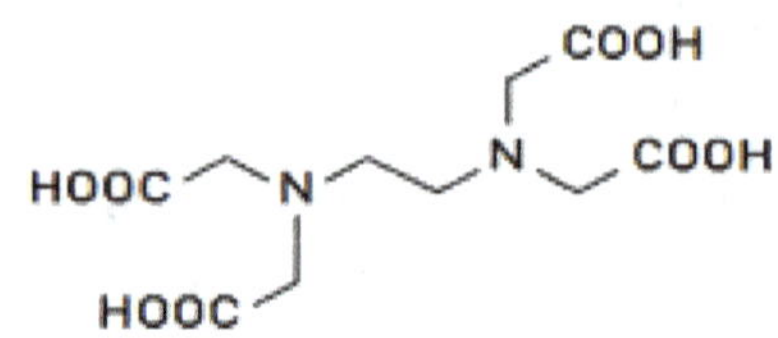

Ethylene diamine tetra acetic acid
EDTA-H_4

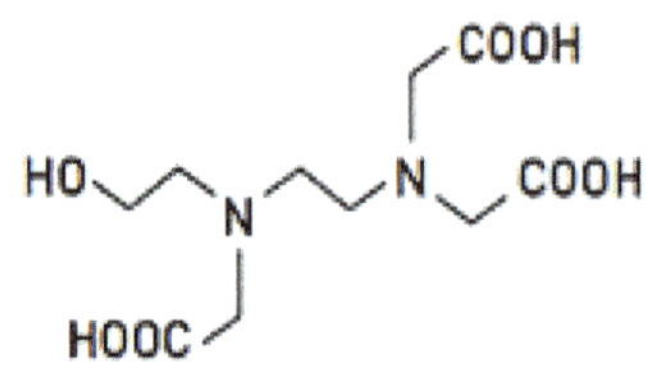

Hydroxyethyl ethylene diamine triacetic acid
HEDTA-H_3

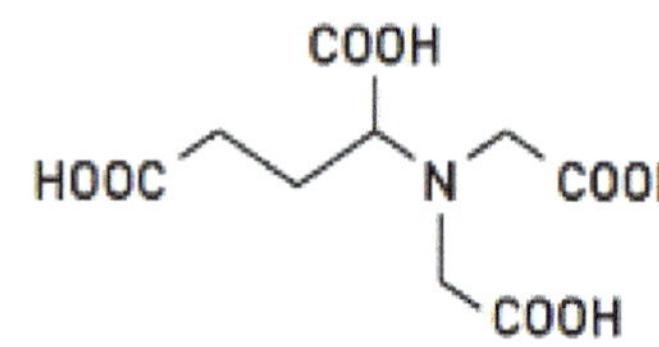

Glutamic acid, N,N-diacetic acid
GLDA-H_4

Figure 1: Molecular structure of the chelating agents [9, 11].

The structure of the DTPA-K_5/ Na_5 (and similarly the others) is the same as DTPA with replacing H_5 with K_5.

Experimental Work

The solubility experiments were carried out using 4 g of barite in 100 ml of solvent. This ratio was obtained based on the ratio of barite solids weight to solvent volume as given in Equation 1. For a barite filter cake or scale formed around the wall of the well, the ratio of barite solids to the solvent is limited by the thickness of the formed cake and the diameter of the wellbore as well as the porosity of the filter cake [12].

$$W_R = \frac{d_h^2 - (d_h - 2th_f)^2 \times \rho_{fc}\,(1 - \varnothing c)}{(d_h - 2th_f)^2} \quad (1)$$

where, d_h (cm) is the diameter of the open hole section; T_{hf} (cm)is the thickness of the filter cake; ρ_{fc} (g/cc) stands for the density of filter cake solids, and ϕ_c (fraction) represents the porosity of the filter cake.

Although the filter cake formed downhole should be very thin to avoid pipe sticking, using a ratio of 4 g/100 ml covers even the worst case of filter cake thickness that reached 1.5 mm [13] for open hole section with a hole size of 8 inches and above (Figure 2).

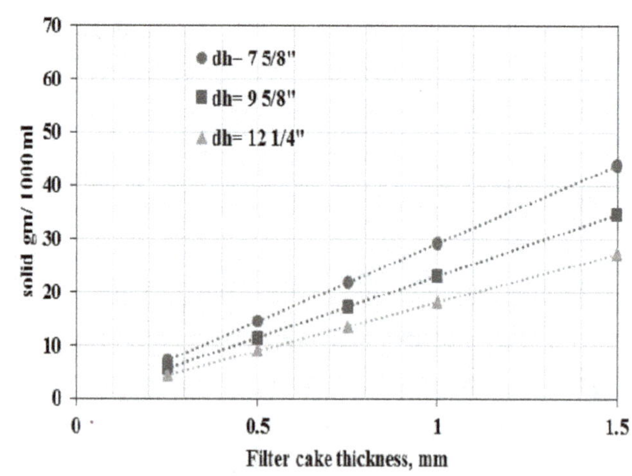

Figure 2: Recommended solid-to-liquid ratio based on filter cake thickness and hole size.

The solubility experiments were conducted at a constant stirring rate (350 rpm) and the solution was mixed using flask connected to reflux condensers to ensure no water loss at high temperatures (Figure 3). The following procedure was followed during the experiments:

a) Mixing 4 g of the industrial barite with 100 ml of a chelating agent (1), for example DTPA-Na. The concentration of the chelating agents ranged between 10 wt.% to 40 wt.%. The experiments were carried out at 95°C (200°F) in a constant time (24 hrs) and at high pH values ranged from 11 to 12.

b) Repeating step (a) using different chelating agents (DTPA-Na_5, DTPA-K_5), (EDTA-Na_2, EDTA-Na_4, EDTA-K_2, EDTA-K_4), (GLDA-Na_4), and (HEDTA-Na_3) at different concentrations.

c) The dissolution rate of barite in the optimal type and at the optimal concentration of chelating agents (obtained from steps a and b) was studied as a function of time and pH. The pH was varied by preparing the powder chelating agent at different concentrations of NaOH and KOH solutions.

d) The weight of the barite at the beginning and the end of the experiment was measured. The solubility of the barite was measured using the following equation:

$$\text{Solubility}\,\% = \frac{dissolved\ barite\,weight}{initial\ barite\,weight} \times 100 \qquad (2)$$

where, the initial barite weight is 4 g in 100 ml of solvent; the dissolved barite weight is the difference between the initial barite (start of the test) and the remaining barite (at the end of the test).

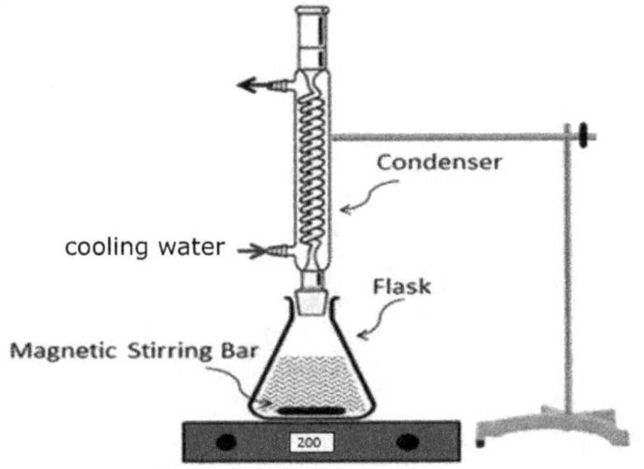

Figure 3: Assembled solubility test setup at high temperatures.

Scanning Electron Microscopy (SEM) Characterization and Analysis

The chemical composition of the industrial barite was determined using scanning electron microscopy (SEM-EDS) to evaluate the purity of barite before the dissolution with the chelating agent. Similarly, the composition of the solids remaining after the dissolution was evaluated to study the change of the chemical composition during the solubility test. The surface area structure of the barite was estimated according the high resolution SEM pictures before and after the dissolution with the chelating agent. The samples were dried for at least 6 hrs and then gold-coated to be viewed using the SEM specimens.

Particle Size Analysis

The particle size distribution of the industrial barite solids was measured using the wet dispersion unit ANALYSETTE 22 Nano Tec plus.

X-Ray Diffraction Analysis

X-ray diffraction spectroscopy was used to measure the mineralogy of the industrial barite (using with PANalytical X-ray, Philips Analytical). The XRD data were used as a confirmation for the EDS analysis results. The XRD was also carried out to determine the mineralogy of the solids precipitated at the end of the solubility test.

Fourier Transform Infrared Spectroscopy (FTIR)

The surface functional groups of industrial barite before and after the dissolution were evaluated using Fourier transform infrared (FPC FTIR Perkin Elmer spectrophotometer) analysis. A pellet was prepared by mixing (1-2 mg) of the barite with 100 mg of KBr. Then, the sample was exposed to a laser beam in the FTIR instrument. The instrument wave number ranged from 4000 to 400 cm^{-1}.

RESULTS AND DISCUSSION

Solubility of Barite in Diethylene Triamine Penta Acetic Acid (DTPA)

As shown in Table 1, the diethylene triamine penta acetic acid (DTPA) is available in a powder form (99%; Mw=393.4) and as an aqueous solution at a concentration of 40 wt.%. The DTPA solution exists either as potassium base or sodium base, viz., diethylene triamine penta potassium (DTPA-K_5) and penta sodium acetic acid (DTPA-Na_5) respectively. The DTPA solution was prepared as an aqueous solution using either NaOH [7] or KOH [8-9]. The hydroxide compounds were used to increase the pH of the distilled water. Since the chelating agent (DTPA-H_5 powder) solubility in the distilled water is limited, the pH of the water must be raised to prepare the aqueous solution. Increasing the pH of the water either by using the NaOH or KOH increased the chelating agent solubility.

The DTPA solution potassium base was tested at variable concentrations ranging from 10 to 40 wt.%. These concentrations were prepared by diluting the 40 wt.% DTPA-K_5. The pH of the DTPA was kept in the range of 11-12. The dissolution capacity of the barite in DTPA potassium base is shown in Figure 4. The results show that the DTPA-K_5 concentration of 20 wt.% at a high pH (11-12) yielded the maximum dissolution rate of barite particles. The test was run for 24 hrs at 200 °F. The maximum removal capacity was about 67 wt.% (equal to 26.8 grams of industrial barite per liter of 20 wt.% DTPA-K_5), as shown in Figure 4 and Figure 5.

Sodium hydroxide was used to increase the pH to 12 to prepare the sodium base of the DTPA-Na_5. The same experimental conditions of 200 °F and 4 g of barite in 100 ml of solvent were applied for this purpose. Various concentrations of DTPA-Na_5 were prepared to optimize the concentration of DTPA-Na_5. The pH of the solution was kept in the range of 11 to 12. The experimental results show that about 15 g of barite particles was dissolved (a solubility percentage of 38 wt.%) at the optimum concentration of DTPA (20 wt.%) as shown in Figure 4 and Figure 5.

The potassium base of the DTPA chelating agent (DTPA-K_5) yielded a higher dissolution rate of barite compared to the sodium base (Figure 4). Therefore, special attention was given here to study the effect of pH on the dissolution of barite in 20 wt.% DTPA-K_5 solution (optimum concentration as shown in Figure 4).

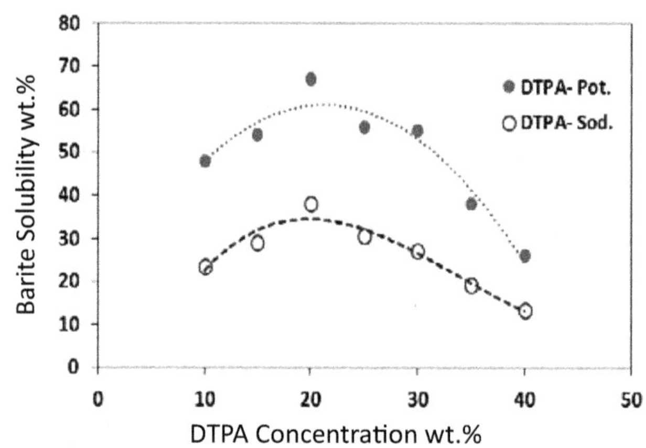

Figure 4: Barite solubility (in percentage) with DTPA-potassium/sodium base (pH=11-12; T= 200 °F; Duration= 24 hr).

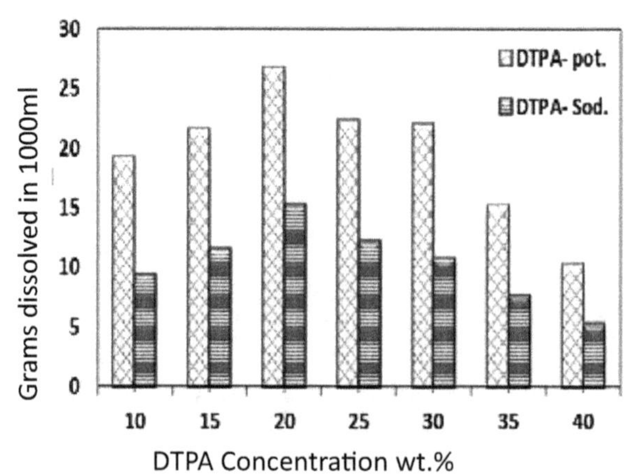

Figure 5: Comparing DTPA potassium/sodium base solubility in grams per liter (pH=11-12; T= 200 °F; Duration= 24 hr).

The solubility of DTPA powder is limited in water, so adding KOH is required to increase the solubility. This also helps to increase the pH. 1.3 M of KOH (Mw=56.11) solution was found to be sufficient to make DTPA at a concentration of 20 wt.% with a pH around 3.5 as shown in Figure 6. The required concentration of KOH needed to prepare a certain pH of the DTPA solution of 20 wt.% is shown in Figure 6. As the pH of the solution reaches a value higher than 12.5, the DTPA becomes fully deprotonated and less sensitive to the concentration of KOH.

The dissolution capacity of industrial barite particles increases by increasing the pH of 20 wt.% DTPA potassium base (Figure 7). The barite solubility reached the maximum value at a pH greater than 11, which is prepared by 3.5 M of KOH. It was noted that increasing the KOH concentration from 3.5 to 4.5 M (to obtain a pH greater than 11) has no effect on the barite dissolution in a 20 wt.% DTPA solution. Higher concentrations of the KOH, more than 4.5 M, have a reverse effect on barite dissolution rate in the 20 wt.% DTPA solution. Therefore, for preparing DTPA penta potassium using DTPA powder, it is recommended that the KOH concentration be kept in the range of 3.5 to 4.5 M to achieve the maximum barite dissolution.

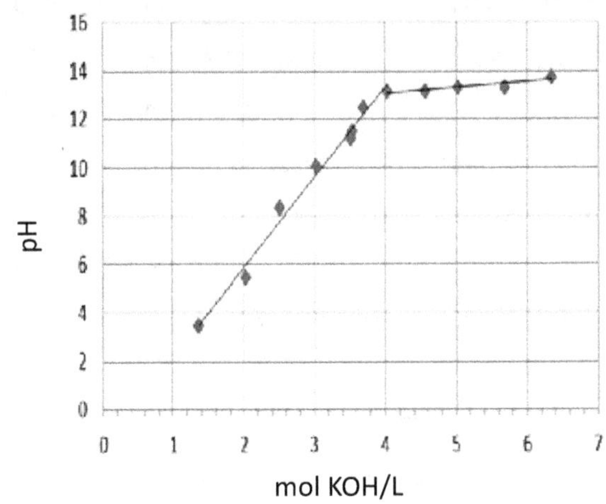

Figure 6: Molarity of KOH to prepare a DTPA solution of 20 wt.%.

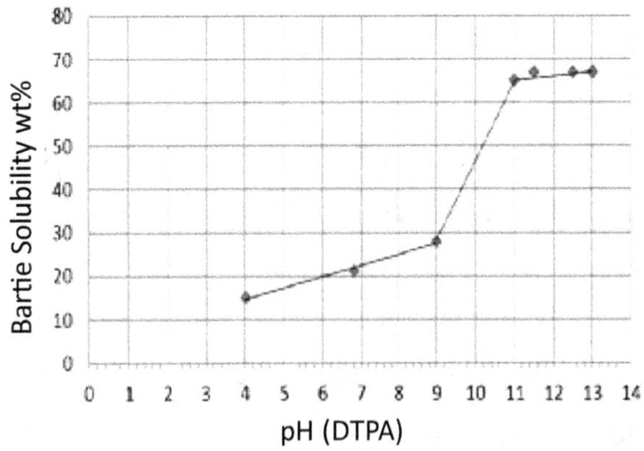

Figure 7: Effect of pH on barite solubility (T= 200 °F; Duration= 24 hr).

The rate of the dissolution of Ba^{2+} from $BaSO_4$ using chelating agents is affected by several resistances based on the concentration of the chelating agent and the pH of the solution [14]. At a low pH, where the chelating agent exists in the acidic form, the presence of hydrogen ions compacts the chelating agent and minimizes the dissolution of Ba^{2+}. The chelating ability is increased by increasing the pH and reaches the maximum for fully deprotonated chelating agent. In our case, at a high pH, higher than 11, the dissolution is affected by the concentration of DTPA in the solution. At a low concentration of DTPA, lower than 15 wt.%, the driving force between the bulk concentration of DTPA in solution and to the one on the surface of barite is low. Therefore, the rate of dissolution is mainly controlled by the mass transport of DTPA. However, at high concentrations, this effect is negligible compared to the effect of the adsorption of DTPA on the surface of barite and the solution complexation reaction thereafter. Thus the DTPA is first adsorbed onto the surface followed by solution complexation reaction between DTPA and $BaSO_4$ according to the following equation:

$$HY^{-4} + OH^{-} \rightleftharpoons Y^{-5} + H_2O \tag{3}$$

$$BaSO_4 + Y^{-5} \rightleftharpoons BaY^{-3} + SO_4^{-2} \quad (4)$$

$$BaSO_4 + HY^{-4} + OH^- \rightleftharpoons BaY^{-3} + SO_4^{-2} + H_2O \quad (5)$$

A higher concentration of DTPA, greater than 25 wt.%, leads to a decrease in the rate of dissolution due to increasing the competitive adsorption of DTPA on the surface of barite, thereby decreasing the rate of desorption of Ba^{2+} into the solution (increasing desorption resistance of Ba^{2+}). Moreover, when the concentration is increased by more than 20 wt.% of DTPA, the viscosity rises and the reaction is retarded [15].

The stability of the solution after the solubility test was monitored for one hour after the test. The stability test gives another confirmation that 20 wt.% DTPA-K5 is the most stable concentration because it was able to hold the soluble barite in the solution. Figure 8 shows that there was a precipitation of barite solids in the solution after 1 hr. This precipitation increases with increasing the DTPA concentration. The concentration range of 20 to 25 wt.% was the most stable with time. Monitoring these solutions for longer times up to 48 hrs showed that the 20 wt.% yielded constant stability.

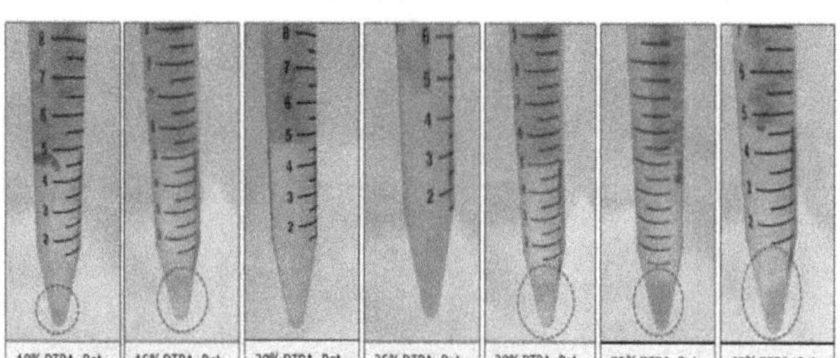

Figure 8: Stability of DTPA solution with time.

The above results were conducted based on 24-hr solubility test. Therefore, the effect of time on the dissolution rate of barite in 20 wt.% DTPA-K5 was tested here. The relationship between the industrial barite solubility in the 20 wt.% DTPA penta potassium and time indicated that there was a significant increase in the percentage of barite solubility at the time period up to 24 hrs. The maximum value of barite solubility was 67 wt.% at 24 hrs. Afterwards, the dissolution rate of barite in DTPA remained almost the same in the range of 67 wt.% up to 70 wt.% (Figure 9). Based on this result, 24 hrs can be taken as the optimum time to achieve good barite solubility. Increasing the time for more than 24 hrs is not recommended due to the minor impact on the barite dissolution beyond this time.

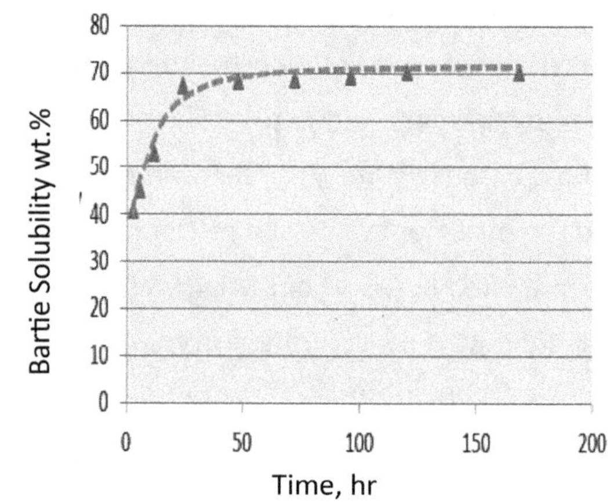

Figure 9: Relationship between time and barite dissolution in 20 wt.% DTPA-K_5 (T= 200°F).

Solubility of Barite in Ethylene Diamine Tetra Acetic Acid (EDTA)

The EDTA sodium base solutions can be prepared using the NaOH from the EDTA powder ($C_{10}H_{16}N_2O_8$). EDTA tetra sodium (EDTA-Na_4) was available at 39 wt.% aqueous solution. The results show that the maximum dissolution capacity of the industrial barite solids was around 26 wt.% using 20 wt.% of EDTA-Na_4 as shown in Figure 10. The pH of the EDTA-Na_4 solution was 12 and the test was conducted for 24 hrs at 200 °F.

The other type of EDTA sodium base was EDTA disodium (EDTA-Na_2 98-100% powder; $C_{10}H_{14}N_2Na_2O_8 .2H_2O$). The disodium EDTA solution was prepared using distilled water and NaOH. The solubility of EDTA disodium (EDTA-Na_2) in water was limited, so the pH of the solution was increased by NaOH. Where EDTA disodium becomes soluble, the lowest pH value is around 8.5. The dissolution of the solids was very small in EDTA disodium at concentrations ranging from 10 to 30 wt.%. The maximum dissolution capacity of the solids was around 6 wt.% using 20 wt.% of EDTA-Na2 during 24 hrs at 200 °F.

Di potassium EDTA-K_2 (99% Powder) was used to prepare different concentrations of EDTA-K_2 to evaluate the barite dissolution rate. The di potassium EDTA 20 wt.% is soluble in water at a pH around 5. In this condition, the solubility of barite was limited in EDTA-K_2; the barite solubility reached 10 wt.% as the maximum value using 20 wt.% EDTA-K_2 in 24 hours.

The potassium base EDTA-K_4 was prepared using EDTA powder ($C_{10}H_{16}N_2O_8$) and the KOH water solution. The results of EDTA-K_4 potassium base is shown in Figure 10. 20 wt.% of EDTA-K_4 dissolved 64 wt.% of the barite. Altering the base of EDTA from sodium to potassium using KOH enhanced the solubility of industrial barite by 38% from 26 to 64 wt.% (Figure 10). The capacity of barite dissolution reached 25.6 g/L by using potassium base EDTA compared to sodium base (Figure 11).

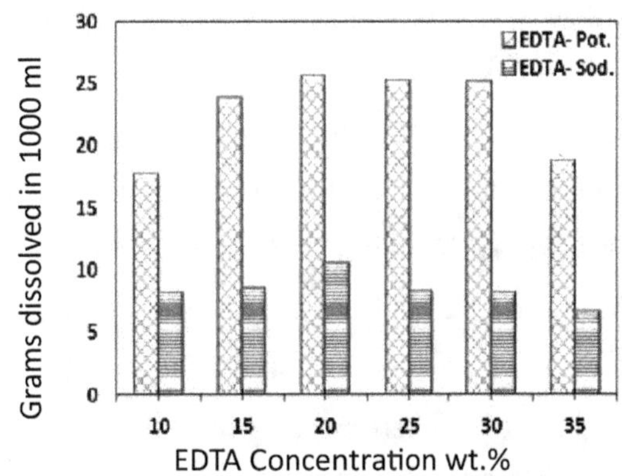

Figure 11: Comparing EDTA potassium/sodium base solubility in grams per liter (pH=11-12; T= 200 °F; Duration= 24 hr).

1.7 M of KOH (Mw=56.11) solution is sufficient to make a concentration of 20 wt.% of EDTA powder soluble in distilled water at a pH of 4.8 as shown in Figure 12. The required concentration of KOH needed to prepare an EDTA solution 20 wt.% with a certain pH is shown in Figure 12. Using KOH solution at concentrations ranging from 4 to 5 M is sufficient to prepare 20 wt.% EDTA with pH ranging from 11 to 12.5. From the point of view of the dissolution of barite, it is not recommended that the concentration of KOH solution be increased to more than 5 M to prepare the EDTA solution.

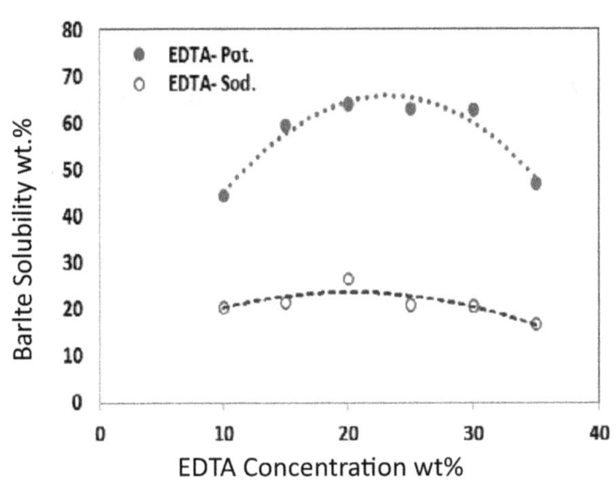

Figure 10: Barite solubility (%) with EDTA- potassium/ sodium base (pH=11-12; T= 200°F; Duration= 24 hr).

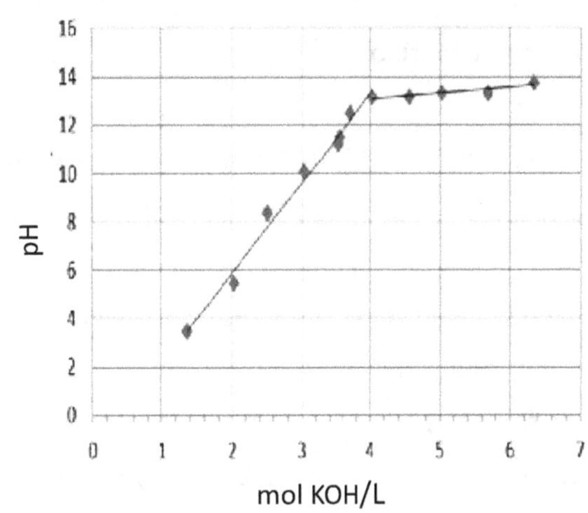

Figure 12: Molarity of KOH to prepare a 20 wt.% EDTA solution.

Hydroxyethyl Ethylene Diamine Triacetic Acid (HEDTA)

The dissolution rate of industrial barite was evaluated using sodium base hydroxyethyl ethylene diamine triacetic acid (HEDTA-Na_3) at high pH values. The results show that the dissolution of barite particles in HEDTA was low. The optimum concentration of HEDTA is 20 wt.%, where 6.5 g of barite was soluble in one litter of 20 wt.% HEDTA-Na_3.

Glutamic Acid, N, N-Diacetic Acid (GLDA)

Two types of GLDA sodium base high pH (11-12) and low pH (3.5) were used to evaluate the dissolution rate of barite. It was found that the barite gave zero solubility in the GLDA at both low and high pH values.

Characterization of Barite before and after Solubility Using DTPA-K_5

4.5.1 SEM, XRD, and Particle Size Analysis

The solubility experiments showed that DTPA-K_5 and EDTA-K_4 were the most effective chelating agents in dissolving barite. Several analyses were conducted to understand the mechanism of the dissolution of barite in DTPA-K_5. The SEM-EDS analysis for the barite was conducted in different spectrum ranges to cover the variable sites of the same sample. Figure 13 shows the chemical composition of the industrial barite used in this work before and after the dissolution in DTPA-K_5. The data presented that the main elements are barium and sulfur which obtained form barium sulfate ($BaSO_4$). Based on EDS data, the impurities including (Si, Al, Ca, etc.) was ranged from 5 to 7 wt.% in average over the whole spectrum range. Although the EDS result showed that the purity of the barite is around 94%; besides, the XRD considered the purity of the barite as 99-100 wt.%. The impurities were not observed by XRD. An untreated sample demonstrated 10 major peaks related to pyrite crystalline phases. These peaks appeared at 2θ values of 20.46 (011), 22.79 (111), 24.87 (002), 25.86 (210), 26.85 (102), 28.75 (211), 31.53 (112), 32.81 (020), 42.59 (113), and 42.93 (122). Upon treatment, the intensity of these peaks reduced by 50% and became wider, which lead to more amorphous phases.

The particle size results indicated that the industrial barite used in this study had a narrow distribution with a mean particle size ranged from 30 to 40 micron. The 200 μ scale high resolution image of SEM in Figure 14 shows that the majority of the barite particles have a sharp edge and most of the particles have a size of 30 to 50 micron. This observation about the particle size of the industrial barite using SEM confirmed the collected data from the particle size distribution analysis.

The lower scale picture (5 μ) focuses on one particle to display the size of the pores in the barite particles; in addition to, the pore size of the barite particles was found to be around 2-3 μ in length and 1-2 μ in width. The pore size on the surface of the barite particles increased due to the dissolution of DTPA-K_5 which increased the area of the reaction and the dissolution rate. The composition of the solid remaining after the dissolution was analyzed to evaluate the precipitated solids. The EDS showed that the precipitated solids contained only barium, which indicated that no new composition formed and the precipitate was barite (Figure 13). The appearance of potassium was due to the base of the DTPA (potassium base). The XRD results of the precipitated showed that there is quartz (SiO_2) formed during the dissolution test.

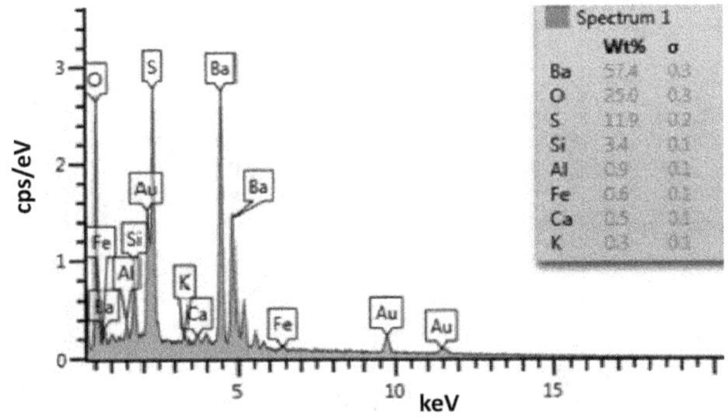

EDS before the dissolution by DTPA-K_5

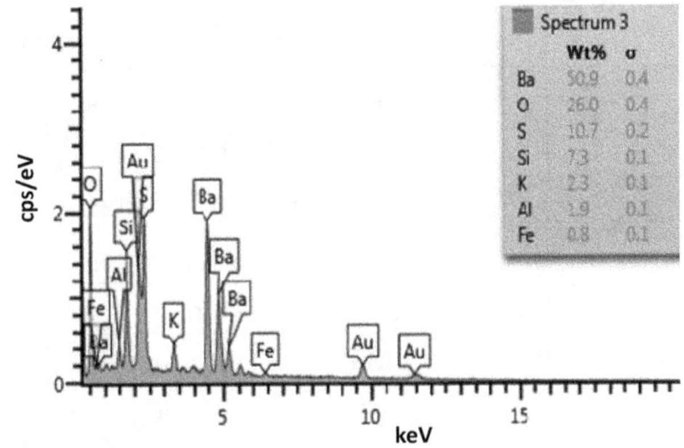

EDS after the dissolution by DTPA-K_5

Figure 13: SEM (EDS) analysis of barite solids before and after the dissolution in DTPA-K_5.

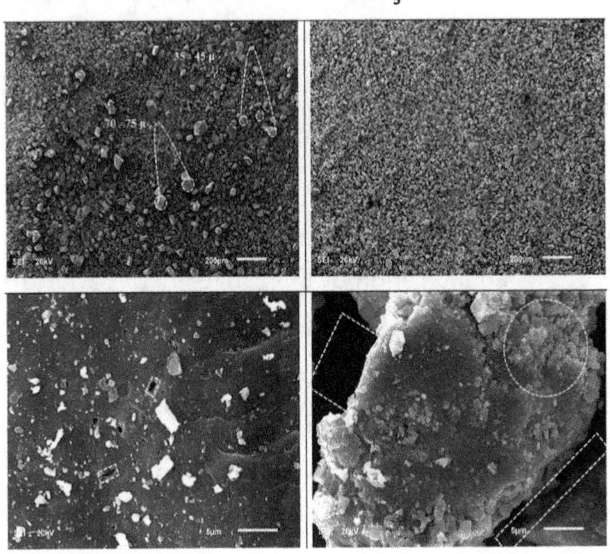

SEM image of barite particles before the dissolution by DTPA-K_5 200 µ, and 5 µ

SEM image of barite particles after the dissolution by DTPA-K_5 200µ, and 5 µ

Figure 14: SEM images of barite particles before and after dissolution with DTPA-K_5.

4.5.2 FTIR Analysis

The upper section of Figure 15 shows the FTIR spectroscopy for the barite solids before the dissolution, whereas the FTIR for the solids remaining after the solubility with DTPA-K_5 is shown in the bottom section of the figure. Table 4 shows that the major FTIR absorption bands for barite solids before dissolution (BSB), and barite solids after the dissolution (BSAD) can be divided into seven groups depending on the different frequencies of the band position. The comparisons between the FTIR peaks for BSB and BSAD and possible assignments for each peak are presented in Table 4. There are four fundamental vibrational modes of sulfate bands called nondegenerate (v_1), one doubly degenerate (v_2), and two triply degenerate (v_3 and v_4) [16–21].

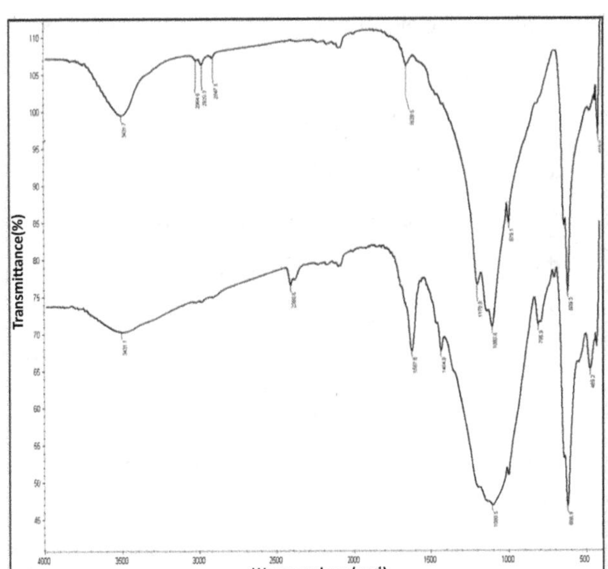

Figure 15: FTIR analysis for barite particles before and after the dissolution in DTPA-K_5.

Table 2: Infrared band positions of the studied barite.

	Wave number (cm^{-1})		Assignments
Peaks group	**Before**	**After**	
Group A	3431.7	3431.1	H-O-H stretching
Group B	2964.6	-------	CH stretch
	2920.3	-------	CH stretch
	2847.5	-------	CH stretch
		2360.6	KBr disc
Group C	1628.5	-------	H-O-H bending
	-------	1587.6	of water
	-------	1404.6	N-H amine group
Group D	1178.9	-------	(v_3) sulfate vibrational modes
	1083.8	-------	(v_3) sulfate vibrational modes
	-------	1080.5	(v_3) sulfate vibrational modes
	979.1		(v_1) sulfate vibrational modes
Group E	-------	795.9	Si-O-Si stretching
Group F	609.3	606.9	(v_4) sulfate vibrational modes
Group L	419.9	-------	"Al-OH" translation
		469.2	(v_2) sulfate vibrational modes

FTIR spectroscopic analyses for the BSB and BSAD solid materials showed that the dissolution of barite particles in DTPA-K_5 affected the bands position and the absorbance intensity. Group-A shows two bands at 3431.7 cm^{-1} in BSB and 3431.1 cm^{-1} in BSAD, which signifies the presence of O–H stretching mode of hydroxyl groups and absorbed water [22,23]. The peak at 3431.1 cm^{-1} in BSAD becomes broader after the dissolution and the absorbance intensity of this peak was decreased when the barite was dissolved in DTPA-K_5 (potassium hydroxides base). In group-B, the peaks at 2964.6, 2920.3, and 2847.5 cm^{-1} appeared in BSB with a very low intensity, which are related to CH stretch [23,24]. These peaks are completely disappeared after the dissolution. The peak at 2360.6 cm^{-1} in BSAD was appeared which is not diagnostic of any mineral in the used barite sample, and it appeared due to the use of KBr in the sample preparation for FTIR [21].

On the other hand, in group-C, the peak at 1628.5 cm^{-1} in BSB is subsequent to a slight amount of water being present in barite crystal [19]. The intensity of this peak increased and shifted to 1587.6 cm^{-1} in BSAD as an indication of the capability of barite to absorb water due to the dissolution of barite solids in DTAP [23,25]. The new peak at 1404.6 cm^{-1} was found only in BSAD, which indicated the presence of N-H bending in the amine group created by using the DTPA.

In group-D, the peaks at 1178.9 and 1083.8 cm^{-1} in BSB are related to the third vibrational modes (v_3) of sulfate [17], and the peak at 979.1 cm^{-1} is assigned to the (v_1) vibrational modes of sulfate [18]. These peaks were diminished and a new peak appeared at 1080.5 cm^{-1} in BSAD as a result of the dissolution of barite in DTPA-K_5. The new peak at 1080.5 cm^{-1} is attributed to (v_3) sulfate vibrational mode [17,18]. The decrease of the 1080.5 cm^{-1} peak intensity means that some of the sulfate functional groups were removed due to the dissolution. The new peaks at 795.9 cm^{-1} was related to Si-O-Si group [26]. This peak confirmed the SEM results, where the weight percentage of Si increases after the dissolution of barite in DTPA. The peak at 609.3 cm^{-1}, which is related to the (v_4) vibrational modes of sulfate, was shifted to 606.9 cm^{-1} with the same intensity for the same vibrational mode [17,18]. This was an indication that the v_4 vibrational mode of sulfate was not easy to break during the dissolution of barite. The peak at 419.9 cm^{-1} refers to the presence of the Al-OH translation modes in BSB [27]. Al was seen also in the SEM results as one of the impurities in the industrial barite used in this study. The intensity of this peak was decreased as result of the dissolution of barite in DTPA. The peak at 469.2 cm^{-1} in BSB was seen in low intensity, and, in the BSAD, it was displayed as an indication of (v_2) sulfate vibrational modes [18,19,28].

CONCLUSIONS

Based on the results obtained from evaluating the dissolution rate of the industrial barite particles in different chelating agents at a high temperature (95°C) in the current study, the following conclusions can be drawn from this current study:

- Potassium base DTPA-K_5 and EDTA-K_4 at a concentration of 20 wt.% provided the maximum barite dissolution capacity, viz. 26.8 g/l and 25.6 g/l during 24 hrs respectively.
- The recommended pH of DTPA-K_5 and EDTA-K_4 should be greater than 11 to achieve the maximum solubility of barite.
- At a high pH, the dissolution of barite is controlled by the adsorption of DTPA on the surface and the solution complexation reaction thereafter.
- Using sodium base DTPA, EDTA, and HEDTA at a high or low pH is not recommended for dissolving barite.
- The barite dissolution rate in chelating agents DTPA-K_5 and EDTA-K_4 increases with time up to 24 hrs. The soaking process after 24 hrs has a minor effect on barite dissolution.
- The FTIR results showed the high capability of the DTPA potassium base to break the v_3 and v_1 sulfate vibrational modes; however, the v_4 sulfate vibrational mode was hard to break during the dissolution.
- The original barite consisted of v_1, v_3, and v_4 vibrational sulfate modes. After dissolution, the barium sulfate was found to be in v_2 and v_4 vibrational sulfate modes. The v_2 mode was formed during the dissolution.
- The dissolution capacity of other chelating agents was found to be as follows: DTPA-Na_5 (15.2 g/L), EDTA-Na_4 (10.5 g/L, HEDTA-Na_3 (6.7 g/L), GLDA-Na_4 (zero), EDTA-K_2 (4 g/L), and EDTA-Na_2 (2.4 g/L) during a soaking time of 24 hours at 95°C.

REFERENCES

1. Fink J. K., "Petroleum Engineer's Guide to Oil Field Chemicals and Fluids," Waltham, MA: Gulf Professional Pub., **2012**.
2. Bageri B. S., Al-Majed A. A., Al-Mutairi S. H., Ul-Hamid A., et al., "Evaluation of Filter Cake Mineralogy in Extended Reach and Maximum Reservoir Contact Wells in Sandstone Reservoirs," SPE/IADC163519, Presented at SPE/IADC Drilling Conference, Amsterdam, The Netherlands, **2013**, 5-7 March.
3. Dunn K., Daniel E., Shuler P. J., Chen H. J. et al., "Mechanisms of Surface Precipitation and

Dissolution of Barite: A Morphology Approach.," *J. Colloid Interface Science*, **1999**, *214*, 427–437.
4. Clemmit A. F., Ballance D. C., and Hunton A. G., "The Dissolution of Scales in Oilfield Systems," SPE-14010, Presented at Offshore Eur. Aberdeen, United Kingdom, **1985**, 10-13 Sept.
5. Morris R. L. and Paul J. M., "Method for Removing Sulfate Scale," US Patent, **1990.**
6. Lakatos I., Lakatos-Szabo J., and Kosztin B., "Optimization of Barite Dissolvers by Organic Acids and pH Regulation," SPE 74667, Presented at International Symposium on Oilfield Scale, Aberdeen, United Kingdom, **2002**, 30-31 January.
7. Lakatos I., Lakatos-Szabo J., and Kosztin B., "Comparative Study of Different Barite Dissolvers: Technical and Economic Aspects," SPE 73719, Presented at International Symposium and Exhibition on Formation Damage Control, Lafayette, Louisiana, **2002**, 20-21 February.
8. Putnis C. V., Kowacz M., and Putnis A., "The Mechanism and Kinetics of DTPA-Promoted Dissolution of Barite," *Appl. Geochemistry*, **2008**, *23*, 2778–2788.
9. Putnis A., Putnis C. V., and Paul J. M., "The Efficiency of a DTPA-based Solvent in the Dissolution of Barium Sulfate Scale Deposits," SPE 29094, Presented at SPE International Symposium on Oilfield Chemistry, 14-17 February, **1995**, 773-785.
10. Nasr-El-Din H. A., Al-Mutairi S. H., Al-Hajji H. H., and Lynn J. D., "Evaluation of a New Barite Dissolver: Lab Studies," SPE 86501, Presented at SPE International Symposium on Exhibition and Formation Damage Control, **2004**,18-20 February.
11. "Akzo Nobel Dissolvine Product Guide 10-2012" [Online]. Available: (https://www.akzonobel.com/dissolvine/system/images/AkzoNobel_Dissolvine_Product_Guide_10-2012_tcm108-37894.pdf).
12. Bageri B. S., Mahmoud M. A., Al-Mutairi S. H., and Abdulraheem A., "Filter Cake Porosity and Permeability Profile along the Horizontal Well and," International Petroleum Technology Conference held in Doha, Qatar, **2015**, 6-9 December.
13. Bageri B. S., Mahmoud M., Al-Mutairi S. H., and Abdulraheem A., "Effect of Sand Content on the Filter Cake Properties and Removal During Drilling Maximum Reservoir Contact Wells in Sandstone Reservoir," *J. Energy Resour. Technol.*, **2015**, *138*, 032901.
14. Fred C. and Fogler H., "The Influence of Chelating Agents on the Kinetics of Calcite Dissolution," *J. Colloid Interface Sci.*, **1998**, *204*, 187–197.
15. Mahmoud M. A., Nasr-El-Din H. A., Wolf C. De, and James N., "Stimulation of Carbonate Reservoirs Using GLDA (Chelating Agent) Solutions," SPE-132286, Presented at Trinidad and Tobago Energy Resources Conference, Port of Spain, Trinidad, **2010**, 27-30 June.
16. Ramaswamy V., Vimalathithan R. M., and Ponnusamy V., "Synthesis of Well Dispersed, Elliptical Shaped Barium Sulfate Nanoparticles via Water-chloroform Mixed Solvent," *Arch. Phys. Res.*, **2010**, *1*, 217–226.
17. Povnnennykh A. S., "The Use of Infrared Spectra for the Determination of Minerals," *Am. Mineral.*, **1978**, *63*, 956–959.
18. Dimova M., Panczer G., and Gaft M., "Spectroscopic Study of Barite from the Kremikovtsi Deposit (Bulgaria) with Implication for its Origin," *Geol. Anal. Balk. Poluostrva*, **2006**, *67*, 101–108.
19. Lane M. D., "Mid-infrared Emission Spectroscopy of Sulfate and Sulfate-bearing Minerals," *Am. Mineral.*, **2007**, *92*, 1–18.
20. Myneni S. C. B., Traina S. J., Waychunas G. a., and Logan T. J., "Vibrational Spectroscopy of Functional Group Chemistry and Arsenate Coordination in Ettringite," *Geochim. Cosmochim. Acta*, **1998**, *62*, 3499–3514.
21. Lugwisha E. H. J., "Identification of Clay Minerals of the Eastern Southern Region of Lake Victoria by Ethylene Glycol and Heat: X-Ray Diffraction and Infrared Spectroscopy Studies," *Tanz. J. Sci.*, **2011**, *37*, 167–178.
22. Shawabkeh R. A., "Synthesis and Characterization of Activated Carbo-aluminosilicate Material from Oil Shale," *Journal of Microporous and Mesoporous Materials*, **2004**, *75*, 107–114.
23. Saikia B. J. and Parthasarathy G., "Fourier Transform

Infrared Spectroscopic Characterization of Kaolinite from Assam and Meghalaya, Northeastern India," *J. Mod. Phys.*, **2010**, *01*, 206–210.

24. Mendelovici E., Frost R. L., and Kloprogge J. T., "Modification of Chrysotile Surface by Organosilanes: An IR–photoacoustic Spectroscopy Study," *J. Colloid Interface Sci.*, **2001**, *238*, 273–278.
25. Frost R., Martens W., Williams P., and Kloprogge J., "Raman and Infrared Spectroscopic Study of the Vivianite-Group Phosphates Vivianite, Baricite and Bobierrite," *Mineral. Mag.*, **2002**, *66*, 1063–1073.
26. Palanivel R. and Velraj, G., "FTIR and FT-Raman Spectroscopic Studies of Fired Clay Artifacts Recently Excavated in Tamilnadu, India," *Indian J. Pure Appl. Phys.*, **2007**, *45*, 501–508.
27. Kloprogge J. T., Hickey L., and Frost R. L., "FT-Raman and FT-IR Spectroscopic Study of Synthetic Mg/Zn/Al-Hydrotalcites," *J. Raman Spectrosc.*, **2004**, *35*, 967–974.
28. Simonescu C. M., "Application of FTIR Spectroscopy in Environmental Studies," *Adv. Asp. Spectrosc.*, **2012**, *2*, 49-84.
29. Yang Y., Xiang H. W., Xu Y. Y., Bai L., et al., "Effect of Potassium Promoter on Precipitated Iron-manganese Catalyst for Fischer-Tropsch Synthesis," *Appl. Catal. A: Gen.*, **2004**, *266*, 181-194.

12

Geochemical Characteristics and Hydrocarbon Generation Modelling of Early Triassic to Late Cretaceous Formations within Ogaden Basin, Ethiopia

Samuel Getnet Tsegaye[1*], M. E. Nton[1], O. A. Boboye[1], and Welela Ahmed[2]

[1*] Department of Petroleum Geosciences, Pan African University Life and Earth Sciences Institute (PAULESI), University of Ibadan, Ibadan, Nigeria

[1] Department of Geology, University of Ibadan, Ibadan, Nigeria

[2] School of Earthsciences, Addis Ababa Science and Technology University, Addis Ababa, Ethiopia

ABSTRACT

The major goal of this study is the evaluation of the source rock of Ogaden Basin using Rock-Eval Pyrolysis, Vitrinite reflectance (%Ro), and well log data of the Permo-Triassic and the Mesozoic sequences including Karro, Hamanlei, and Urandab Formations. Burial history and level of thermal maturity of the Bodle-1 well, the Calub-1 well and the Hilala-2 well were modeled based on the calibration of measured %Ro and T_{max} against calculated %Ro model. The calculated Total Organic Carbon (TOC) content from the well log data compared with the measured TOC from the Rock-Eval pyrolysis in the three mentioned wells is shown to match against the shale source rock.

The total organic carbon (TOC) content value ranged from 0.48 to 4.24 wt % while the S1 and S2 values ranged from 0.05 to 0.76 (mg/g) and 8.52 to 27.88 (mg/g) respectively. Hydrogen index (HI) versus oxygen index (OI) plot has shown that the organic matter is predominately made up of type III kerogen (gas prone) while T_{max} and HI indicated mixed type II/III kerogen (25% of Urandab Formation, 40% of Bokh Formation and 15% of Hamanlei and Adigrat Formations). T_{max} was assessed from vitrinite reflectance (Ro) readings for Bodle-1, Calub-1 and Hilala-2 wells showed an average vitrinite reflectance values in the range of 0.5-1.4 VRo%, 0.35-1.08 VRo %, 0.55 – 0.69 VRo % respectively. The developed burial history models have revealed that Bokh Shale hydrocarbon generation has been started during lower Jurassic, and presently hydrocarbon generation has been started within the "oil window" while Urandab Shale is immature to fairly mature.

Keywords: Calub, Vitrinite Reflectance, Kerogen, Bokh, Bodle

***Corresponding author**

Samuel Getnet Tsegaye

Email: stsegaye0121@stu.ui.edu.ng

INTRODUCTION

Ogaden Basin, one of the sedimentary basins of Ethiopia, and on which it has been focused mainly, is located in the southeast of Ethiopia between latitudes 4° and 8°N and longitudes 40° and 48°E. (Figure 1). The basin is located in Somali Plateau, in south-eastern Ethiopia; the name Ogaden is now commonly used for the entire region below about 1,500 m AMSL, an area of some 300,000 km^2 which encompasses most of the Somali Regional State and includes the southeast and southwest portion of Ethiopia [3,39].

In the work of [32], it has been shown that the Ogaden Basin extends into the neighboring countries of Somalia and Kenya. The basin contains extensive wedge of Late Palaeozoic to Tertiary sediments which reaches up to and contains an average thickness of up to 4,000 meters in deeper parts and thickens towards the West. The Ogaden basin has spectacularly maintained a thick sedimentary apron and salient petroleum geological features favorable for petroleum accumulation from the onshore through the continental shelf and to the deep water terrains [1, 2].

In the references [17, 46], it was reported that the Ogaden Basin was initiated in the Late Carboniferous to Permian when rifting occurred widely across Pangaea. The deep Adigrat sandstone and Hamanlei limestone are the main exploration targets in prospective areas within Ogaden basin. Calub-Hilala oil and gas field is located at the central part of Ogaden Basin which is a relatively under-explored field within the basin.

A lot of information about the sediments and sedimentary processes is contained in well logs. Sediments in different paleo environments display characteristic log motifs. As a result, borehole logs are widely used to interpret sedimentary facies [32].

The logs used in this study include Gamma Ray (GR), Porosity logging tools and true resistivity logging suits. Information about the sediments and sedimentary processes from these logs may not be sufficient alone, due to some lithology having similar natural radioactivity and electrical properties. Information from cuttings and cores is therefore often an essential component of any lithological analysis. From the combined core description and wireline log data, it is commonly possible to generate a series of (wireline) log facies. Such log facies may be used to describe and correlate lithology along with source rock maturity evaluation of a cored and logged interval of the wells [13]. This study attempts to identify the maturity history, burial history and thermal history of the source rocks found within Ogaden Basin.

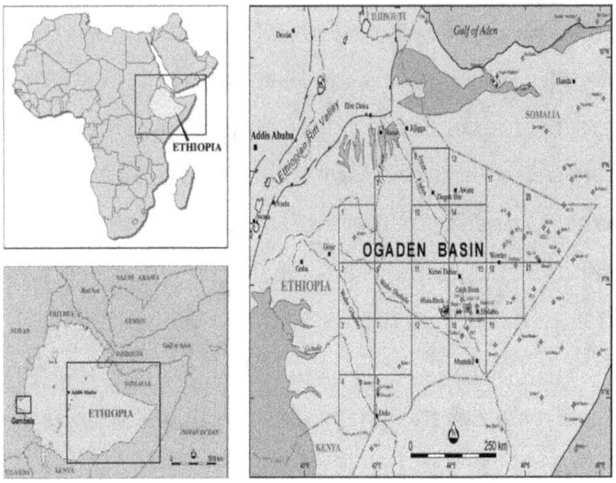

Figure 1: Location Map, Ogaden Basin, Ethiopia (Modified after [24]).

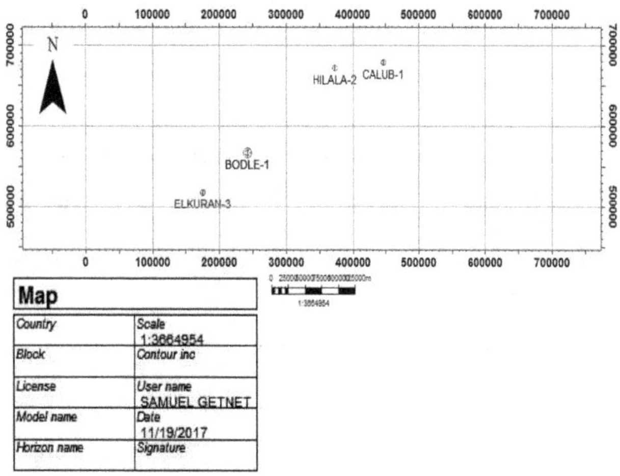

Figure 2: Well location map used for the study plotted using PetrelTM 2009 software.

Geological Setting

The study area is a field situated within an onshore central and southern part of Ogaden Basin, Ethiopia (Figure 1). The map showing the location of the wells used in this study is illustrated in Figure 2. The Ogaden basin is said to have a maximum sediment thickness (at deeper parts) of 7 km at the central and southwestern part of the sub-basin around Bodle deep with an aerial extent of about 75,000 km^2.

The source of the sedimentary fill is mainly the Tethys sea during the Mesozoic times, the cross river channels, and lacustrine depositional environments during the Permo-Triassic. These fills gave rise to the formations found in the basin, which consists of unconsolidated to consolidated sands and overpressured shales to limestone and dolomitic dominated formations. While the sands are fluvial to fluvio marine (channels and barrier bars respectively), the shales are fluvial marine or lagoonal [8, 13, 17, 39].

The formations within the Permo-Upper Cretaceous Ogaden Basin include the Calub, Bokh, Gumburo, Adigrat, Hamanlei, Gebredare, Gorrahei, Mustahil and Ferfer formations. The Karro Formations which are considered to be the oldest formations within the basin includes Calub sandstone, Bokh shale, Gumburo sandstone and Adigrat sandstone which is unconsolidated to slightly consolidated paralic siliciclastic sequence of sandy unit with minor shale intercalations of about 4500 m thick [39].

The Formations from Hamanlei upwards which overlies the Karro formations at the base and mainly westward of the basin are shallow to deep marine origin and composed of thick carbonate with minor amounts of clay and silt. These formations underlie the entire basin and are typically over pressured. The overlying Lower, Middle, and Upper Hamanlei Formations represent an Early Jurassic to Callovian syn-rift marine sequence. In addition the consist of the following lithology: limestone with intercalated shales (in the lowermost section); anhydrite, dolomite and limestones with local oolitic and stromatolitic beds (in the middle section); and Bioclastic Oolitic limestones (in the uppermost section). Adigrat Formation marks the uppermost unit of the Karoo formation complex. Moreover, mainly it consists of about 400 m thick massive continental sands and gravels which are deposited in a terrigenous depositional environment. The sands are yellowish brown or white in color due to the presence of limonite, hematite, and feldspars [6]. However, the Adigrat Formation is fluvio-lacustrine in origin derived from a continental environment, based on lithological character, it is mainly sandstone.

The Permo-Triassic sediments are generally overlain by the marine Mesozoic secessions. The Bodle-1 well is located towards the south of the basin at the Bodle-deep and penetrated the Permo Triassic and the Mesozoic successions. Moreover Calub-1 well drilled on the central part of the basin at Calub field intercepted all the formations within the basin and penetrated up to Calub Formation. The thickness of the Permo Triassic Formations and the Mesozoic Formations decreases towards the center of the basin (Calub-Hilala high) which is the saddle of the Ogaden Basin. The Permo Triassic sediments were not intercepted by Hilala-2 well, but the well has penetrated up to the Mesozoic carbonate secessions. Permo Triassic source rocks are developed within the Permo-Triassic succession of the Bokh Formation, while the Late Jurassic-Early Cretaceous source rocks are developed within the Mesozoic Formation in all the three wells.

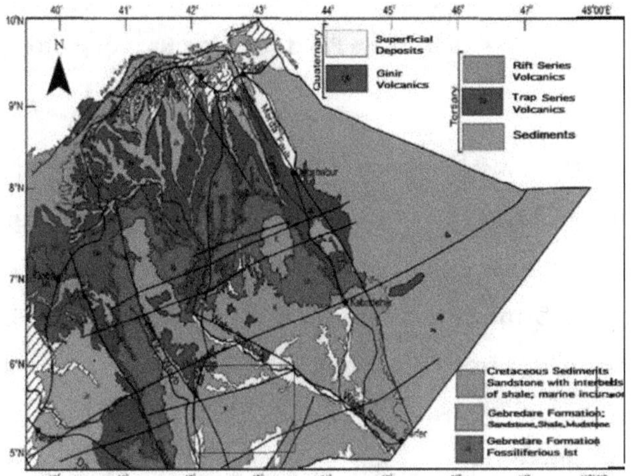

Figure 3: Geological map of Ogaden Basin (Modified after 24).

Materials and Methods

The study depends on geochemical data (Rock-Eval pyrolysis and %Ro) of core and cutting samples from three wells Bodle-1, Calub-1, and Hilala-2. A total of Fourteen (14) representative samples constituting Seven (7) Shale, Six (6) limestone, and one (1) Sandstone samples were selected for TOC/Rock Eval-Pyrolysis analysis (Table 2). While selecting the samples for the organic geochem analysis, an emphasis was given to stratigraphically constrain to the upper and lower boundaries of Permo-Triassic deposits of lithostratigraphic section of Ogaden Basin; the core and cutting samples were used to: analyze source rock maturity for interpretation of source rock facies, determine kerogen types and establish a hierarchal classification of depositional environment based on the organic matter type and delineate and plot oil generating window and recognize associations of intervals with reservoir-quality porosity and permeability values and specific facies and/or boundaries. The samples were sent for analysis to the StratoChem Services (SCS) Company to determine the Rock-Eval pyrolysis and % Ro results. Well log data (from the three wells) (Gamma-ray, Density, Neutron, ILD, PEF and Sonic logs) was obtained from the Ministry of Mines, Petroleum and Natural Gas, Addis Ababa, Ethiopia for the Permo-Triassic and Mesozoic formations of Ogaden Basin. An attempt was made to reconstruct basin modeling and thermal maturity histories, and as reported by [32], the mentioned modeling has become an important tool in the study of burial history and thermal evaluation of sedimentary basins and has been used by many authors.

One-dimensional modeling of burial history and thermal maturity was performed on this present study for the three wells at three different well locations (Figure 2) using PetroMod1DTM Express (version 1.1) of Integrated Exploration Systems GmbH (IES), Germany. The well locations were chosen because (1) they were drilled to a depth that penetrated a significant part of the geologic section of interest, (2) they represent different geologic settings within the province, and (3) they have measured vitrinite reflectance.

Lithology of the stratigraphic units were also interpreted from well logs data using Petrel™ 2009 software and were generalized for modeling purposes (Tables 2 to 4).

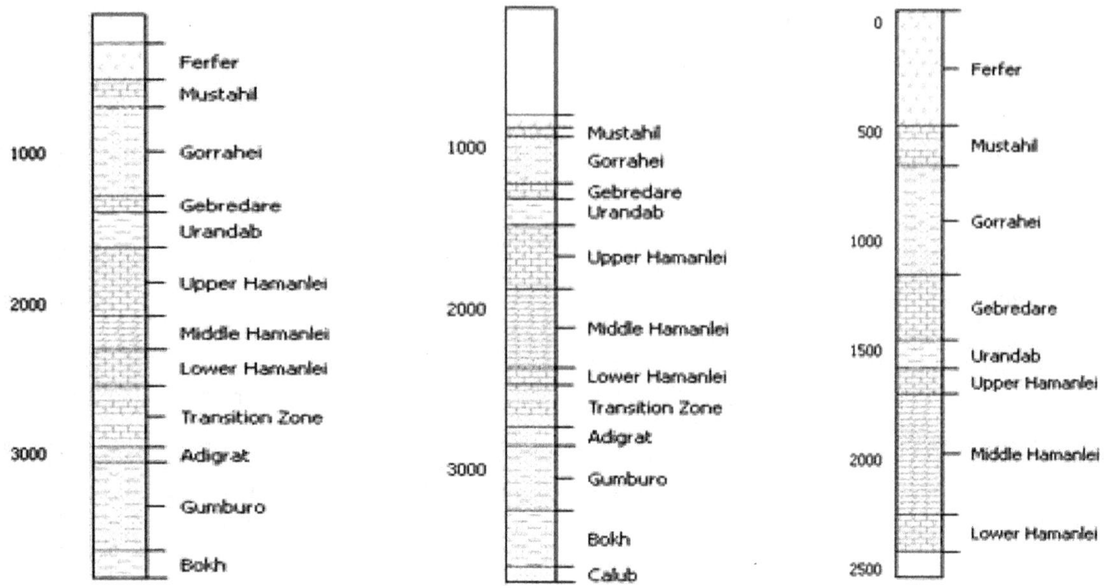

Figure 4: Simplified stratigraphic charts through East-Southwest of the Calub-Hilala Field (Ogaden Basin), plotted using PetroMod 1D software.

RESULTS AND DISCUSSIONS

A- Vitrinite Reflectance Values (VRo %)

Vitrinite Reflectance (%Ro) is the most widely used method for measuring the thermal maturity of sedimentary strata. The Permo-Triassic shale samples were taken mainly from Bokh shale from Bodle-1, Calub-1, and Hilala-2 wells showed an average vitrinite reflectance values in the range of 0.5-1.4 VRo%, 0.35-1.08 VRo %, 0.55-0.69 VRo % respectively, indicating that they fall within the fairly mature to thermally mature zone (Table 1). Most samples from the three wells showed a Vitrinite Reflectance (Ro) value between 0.5 and 1.4 VRo % and this suggests that the samples are within the range of immature to oil generating window phases. Vitrinite reflectance value greater than 0.6 % VRo is considered mature, while values greater than 1.3 % VRo indicates gas window maturity [26, 35].

The Ro and Tmax result at certain depths showed lower (suppression) values and results didn't increase with depth and were considered as anomalies. The observed anomalies were at measured values of Ro within the depth range of 3115 m on a sample taken from Bodle-1 well. A similar effect was observed at measured values of T_{max} at depth ranges of 1364 m and 2775 m on samples from Urandab and Adigrat Formations at Bodle-1 and Calub-1 wells respectively. The results demonstrated that depending on the anoxic or oxic nature of the depositional conditions different initial hydrogen contents and oxygen contents and slightly different initial reflectivity can be recorded [15]. In addition, vitrinite macerals with different initial hydrogen contents have different reaction kinetics, with hydrogen-poor, oxygen-rich vitrinite measuring at an enhanced rate compared to vitrinite with a higher hydrogen content and lower oxygen content. These variations in the hydrogen content and the thermostability of vitrinite macerals caused the Ro and the Tmax suppression and enhancement anomalies (Table 1, Table 2).

Table 1: Vitrinite reflectance measurement of Permo-Cretaceous stratigraphic levels in the Bodle-1, Calub-1 and Hilala-2 wells (Result summarized with increasing depth per well (Bodle-1, Calub-1, and Hilala-2 wells)).

Well	Formation	Depth (m)	Measured vitrinite reflectance values
Bodle-1	Gorrahei Fm	1134	0.50
Bodle-1	Gorrahei Fm	1268	0.58
Bodle-1	Gorrahei Fm	1588	0.67
Bodle-1	Gebredarre Fm	1951	0.62
Bodle-1	M Hamanlei Fm	2716	0.99
Bodle-1	M Hamanlei Fm	2807	0.96
Bodle-1	M Hamanlei Fm	2896	1.09
Bodle-1	M Hamanlei Fm	3264	1.35
Bodle-1	M Hamanlei Fm	3264	1.40
Bodle-1	M Hamanlei Fm	2716	0.99
Bodle-1	M Hamanlei Fm	2807	0.96
Bodle-1	M Hamanlei Fm	2896	1.09
Bodle-1	M Hamanlei Fm	3115	0.82
Bodle-1	M Hamanlei Fm	3264	1.35
Bodle-1	M Hamanlei Fm	3264	1.40
Calub-1	Gorrahei Fm	243.8	0.35
Calub-1	Gorrahei Fm	795.5	0.39
Calub-1	Urandab Fm	1370	0.71
Calub-1	Urandab Fm	1370	0.71
Calub-1	Urandab Fm	1378	0.48
Calub-1	U Hamanlei Fm	1530	0.64
Calub-1	U Hamanlei Fm	1801	0.64
Calub-1	M Hamanlei Fm	2350	0.74
Calub-1	Adigrat Fm	2734	0.69
Calub-1	Adigrat Fm	2792	0.61
Calub-1	Gumburo Fm	2914	1.08
Calub-1	Bokh Fm	3249	0.98
Calub-1	Bokh Fm	3417	1.07
Hilala-2	Urandab Fm	1619	0.55
Hilala-2	U Hamanlei Fm	1640	0.55
Hilala-2	U Hamanlei Fm	1650	0.55
Hilala-2	M Hamanlei Fm	2338	0.69
Hilala-2	M Hamanlei Fm	2342	0.65

Table 2: TOC / RockEval pyrolysis analysis data (Results summarized with increasing depth per well (Bodle-1, Calub-1, and Hilala-2 wells)).

S/N	Well	Sample ID	Depth (m)	Formation	TOC (Wt %)	S1	S2	S3	T_{MAX}	HI	OI	PI
1	Bodle-1	BoW2	1322	Gebredare	0.34	0.05	0.46	1.02	431	137	304	0.1
2	Bodle-1	BoW3	1364	Urandab	0.60	0.06	0.18	0.61	344	321	1089	0.25
3	Bodle-1	BoW4	2489	L. Hamanlei	0.5	0.08	1.10	0.80	437	219	159	0.07
4	Bodle-1	BoW9	3717	Bokh	1.38	0.06	0.22	0.66	415	16	48	0.21
5	Bodle-1	BoW11	3844	Bokh	4.24	0.76	27.88	1.16	427	658	27	0.03
6	Calub-1	CW8	1290	Gebredare	0.50	0.05	0.29	0.39	440	199	267	0.15
7	Calub-1	CW9	1468	Urandab	0.66	0.04	0.75	2.57	428	114	391	0.05
8	Calub-1	CW13	2625	Transition	0.40	0.04	0.14	0.34	413	326	791	0.22
9	Calub-1	CW14	2775	Adigrat	0.30	0.17	0.31	0.38	397	246	302	0.35
10	Calub-1	CW16	3623	Bokh	0.71	0.48	0.22	0.50	422	31	71	0.69
11	Hilala-2	HW6	1574	Gebredare	0.65	0.40	0.19	0.45	423	29	69	0.68
12	Hilala-2	HW7	1623	Urandab	1.75	0.23	8.52	1.09	429	587	62	0.03
13	Hilala-2	HW8	1628	M.Hamanlei	0.25	0.04	0.11	0.32	426	44	127	0.27
14	Hilala-2	HW12	2214	L.Hamanlei	0.78	0.08	0.20	1.23	427	26	158	0.29

Table 3: Summary of Paleo-Water Depth and Paleo Heat Flow inputs estimation used to develop burial history and maturity history, models.

Age/ Stratigraphic Unit	Formation	Lithology	Depositional Environment	Estimated Paleo Water Depth (PWD) (m)	Estimated Heat Flow (mW/m^2)	Estimated Thermal History (°C)
Albian	Ferfer	Gypsum, Anhydrite	Outer-Inner shelf	<50	74	21.35
Aptian	Mustahil	Shale, Sand-stone	Outer-Inner shelf	40	76	20.65
Neocomian-Barremian	Gorrahei	Shale, Anhydrite, Sandstone, Limestone	Inner shelf	50	79	18.85
Upper Kimmerdgian	Gebredare	Pelletal, Bioclastic, locally oo-litic	Outer-inner shelf	80	81	18.44
Lower Oxfordian-Lower Kimmerdgian	Urandab	Shale, Marl, Limestone	Basinal-outer shelf	About 160	83	17.98
Bathonian-Callovian	U.Haman-lei	Grainstone, Oolicitc, Peletal, Wackstone	Inner-outer shelf	90	85	17.61
Lias-Bathonian	M.Haman-lei	Dolomite, Anhydrite, Limestone	Supratidal -inner shelf	5-20	86	15.88
Lias	L.Haman-lei	Limestone locally oo-litic	Inner shelf	30-40	89	16.12
Upper Triassic to Lias	Adigrat	Sandstone, shale	Fluvial-marginal marine	0-2	91	16.68
Middle Trias-sic	Gumburo	Sandstone, shale	Fluvial	0-1	91.5	13.97
Lower Triassic	Bokh	Shale	Lacustrine with marine influence	0-5	92	11.37
Permian	Calub	Sandstone	Glacial influence	0	94	5.00

Table 4: Data generated from PetroMod and used to model burial-history curves for the Calub location using Calub-1 well around the central part of Ogaden Basin. Heat flow and thermal gradient used to calibrate model are given.

Layer	Top (m)	Base (m)	Thickness (m)	Eroded (m)	Depo. From [Ma]	Depo to [Ma]	Eroded from [Ma]	Eroded to [Ma]	Lithology	PSE
Ferfer	783.3400	859.0000	75.6600		112.0000	99.6000			Gypsum	Overburden Rock
Mustahil	859.0000	921.0000	62.0000		125.0000	112.000			Sand&Lime	Overburden Rock
Gorrahei	921.0000	1203.5100	282.5100		145.5000	125.000			Sandshaly	Reservoir Rock
Gebredare	1203.5100	1301.7500	98.2400	286.0000	153.0000	150.800	150.8000	145.000	Limestone(Shaly)	Reservoir Rock
Urandab	1301.7500	1463.2100	161.4600		158.0000	153.000			Shale	Source Rock
Upper Hamanlei	1463.2100	1864.0200	400.8101		164.7000	161.200			Limestone (Micrite)	Reservoir Rock
Middle Hamanlei	1864.0200	2349.0200	485.0000		196.5000	164.700			Dolomite (Typical)	Reservoir Rock
Lower Hamanlei	2349.0200	2459.9500	110.9299		198.0000	196.5000			Limestone (Ooid Grainstone)	Reservoir Rock
Transition Zone	2459.9500	2721.6499	261.7000		205.0000	203.0000	203	198	Sand&Lime	Seal Rock
Adigrat	2721.6499	2840.0000	118.3501		228.7000	205.0000			Sandshaly	Reservoir Rock
Gumburo	2840.0000	3245.3201	405.3201		245.9000	228.7000			Sandstone	Reservoir Rock
Bokh	3245.3201	3588.5901	343.2700		251.0000	245.900			Shalesand	Source Rock
Calub	3588.5901	3691.3999	102.8098		270.6000	251.000			Sandstone	Reservoir Rock
						270.600				

Table 5: Data generated from PetroMod and used to model burial-history curves for Bodle deep using Bodle-1 well towards the South-western part of Ogaden Basin. Heat flow and thermal gradient used to calibrate model are given for each location.

Layer	Top (m)	Base (m)	Thickness (m)	Eroded (m)	Depo. From [Ma]	Depo to [Ma]	Eroded from [Ma]	Eroded to [Ma]	Lithology	PSE
Ferfer	310.7800	551.0000	240.2200		112.0000	99.6000			Gypsum	Overburden Rock
Mustahil	551.0000	735.0000	184.0000		125.0000	112.0000			Sand&Lime	Overburden Rock
Gorrahei	735.0000	1325.0000	590.0000		145.5000	125.0000			Sandshaly	Reservoir Rock
Gebredare	1325.0000	1439.0000	114.0000	286.0000	153.0000	150.8000	150.800	145.000	Limestone(Shaly)	Reservoir Rock
Urandab	1439.0000	1673.0000	234.0000		158.0000	153.0000			Shale	Source Rock
Upper Hamanlei	1673.0000	2125.0000	452.0000		164.7000	161.2000			Limestone (Micrite)	Reservoir Rock
Middle Hamanlei	2125.0000	2352.0000	227.0000		196.5000	164.7000			Dolomite (Typical)	Reservoir Rock
Lower Hamanlei	2352.0000	2594.0000	242.0000		198.0000	196.5000			Limestone (Ooid Grainstone)	Reservoir Rock
Transition Zone	2594.0000	3000.0000	406.0000		205.0000	203.0000	203	198	Sand&Lime	Seal Rock
Adigrat	3000.0000	3095.0000	95.0000		228.7000	205.0000			Sandshaly	Reservoir Rock
Gumburo	3095.0000	3687.0000	592.0000		245.9000	228.7000			Sandstone	Reservoir Rock
Bokh	3687.0000	3875.6499	188.6499		251.0000	245.9000			Shale	Source Rock
						251.0000				

Table 6: Data generated from PetroMod and used to model burial-history curves of Hilala trough using Hilala-2 well around the central part of Ogaden Basin. Heat flow and thermal gradient used to calibrate model are given for each location.

Layer	Top (m)	Base (m)	Thickness (m)	Eroded (m)	Depo. From [Ma]	Depo to [Ma]	Eroded from [Ma]	Eroded to [Ma]	Lithology	PSE
Ferfer	0.0000	510.0000	510.0000		112.0000	99.6000			Gypsum	Overburden Rock
Mustahil	510.0000	690.0000	180.0000		125.0000	112.0000			Sand&Lime	Overburden Rock
Gorrahei	690.0000	1174.0000	484.0000		145.5000	125.0000			Sandshaly	Reservoir Rock
Gebredare	1174.0000	1451.5800	277.5800	286.0000	153.0000	150.8000	150.8000	145.00	Limestone(Shaly)	Reservoir Rock
Urandab	1451.5800	1578.3900	126.8101		158.0000	153.0000			Shale	Source Rock
Upper Hamanlei	1578.3900	1688.0000	109.6100		164.7000	158.0000			Limestone (Micrite)	Reservoir Rock
Middle Hamanlei	1688.0000	2221.1499	533.1499		196.5000	164.7000			Dolomite (Typical)	Reservoir Rock
Lower Hamanlei	2221.1499	2386.9399	165.7900		198.0000	196.50			Limestone (Ooid Grainstone)	Reservoir Rock

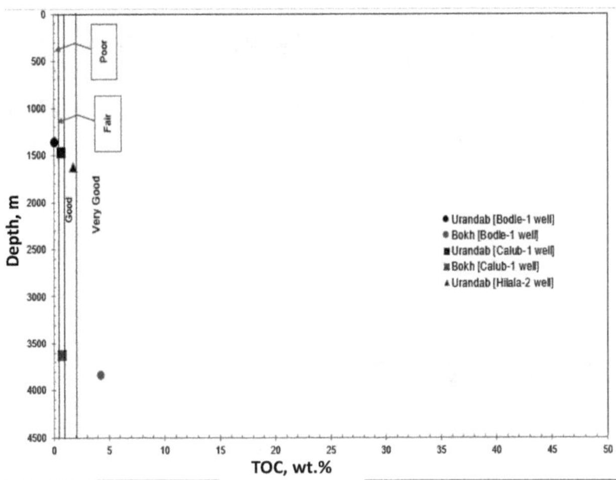

Figure 5: TOC vs. Depth plot of the source (shale) formations (Bokh Shale and Urandab Shale) from three wells.

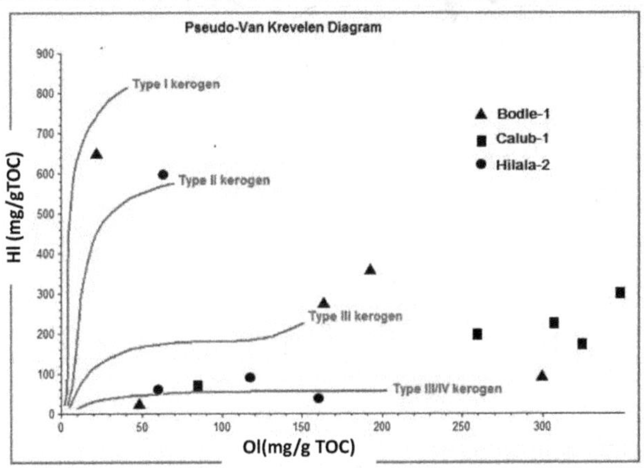

Figure 7: Rock-Eval hydrogen index (HI) versus oxygenindex (OI), showing kerogen quality of the Mesozoic source Rocks in Bodle-1, Calub-1 and Hilala-2 wells; Most samples fall in the zone of Kerogen type III organic matter (Modified after [43]).

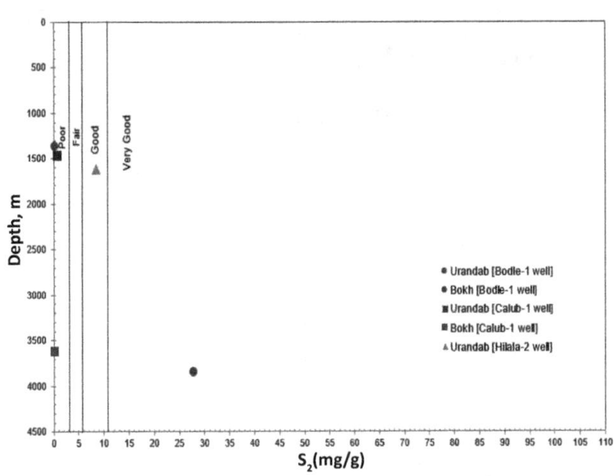

Figure 6: Depth vs. S2 yield plot of the source (shale) formations (Bokh Shale and Urandab Shale) from three wells.

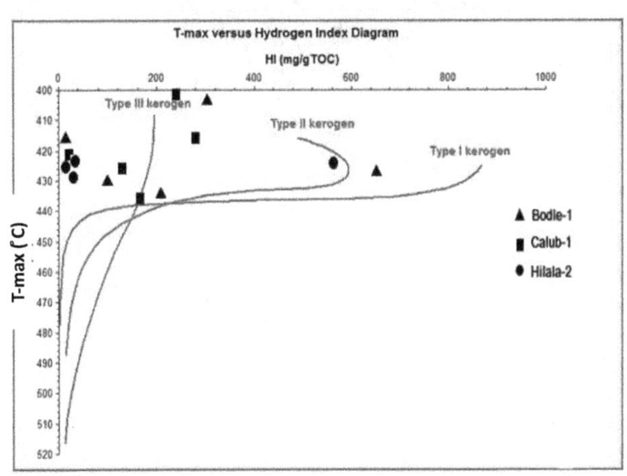

Figure 8: Rock-Eval hydrogen index (HI) versus Tmax, showing kerogen quality of the Permo-Cretaceous source Rocks in Bodle-1, Calub-1 and Hilala-2 wells. Most samples fall in the zone of Kerogen type III organic matter (Modified after [43]).

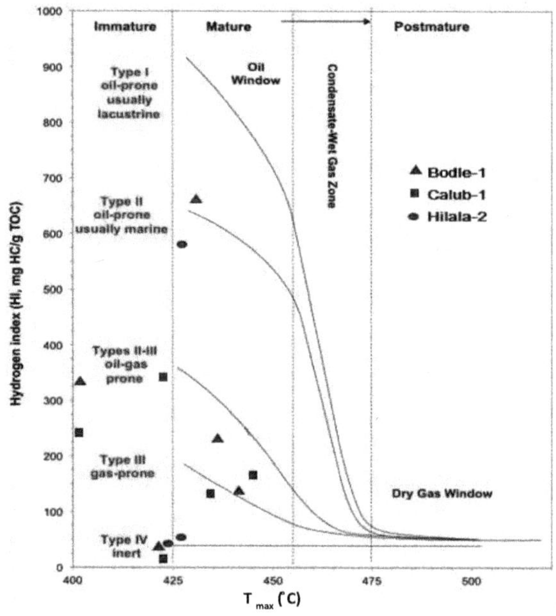

Figure 9: Rock-Eval hydrogen index (HI) versus T_{max}, showing kerogen type and maturity (Modified after [43]).

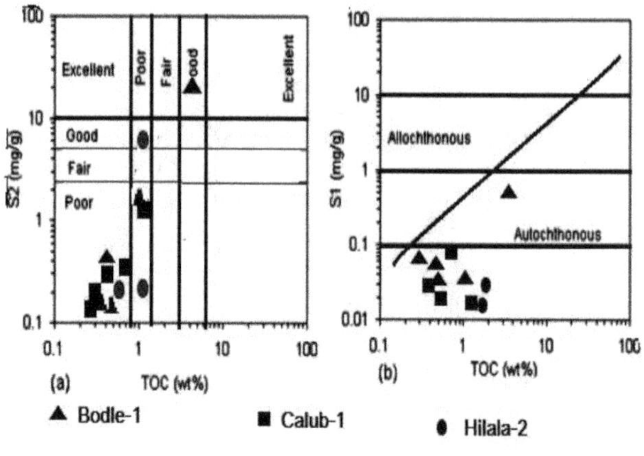

Figure 10: S1, S2 vs TOC yield plot of samples taken from three wells (Modified after [43]). Plot revealed that, most source rocks are (autochthonous) indicating that the oil produced from the studied wells are in-situ deposits with an in-situ source rock.

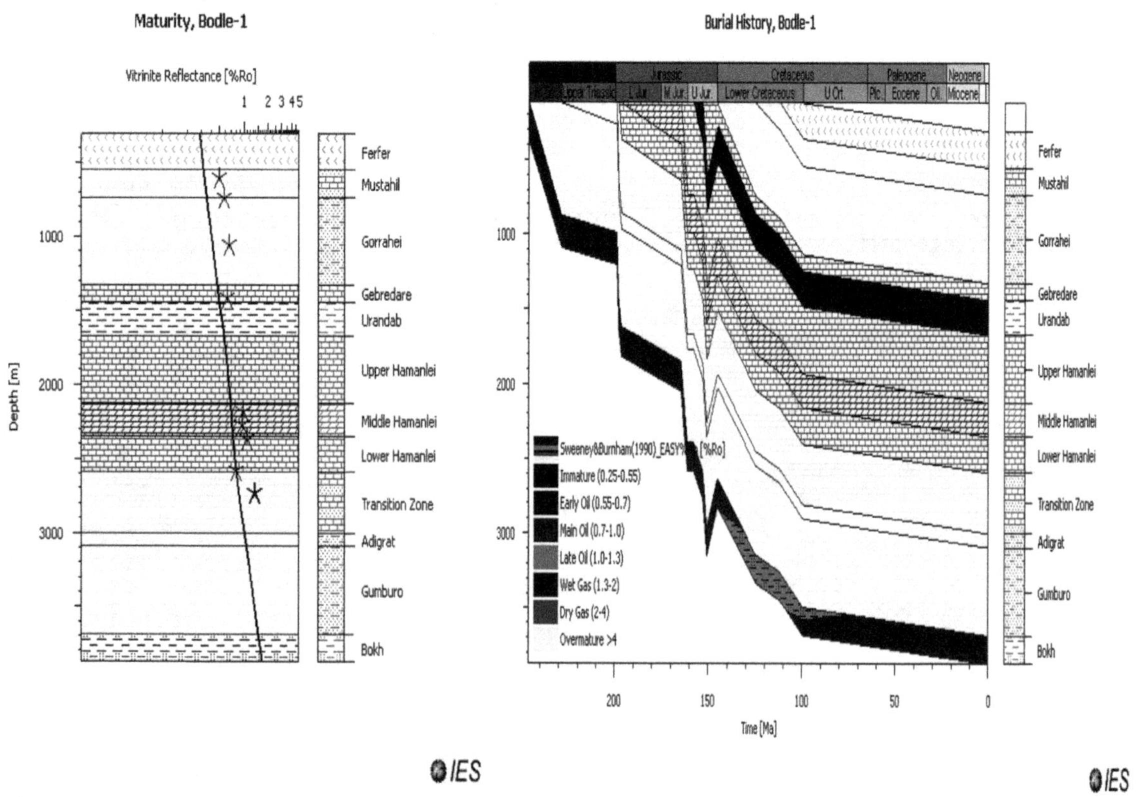

Figure 11: Correlation of measured and modeled vitrinite reflectance data and burial history (with maturity overlay) of Bodle -1 well.

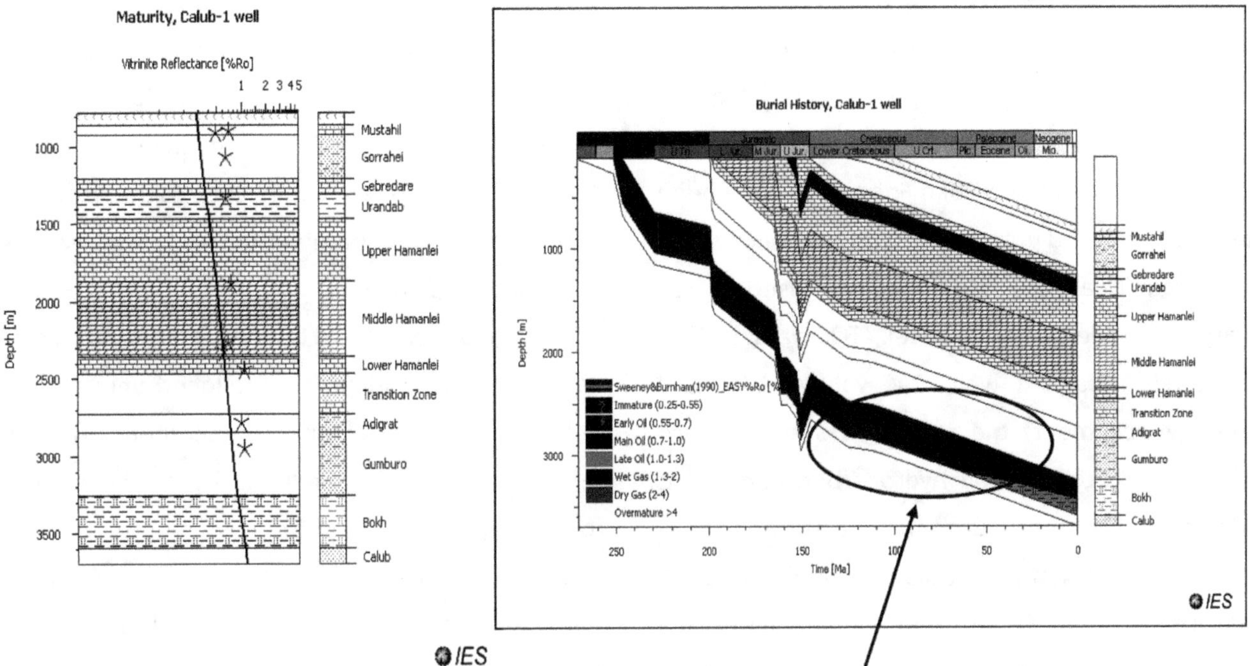

For Bokh shale in Calub-1 well, hydrocarbon generation started during the lower Jurassic and presently within the late oil generating window, whereas Urandab shale at calub-1 well is immature.

Figure 12: Correlation of measured and modeled vitrinite reflectance data and burial history (with maturity overlay) of Calub -1 well.

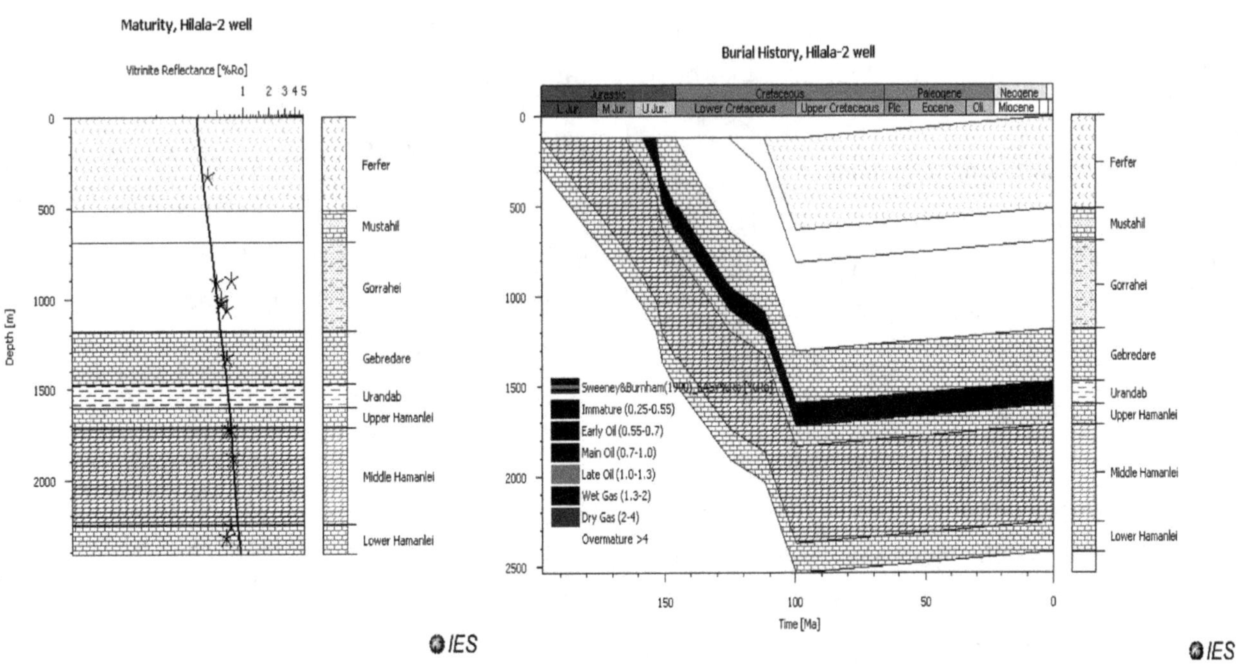

Figure 13: Correlation of measured and modeled vitrinite reflectance data and burial history (with maturity overlay) of Hilala -2 well.

B- Source Rock Characteristics

Table 2 shows the rock Eval-Pyrolysis/ TOC reading for the fourteen (14) samples from Bodle-1, Calub-1, and Hilala-2 wells. The Permo-Triassic source beds within the Bokh Formation have been identified as good potential source rock because of total organic carbon (TOC) values and based on [35] marginal models, (Bodle-1 well, TOC values, 1.21 – 3.1wt%), (Calub-1 well; TOC values, 1.04-3.62 wt%), which meets the standard as source rock with fairly good to good hydrocarbon generative potential. The maximum present-day hydrogen index suggests that gas would be expelled at peak maturity. S2 gives the amount of oil and gas which can be generated from its present stage of thermal maturation to the graphite stage. The TOC content and the pyrolysis S2 yield values show that the Eocene source rocks are good to very good source rocks with poor to fair generative potential (Figure 6). The kerogen typically has a low hydrogen index (HI), ranging from 16 mg HC/g TOC to 212 mg HC/g TOC, suggesting Type III and Type III/IV kerogen (Figures 5 and 6).

Temperature maximum (T_{max}) value of the studied samples ranges from 344 to 440 °C (Table 2). T_{max} values less than 435 °C are indicative of rocks that are thermally immature with respect to oil generation; whilst values in the range of 435-460 °C are indicative of rocks at ideal condition for oil generation. In addition, the values greater than about 465 °C are indicative of rocks that are beyond the stage of effective petroleum generation [34, 35]. The hydrogen index of studied oil shales ranged from 115 to 1187 mg HC/g TOC indicating oil and gas prone source rocks (Table 2). T_{max} values range from 412 to 430 °C for Bodle-1 well, 416 to 434 °C for Calub-1 well, 406 to 459 °C for Hilala-2 well which are in reasonably good agreement with vitrinite reflectance data, indicating that most samples have entered the early maturity stage. The Permo-Triassic source unit is effectively has good generative potential within the Bokh formation based on TOC content, thermal maturity and widespread distribution.

The plot of S1 versus TOC can be used to discriminate between non indigenous (allochthonous) and indigenous hydrocarbons (autochthonous). This relation shows that the majority of the studied rock samples for the Cretaceous and Miocene source rocks were characterized by indigenous hydrocarbons (autochthonous) indicating that the oil produced from the studied wells are in-situ deposits with an in-situ source rock (Figure 10). The organic matter type plots revealed that most samples lay on the region of Kerogen type II to III (Figures 8 to 10). Moreover, the result goes in constant with the work of [44,45] that most Ethiopian shales can be grouped under Type II and rarely Type III kerogen.

Heat Flow History

For heat flow and/or basement response function and prediction, a one-dimensional model (Airy compensation without lateral heat transfer) model was adopted from [41]. The models for the three wells assumed that the heat flow varies from 74 mW/m^2 to 94 mW/m^2 and the Paleo heat flow value generally decreases to present-day heat flow measurements. The model, using the best match of the measured and modeled vitrinite reflectance data, suggests that the present-day heat flow in the studied wells is 74 mW/m^2 to 94 mW/m^2. The value of heat flow in the three wells which is greater than 80 mW/m^2 suggests that Ogaden basin be within the geothermally active areas during depositions of the Calub, Bokh, Gumburo and Adigrat formations. Reduction in the heat flow value might be associated with rapid sedimentation that took place within the basin mainly during Jurassic periods during deposition of Hamanlei formation which were followed by a marine depositional environment where a rapid sedimentation were recorded (Figures 11, 12, and 13) In addition, rapid sedimentation may cause a reduction in heat flow and thermal maturity.

The Maturity of Organic Matter

In the reference [42], it is Shown that T_{max} can be a useful indicator of maturity; Moreover, 435°C ± 10% marks the transition from immature to mature organic matter. Values in the range of 435 to 460°C ± 10% represent the peak oil generation (the oil window). Values of 455 to 460°C ± 10% represent the transition from oil to wet gas generation [42]. T_{max} values that range from 339 °C with an average value of 395 °C indicate that source interval is immature [31]. The Hydrogen Index (HI, defined under source rock evaluation) is also used as a maturity indicator. HI will decrease with increasing maturity as labile hydrogen-rich compounds leave the kerogen nucleus. From the analyses result (Table 2) most samples showed a Tmax reading of averagely > 425 °C, which reveals that the source rocks within the formations have entered into early maturity phase.

Hydrocarbon Generation Phases

The one-dimensional basin modeling of Bodle-1, Calub-1, and Hilala-2 wells was performed using [41] kinetic models to infer hydrocarbon generation potential of the Permo-Cretaceous organic-rich shaly interval. A reasonable correlation between measured vitrinite reflectance values was established.

In Bodle-1 well, the predicted oil-generative window from the Permo Triassic source unit (Bokh shale) occurred in Lower Jurassic and during upper cretaceous wet gas generation have started from the source interval while liquid hydrocarbon is expected from the source unit within the Upper Jurassic and Lower Cretaceous (Figure 11). Hydrocarbon generation from Bokh shale in well Bodle-1 has been started during the lower Jurassic and presently, it is capable of generating wet gas (condensate), and this is consistent with the result of [1, 4, 24, 25, 39]. Urandab shale in well Bodle-1 started oil generation during the Oligocene and presently has the potential to generate.

For Bokh shale in well Calub-1, hydrocarbon generation started during the lower Jurassic and presently within the late oil generating window, whereas Urandab shale at well calub-1 is fairly mature. Initial oil generation form the Triassic source interval in well Calub-1 started during Oligocene Epoch (32.84 Ma), and late hydrocarbon generation reached during Miocene (19.63 Ma; Figure 11).

In Hilala-2 well, only Urandab shale was penetrated. Moreover the formation is presently within the early oil generating window. In addition, hydrocarbon generation has started during the lower Cretaceous. In conclusion, the models developed (Figure 13) have shown that Urandab and Bokh shale have the potential to generate hydrocarbon, and this is consistent with previous studies. Oil and gas generated by the source rocks are expected to have charged reservoir rocks specially Adigrat, Gumburo and Hamanlei formations. Furthermore, a small amount of hydrocarbon may exist within the Gorrahei and Gebredare formations.

Early Eocene source rocks entered the oil window phase in Miocene while hydrocarbon generation window from the mid-Eocene source rocks occurred in the Pleistocene. Liquid hydrocarbon is expected from the Eocene source units (Figure 13). Geological factors responsible for the variation in the hydrocarbon generation time of Eocene source beds in the studied wells are influenced by different criteria such as overburden thickness, lithological variation of the overburden rock, thermal maturity, and the basal heat flow. Petroleum generation within the Ogaden Basin began in the Permo-Triassic and continues today.

Implications for Hydrocarbon Exploration

The models developed (Figures 11, 12 and 13) revealed that Urandab and Bokh shale have the potential to generate hydrocarbon and this is consistent with previous studies. Oil and gas generated by the source rocks are expected to have charged reservoir rocks specially Adigrat, Gumburo and Hamanlei formations. Furthermore, a little hydrocarbon appearance is expected within the Gorrahei and Gebredare formations.

Possible hydrocarbon migration is expected along fault breakouts and unconformities to adjoining prolific reservoir rocks (Adigrat sandstone). The presence of hydrocarbon in the studied well suggests the viability of the Permo–Triassic, and the lower Jurassic sourced play in central part of the basin (Calub-Hilala Field). Hydrocarbon drilling activities should focus on identifications of productive reservoir sand bodies, and carbonate reservoir rocks since as reported by [17], the petroleum resources in the East African region are estimated to be 2 billion barrels of oil in place and 3tcf of natural gas. Additional resources continue to be firmed up by aggressive exploration programs in the region.

CONCLUSIONS

This study has established the importance of source rock evaluation studies and basin modeling in determining the hydrocarbon generation potentials of central Ogaden Basin. Investigation of Permo-Triassic and Cretaceous formations indicates that most shale of the Bokh formation and the Urandab formation as observed in the wells Bodle-1, Calub-1 and Hilala-2 are good source rocks in the central part of Ogaden basin. The maximum present day HI suggests that gas should be expelled at peak maturity. Kerogen type and total organic carbon data from Rock-Eval pyrolysis indicate that most shale units of the Permo-Triassic Bokh shale contain mainly type II kerogen with low to fair hydrogen index values (55-200 mg HC/g TOC).

Numerical modeling of the three wells in the central Ogaden Basin indicates that the Permo-Triassic Bokh formation and the Cretaceous Urandab formation source rocks entered the oil window stage for significant hydrocarbon generation and capable of charging the interbedded Triassic (Adigrat Sandstone) and lower Jurassic (Hamanlei Carbonate) reservoir bodies. The Triassic sediments are effective source rocks for the oil in the central part of Ogaden basin. This study presents information that improves understanding of the Permo-Cretaceous sourced facies. Because of most favorable maturity, the Permo-Triassic source rock (Bokh shale) is expected to have been responsible for oil and condensate discoveries in the central part of Ogaden basin.

REFERENCES

1. Abate E., Bruni P. and Sagri M. "Geology of Ethiopia: A Review and Geomorphological Perspectives," *Springer Netherlands*, **2015**, 33-64.
2. Astin T. R. and Worku T., "The Karoo Sediments: Late Palaeozoic to Early Jurassic of Ogaden Basin," *Ethiopia*, **1991**, *76*, 7-21.
3. BEICIP F., "The Petroleum Potential Of Ethiopia, Beicip-Franlab Consultants Report For Ministry Of Mines And Energy," Ethiopia, 1st ed, *Addis Ababa, Ethiopia*, **1985**.
4. Beksa A. K., "Microfacies Analysis and Paleontology of Carbonate Rocks in Shanan Section," *South Eastern Ethiopia*, **2017**.
5. Boboye O. A. and Okon E. E., "Sedimentological and Geochemical Characterization of the Cretaceous strata of Calabar Flank," *Southeastern Nigeria*, **2014**.
6. Clift W. O., "Sedimentary History of the Ogaden district, Ethiopia," *Rep. 20th Int. Geol. Cong. Petrol. Gad Sympos.*, **1956**, 78-112,
7. Corti G., "Control of Rift Obliquity on the Evolution and Segmentation of the Main Ethiopian Rift," *Nature Geosci.*, **2008**, *1*(4), 258-262.
8. Dawit L. E., "Adigrat Sandstone in Northern and Central Ethiopia: Stratigraphy, Facies, Depositional Environments and Palynology," Doctoral Thesis, Technische Universität Berlin, Fakultät VI-Planen Bauen Umwelt, **2010**.
9. Erik F., "Microfacies of Carbonate Rocks: Analysis, Interpretation and Application," 1st ed., *Institute of Palaeontology*, Loewenichstrasse 28, Germany, **2004**.
10. Gani M. G., Abdelsalam S. and Gani M. R., "Stratigraphic and Structural Evolution of the Blue Nile Basin, North-Western Ethiopian Plateau," *Geological Journal*, **2009**, *44*, 30-56.
11. Gluyas J. and Swarbick R., "Petroleum Geoscience," *Oxford: Blackwell Publishing Co*, **2009**, 40-252.
12. Hallam A., "Jurassic Environments," (Cambridge Earth Science Series) Reissue Edition, University of Oxford, UK, **2009**.
13. Hankel O., "Early Permian to Middle Jurassic Rifting and Sedimentation in East Africa and Madagascar," Springer, **1994**, *83*(4), 703-710.
14. Hantschel T. and Kauerauf A. I., "Fundamentals of Basin and Petroleum Systems Modelling," Springer, **2009**.
15. Hao F. and Chen J., "The Cause and Mechanism of Vitrinite Reflectance Anomalies," *Journal of Petroleum Geology*, **1992**, *15*(4), 419-434.
16. Haq B. U., Hardenbol J. and Vail P. R., "Chronology of the Fluctuating Sea Level Since the Triassic," *Science*, **1987**, *235*, 1156-1167.
17. Hunegnaw A., Sage L. and Gonnard R., "Hydrocarbon Potential of the Intracratonic Ogaden Basin,"*Journal of Petroleum Geology,* **1998**, *21*(4), 401-425.
18. Gluyas J. and Swarbrick R., "Petroleum Geosciences," Amazon, 1st Ed. UK, **2004**.
19. Martin B. and Geleta S., "A Planktonic Marker and Callovian-Oxfordian Fragmentation of Gondwana: Data from Ogaden Basin," *Palaeo*, **1992**, *104*, 177-184.
20. Mascle J., Moungenot D., Blarez E., Marinho M., and et al., "African Transform Continental Margins: Examples from Guinea, *Ivory Coast and Mozambique*," *Geological Journal*, **1987**, *22*, 537-561.
21. Maurice E. T., "Sedimentary Petrology: An

Introduction to the Origin of Sedimentary Rocks," 3rd ed. *Department of Geological Sciences: University Of Durham*, **2001**.
22. Mengesha T., Tadiwos C. and Workneh H., "Explanation of the Geological Map of Ethiopia," 2nd *ed. Addis Ababa*, Ethiopia, **1996**.
23. Millard F. C. and Philip D. R., "Evolution of the Conjugate East African-madagascan Margins and the Western Somali Basin," *Model. J. Petrol. Geol.*, **1988**.
24. Ministry of Mines of Ethiopia., "Joint Ogaden Basin Study Project" *Petroleum Licensing & Administration*, **2005**.
25. Ministry of Mines of Ethiopia., "Petroleum Exploration in Ethiopia," *Petroleum Licensing & Administration*, **2011**.
26. Monson B., "Reflectance Measurement Manual," Department of Geology, Queen University of Belfast: Belfast, **1995**.
27. Morrison J., Birse A., Samuel M. A., Richardson S.M., and et al., "The Cretaceous Sequence Stratigraphy of the Socotran Platform," Republic of Yemen, *Marine and Petroleum Geology*, **1997**.
28. Moungenot D., Recq M., Virlogeux P., and Lepvrier C., "Seaward Extension of the East African Rift," Nature, **1986,** *321*, 509-603.
29. Nabil M. A. and Mohammed A. A., "Petroleum Source Rocks Characterization and Hydrocarbon Generation of the Upper Jurassic Succession in Jabal Ayban Field, Sabatayn Basin, Yemen," *Egyptian Journal of Petroleum*, Open Access, **2018**.
30. Nick L., "Petroleum Resources in East Africa," , *EAC.*, **2010**, 68-71.
31. Nton M. E. and Ekom J. C., "Hydrocarbon Generative Potential of Cretaceous Sediments of Kanadi-1 Well," Bornu Basin, Nigeria, **2016**.
32. Oladotun A., Oluwajana, Olugbenga A. and Ehinola, "Modelling Hydrocarbon Generation Potentials of Eocene Source Rocks in the Agbada Formation, Northern Delta Depobelt, Niger Delta Basin, Nigeria," *Journal of Petroleum Exploration and Production Technology,* **2017,** *7*(2), 379-388.
33. Paolo B., "Landscapes and Landforms of Ethiopia," *World Geomorphological Landscapes*, **2015**, 326-340.
34. Peters K. E., "Assoc. Petroleum Geologists Bull," **1986**.
35. Peters K. E. and Cassa M. R., "Applied Source Rock Geochemistry; the Petroleum System-from Source to Trap Magoon," *Am Assoc Pet Geol Bull.*, **1994,** 93-120.
36. Phan K., "Magnetic Data Processing for the Purpose of Hydrocarbon Exploration," Institute of Geological Sciences, *NCNST*, **1999**.
37. Premanand M. and Sagar N., "Using Remote Sensing and GIS Technologies as an Aid for Hydrocarbon Exploration in the Assam-Arakan Fold-thrust Belt," *Geology and Mineral Resource*, Map India Conference, **2003.**
38. Proeh T. S., ''Pore Pressures, Fracture Gradients, and Drilling Economics," *SPE/IADC Drilling Conference*, Dallas, Texas, **1994**.
39. Purcell P. G., "The Geology and Petroleum Potential of the Ogaden Basin, Ethiopia,", Unpublished Report, **1979.**
40. Shalaby M. R., Abdullah W. H., and Abu S., "Burial History, Basin Modelling and Organic Geochemistry in the Western Desert," Egypt, *Geological Society of Malaysia Bulletin*, **2008.**
41. Sweeney J. J. and Burnham A. K., "Evaluation of a Simple Model of Vitrinite Reflectance Based

on Chemical Kinetics," *Assoc Am Pet Geol Bull.*, **1990**, *74*(10), 1559-1570.

42. Tissot B. P., Pelet R., and Ungerer P. H., "Thermal History of Sedimentary Basins, Maturation Indices and Kinetics of Oil and Gas Generation," *AAPG Bulletin*, **1987**, 71, 1445-1466.
43. Van Krevelen D. W., "Graphical-statistical Method for the Study of Structure and Reaction Processes of Hydrocarbons and Coal," *Fuel*, **1950**, *29*, 228-269.
44. Wolela A., "Fossil Fuel Energy Resources of Ethiopia," *Bulletin of the Chemical Society of Ethiopia*, **2007**, 22, 67-84.
45. Wolela A., Alula H. G., and Belay W., "Sedimentology, Depositional Environments and Hydrocarbon Potential of the Yayu Basin," *Ministry Of Mines and Energy*, Report, Petroleum Operations Department, **2002.**
46. Worku T. and Astin T. R., "The Karoo Sediments: Late Paleozoic to Early Jurassic of Ogaden Basin, Ethiopia," *Sedimentary Geology*, **1992**, *76*, 7-21.

13

Mathematical Modeling of Carbon Dioxide Removal from the CO_2/CH_4 Gas Mixture Using Amines and Blend of Amines in Polypropylene: A Comparison between Hollow Fiber Membrane Contactor and Other Membranes

Mohammad Reza Talaghat* and Ahmad Reza Bahmani

Department of Chemical Engineering, Petroleum and Gas, Shiraz University of Technology, Shiraz, Iran

ABSTRACT

In this work, a mathematical model is established to describe the removal of CO_2 from gaseous mixtures including CH_4 and CO_2 in a polypropylene hollow fiber membrane contactor in the presence of conventional absorbents such as monoethanolamine (MEA), methyldiethanolamine (MDEA), and a blend of them. Modeling was performed in axial and radial directions under the fully-wet condition for countercurrent gas-liquid flow arrangement. Both of axial and radial diffusions have been considered in three segments, including shell, membrane, and tube. To evaluate the model, the results of this model were compared with the experimental data and the results of COMSOL software and the results were in agreement with the experimental data and COMSOL outputs. In addition, the effect of various parameters on the removal percentage of carbon dioxide from gas mixtures was studied. It was found out that the CO_2 removal percentage is the best by using MEA solution as the absorbent. This modeling shows that the removal of CO_2 increases by adding MEA into MDEA solution. In this study, the factors that influence the removal percentage of CO_2 from gaseous mixture were investigated. The CO_2 removal efficiency increased with an increase in the liquid flow rate, number of fibers, membrane length, porosity-to-tortuosity ratio, and solvent concentration. The results show that increasing gas flow rate reduces CO_2 removal due to decreasing the contact time. Finally, the performance of this membrane was compared with other membranes such as polyvinyl difluoride (PVDF) and polytetrafluoroethylene (PTFE). The results show that the percentage of CO_2 removal by the polypropylene HFM is higher than that of the PVDF and PTFE hollow fiber membranes in the presence of MEA as the absorbent.

Keywords: Hollow Fiber Membrane, CO_2 Removal, Separation, Modeling, Blend of Amines.

INTRODUCTION

The emissions of greenhouse gases in the atmosphere are generally made by industries. In this regard, fossil fuel (coal, oil, and natural gas) burning is the main source of commercial energy supplies [1]. The flue gases generally consist of nitrogen, oxygen, water vapor, carbon dioxide, sulfur dioxide, nitrogen oxides, and a small quantity of hydrochloric acid [2]. The major parts of the effluent gases are N_2, H_2O, and CO_2 [2,3]. Thus, an important step in

***Corresponding author**

Mohammad Reza Talaghat

Email: talaghat@sutech.ac.ir

many industrial processes is the removal of carbon dioxide from the process of gas streams because of technical and economical or environmental reasons. The conventional technology to capture CO_2 on a large scale is the absorption-desorption process. In this process, aqueous solutions such as monoethanolamine (MEA), diethanolamine (DEA), N-methyldiethanolamine (MDEA) are frequently used as absorbents [3]. In many chemical and petroleum refining processes, the gas absorption operation has been carried out using Venturi scrubbers, packed columns, and plate columns. These processes suffer a number of shortcomings such as flooding, channeling, entraining, foaming, and high capital and operating cost. In recent years, membrane-based technology is widely used in purification, concentration, and fractionation of fluid mixtures. In recent years, membrane gas absorption technology has been considered as one of the promising alternatives to conventional techniques for CO_2 capture due to its favorable mass transfer performance. As a hybrid approach of chemical absorption and membrane separation, it exhibits a number of advantages such as operational flexibility, compact structure, high surface-area-to-volume ratio, linear scale up, modularity, and predictable performance. A schematic drawing of membrane gas absorption process is shown in Figures 1 and 2.

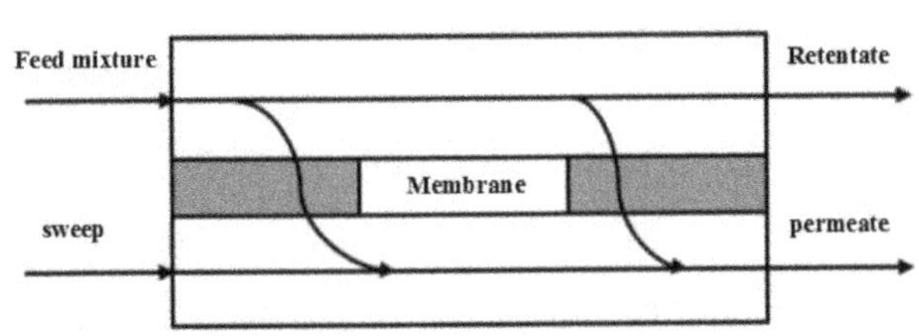

Figure 1: A general membrane process.

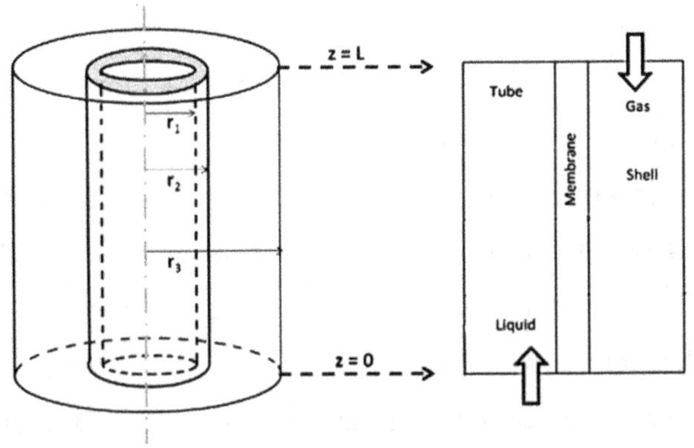

Figure 2: A schematic diagram of the hollow fiber membrane contactor.

The feed gas passes through the shell side of the hollow fibers, and the liquid absorbent flows countercurrent on the other side. Instead of depending on the membrane selectivity, the liquid flowing in the hollow fiber membrane contactor provides the selectivity and the porous, unselective membrane only acts as the contacting interface of the liquid and gas phases. The gasses diffuse through the membrane pores to the other side of the membrane where they are absorbed in the absorbent and taken away from the contactor. Although the membrane contactor offers many advantages over conventional contacting equipment, the resistance of additional mass transfer is introduced due to the existence of membrane phase. The membrane pores can be theoretically filled with either gas for the hydrophobic membrane or liquid for the hydrophilic membrane, corresponding to the non-wetting mode and overall-wetting mode respectively. For membrane gas absorption systems, it is essential to avoid a strong increase in mass transfer resistance in a liquid filled membrane pore compared to a gas-filled pore. However, in practical application, the aqueous solutions with organic absorbents can penetrate into partial pores of the hydrophobic membrane, and the membrane contactor is operated under partial wetting mode. The wetting phenomenon of the membrane leads to an increase in the overall mass transfer resistance and the deterioration of membrane performance [4].

Polymeric membranes in the form of sheets or fine hollow fibers can be used to separate gas and liquid mixtures due to differences in permeation rates through the material. This makes membrane process very useful for a wide range of separations. One of the main challenges of membrane gas absorption technology is the membrane wetting by absorbent over prolonged operating time, which may significantly decrease the mass transfer coefficients of the membrane module. The characterization results from several types of research showed that the absorbent molecules diffused into the membrane polymer such as polypropylene during the contact with the membrane, resulting in the swelling of the membrane. In addition, the effects of operating parameters such as immersion time and absorbent type on the membrane wetting are important [5-9].

For the membrane gas absorption process, the membrane does not provide the selectivity. It only acts as the gas-liquid contact interface. Generally, hydrophobic porous membranes can be used as a gas-liquid membrane contactor. The most widely used hydrophobic polymers include PP, PVDF, PTFE, and PE membranes. PP, PTFE, and PE membranes are usually prepared by stretching and thermal methods, while PVDF asymmetric membranes are prepared via phase inversion method. PP and PE hollow fiber membranes are widely used in industries due to their low cost and commercial availability in various sizes, but their hydrophobicity is lower than membranes based on PTFE, PVDF, and other fluorine-containing polymers [4].

The absorption of carbon dioxide using the chemical absorbent in microporous membrane was first studied by Qi and Cussler [10]. Karoor and Sirkar investigated the absorption of CO_2 and SO_2 in a microporous polypropylene hollow fibers using water as solvent [11]. Dindore et al. [12] investigated the chemical absorption of carbon dioxide using potassium carbonate in polypropylene hollow fiber membrane contactor. The absorption of CO_2 in different absorbents using the hollow fiber membrane contactors was reported by Wang et al. [13]. Al-Marzouqi et al. [14] developed a steady state mathematical model for micro porous hollow fiber membrane contactor to analyze the absorption of CO_2 using chemical absorbent.

In this study, the separation of CO_2 from CH_4 in a polypropylene hollow fiber membrane (HFM) is modeled in two dimensions using amine and a blend of amines. The model is based on fully wet conditions where the solvent completely filled the membrane pores. The effects of different parameters and membrane features on the percentage of CO_2 removal will be considered.

MATHEMATICAL MODELING

The mathematical modeling was performed in axial and radial directions to describe the removal of CO_2 from gaseous mixture in a hollow fiber membrane under the fully-wet condition in the presence of amine and a blend of amines. The model was based on the complete wetting condition in which the solvent completely filled the membrane pores. The schematic diagram for the hollow fiber membrane contactor is shown in Figure 2. The hollow fiber membrane contactor consists of three segments: tube side, membrane, and shell side. The liquid flows in the tube side, and the gas mixture flows in the shell side.

The following assumptions have been made in the mathematical model:

1) The hollow fiber membrane contactor is under steady state and isothermal conditions.

2) The Henry's law is applicable.

3) A complete wetting condition in which the solvent completely fills the membrane pores is considered.

Subject to these assumptions, the equation for absorption of CO_2 in the shell side of the hollow fiber membrane contactor in cylindrical coordinate is obtained using Fick's law:

$$D_{CO_2-shell}\left[\frac{\partial^2 C_{CO_2-shell}}{\partial r^2}+\frac{1}{r}\frac{\partial C_{CO_2-shell}}{\partial r}+\frac{\partial^2 C_{CO_2-shell}}{\partial z^2}\right]= V_{z-shell}\frac{\partial C_{CO_2-shell}}{\partial z} \quad (1)$$

Radial velocity profile in the outside of hollow fiber is given by Equation 2 [15]:

$$V_{z-shell}=2\bar{u}_{gas}\left[1-\left(\frac{r_2}{r_3}\right)^2\right]\left[\frac{\left(\frac{r}{r_3}\right)^2-\left(\frac{r_2}{r_3}\right)^2+2\ln\left(\frac{r_2}{r}\right)}{3+\left(\frac{r_2}{r_3}\right)^4-4\left(\frac{r_2}{r_3}\right)^2+4\ln\left(\frac{r_2}{r_3}\right)^2}\right] \quad (2)$$

The boundary conditions are:

$$z=L \qquad C_{CO_2-shell}=C_{CO_2-shell,initial} \quad (3)$$

$$r=r_2 \qquad C_{CO_2-shell}=\frac{C_{CO_2-membrane}}{m} \quad (4)$$

$$r=r_3 \qquad \frac{\partial C_{CO_2-shell}}{\partial r}=0 \quad (5)$$

where, $D_{CO_2\text{-shell}}$ is the diffusion coefficient of CO_2 in shell ($m^2.s^{-1}$), and $C_{CO_2\text{-shell}}$ is the concentration of CO_2 in the shell ($mol.m^{-3}$); $C_{CO_2\text{-shell,initial}}$ is the inlet CO_2 concentration in shell ($mol.m^{-3}$), and $C_{CO_2\text{-membrane}}$ represents the concentration of CO_2 in the membrane ($mol.m^{-3}$); $V_{z\text{-shell}}$ stands for the axial gas velocity in shell ($m.s^{-1}$); r_1 and r_2 represent the inner and outer tube radius respectively, and r_3 is the inner shell radius (m); *m* is Henry's law constant, and L is the membrane length (m). The kinetic parameters for CO_2 and amines are given in Tables 1 to 6.

Within the membrane pore, the steady-state continuity equation for CO_2 is:

$$D_{CO_2-membrane}\left[\frac{\partial^2 C_{CO_2-membrane}}{\partial r^2}+\frac{1}{r}\frac{\partial C_{CO_2-membrane}}{\partial r}+\frac{\partial^2 C_{CO_2-membrane}}{\partial z^2}\right]+R_{CO_2}=0 \quad (6)$$

$$D_{solvent-membrane}\left[\frac{\partial^2 C_{solvent-membrane}}{\partial r^2}+\frac{1}{r}\frac{\partial C_{solvent-membrane}}{\partial r}+\frac{\partial^2 C_{solvent-membrane}}{\partial z^2}\right]+R_{solvent}=0 \quad (7)$$

The boundary conditions are given by the following equations:

$$r=r_2 \qquad C_{CO_2-membrane}=mC_{CO_2-shell} \quad (8)$$

$$\frac{\partial C_{solvent-membrane}}{\partial r}=0$$

$$r=r_1 \qquad C_{CO_2-membrane}=C_{CO_2-tube} \quad (9)$$

$$C_{solvent-membrane}=C_{solvent-tube}$$

Within the tube side, the steady-state continuity equation for CO_2 and absorbent is defined by:

$$D_{CO_2-tube}\left[\frac{\partial^2 C_{CO_2-tube}}{\partial r^2}+\frac{1}{r}\frac{\partial C_{CO_2-tube}}{\partial r}+\frac{\partial^2 C_{CO_2-tube}}{\partial z^2}\right]+R_{CO_2}=V_{z-tube}\frac{\partial C_{CO_2-tube}}{\partial z} \quad (10)$$

$$D_{solvent-tube}\left[\begin{array}{c}\frac{\partial^2 C_{solvent-tube}}{\partial r^2}\\ +\frac{1}{r}\frac{\partial C_{solvent-tube}}{\partial r}\\ +\frac{\partial^2 C_{solvent-tube}}{\partial z^2}\end{array}\right]+R_{solvent} \quad \textbf{(11)}$$

$$=V_{z-tube}\frac{\partial C_{solvent-tube}}{\partial z}$$

The radial velocity profile in the tube is assumed to follow the Newtonian laminar flow [16]:

$$V_{z-tube}=2\bar{u}_{liq}\left[1-\left(\frac{r}{r_1}\right)^2\right] \quad (12)$$

The boundary conditions are given by the following equations:

$$z=0 \qquad C_{CO_2-tube}=0 \quad (13)$$

$$C_{solvent-tube}=C_{solvent-tube,initial}$$

$$r=0 \qquad \frac{\partial C_{CO_2-tube}}{\partial r}=0 \quad (14)$$

$$\frac{\partial C_{solvent-tube}}{\partial r}=0$$

$$r=r_1 \qquad C_{CO_2-tube}=C_{CO_2-membrane} \quad (15)$$

$$C_{solvent-tube}=C_{solvent-membrane}$$

Method of Solution

The set of partial differential equations along with boundary conditions and the reaction rates were solved by numerical method using MATLAB software. In this study, we used finite difference method for the numerical solutions of differential equations. The physical and chemical properties are listed in Tables 1-6 [17].

RESULTS AND DISCUSSION

Today, the gas-liquid hollow fiber membrane contactors have been a subject of great interest. In these processes, the membrane contactor mainly acts as a physical barrier between two phases (gas and liquid) without any varieties in selectivity. Because of a very high surface/volume ratio, the hollow fiber membrane contactors have a great potential for gas absorption. In this work, mathematical model was performed in axial and radial directions for the removal of CO_2 in the hollow fiber membrane contactors in the presence of MEA, MDEA, and a blend of them under the fully-wet condition for a countercurrent gas-liquid flow arrangement. In this study, the effect of various parameters on the rate of the removal of carbon dioxide from gas mixtures were studied, and the results are presented in the following sections. The effect of gas flow rate on outlet CO_2 concentration in the membrane contractor is displayed in Figure 3.

Table 1: Diffusion coefficient of CO_2 and amine in tube and shell at T = 298 K [19-21].

MEA (B)+MDEA (C)+H_2O	
$k_{2,B}\left(m^3mol^{-1}s^{-1}\right)$	8.98
$\frac{k_{2,B}k_{H_2O}}{k_{-1}}\left(m^6mol^{-2}s^{-1}\right)$	1.16×10^{-5}
$\frac{k_{2,B}k_B}{k_{-1}}\left(m^6mol^{-2}s^{-1}\right)$	2.41×10^{-3}
$\frac{k_{2,B}k_C}{k_{-1}}\left(m^6mol^{-2}s^{-1}\right)$	5.31×10^{-4}
$k_{2,C}\left(m^3mol^{-1}s^{-1}\right)$	1.30×10^{-3}

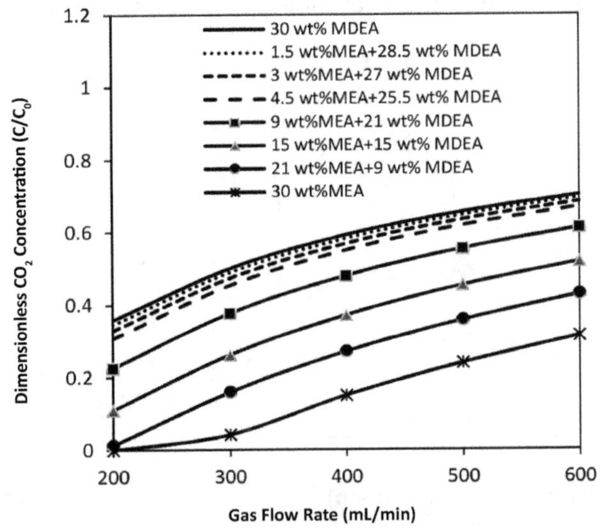

Figure 3: The effect of gas flow rate on outlet CO_2 concentration in the membrane contactor for the absorption of CO_2 (n = 3600, Q_{Liquid} = 500 mL/min, $C_{o, gas}$ = 3.883 mol/m³, L = 0.2286 m, ε/τ = 0.2, T = 313 K).

Table 2: Parameters used in simulation at T=313 K [19-21].

Parameter	Value	Ref.
Inner tube radius(mm)(R_1)	0.11	[21]
Outer tube radius (mm) (R_2)	0.15	[21]
Inner shell radius (mm) (R_3)	0.2645	[21]

Table 3: Kinetic parameters for the blend of amines at T=313 K in tube and shell [19].

Tube Section	
$D_{MEA\text{-}tube}$ (30 wt.% MEA)	8.590×10^{-10}
$D_{MEA\text{-}tube}$ (21 wt.% MEA+9 wt.% MDEA)	8.356×10^{-10}
$D_{MEA\text{-}tube}$ (15 wt.% MEA+15 wt.% MDEA)	8.181×10^{-10}
$D_{MEA\text{-}tube}$ (9 wt.% MEA+21 wt.% MDEA)	7.985×10^{-10}
$D_{MEA\text{-}tube}$ (4.5 wt.% MEA+25.5 wt.% MDEA)	7.821×10^{-10}
$D_{MEA\text{-}tube}$ (3 wt.% MEA+27 wt.% MDEA)	7.769×10^{-10}
$D_{MEA\text{-}tube}$ (1.5 wt.% MEA+28.5 wt.% MDEA)	7.705×10^{-10}
$D_{MDEA\text{-}tube}$ (30 wt.% MDEA)	7.642×10^{-10}
$D_{MDEA\text{-}tube}$ (21 wt.% MEA+9 wt.% MDEA)	5.584×10^{-10}
$D_{MDEA\text{-}tube}$ (15 wt.% MEA+15 wt.% MDEA)	5.565×10^{-10}
$D_{MDEA\text{-}tube}$ (9 wt.% MEA+21 wt.% MDEA)	5.332×10^{-10}
$D_{MDEA\text{-}tube}$ (4.5 wt.% MEA+25.5 wt.% MDEA)	5.221×10^{-10}
$D_{MDEA\text{-}tube}$ (3 wt.% MEA+27 wt.% MDEA)	5.181×10^{-10}
$D_{MDEA\text{-}tube}$ (1.5 wt.% MEA+28.5 wt.% MDEA)	5.141×10^{-10}
$D_{CO_2\text{-}tube}$ (30 wt.% MEA)	11.509×10^{-9}
$D_{CO_2\text{-}tube}$ (21 wt.% MEA+9 wt.% MDEA)	6.922×10^{-9}
$D_{CO_2\text{-}tube}$ (15 wt.% MEA+15 wt.% MDEA)	4.484×10^{-9}
$D_{CO_2\text{-}tube}$ (9 wt.% MEA+21 wt.% MDEA)	3.102×10^{-9}
$D_{CO_2\text{-}tube}$ (4.5 wt.% MEA+25.5 wt.% MDEA)	2.346×10^{-9}
$D_{CO_2\text{-}tube}$ (3 wt.% MEA+27 wt.% MDEA)	2.760×10^{-9}
$D_{CO_2\text{-}tube}$ (1.5 wt.% MEA+28.5 wt.% MDEA)	2.047×10^{-9}
$D_{CO_2\text{-}tube}$ (30 wt.% MDEA)	1.960×10^{-9}
Shell Section	
$D_{CO_2\text{-}shell}$ (CH_4/CO_2)	1.9968×10^{-5}
$D_{CO_2\text{-}shell}$ (N_2/CO_2)	1.2498×10^{-5}

Table 4: The diffusion coefficient of CO_2 and amines.

Tube Section	
$D_{AMP\text{-}tube}$	1.18×10^{-9}
$D_{MEA\text{-}tube}$	1.51×10^{-9}
$D_{DEA\text{-}tube}$	1.47×10^{-9}
$D_{MDEA\text{-}tube}$	1.44×10^{-9}
$D_{CO_2\ tube}$ (aqueous AMP solution)	5.67×10^{-10}
$D_{CO_2\ tube}$ (aqueous MEA solution)	9.32×10^{-10}
$D_{CO_2\ tube}$ (aqueous DEA solution)	6.32×10^{-10}
$D_{CO_2\ tube}$ (aqueous MDEA solution)	6.21×10^{-10}
Shell Section	
$D_{CO_2\ shell}$ (CH_4/CO_2)	1.18×10^{-5}
$D_{CO_2\ shell}$ (N_2/CO_2)	1.18×10^{-5}

Table 5: Rate expression of amines.

Aqueous amine solution	R_B (mol.m^{-3}s^{-1})	R_C (mol.m^{-3} s^{-1})	Reference
Single amines MEA (B):	$\dfrac{K_{2,B}C_{CO_2}C_B}{1+\dfrac{1}{\left(\left(\dfrac{K_{H_2O}}{K_{-1}}\right)C_{H_2O}\right)}+\left(\left(\dfrac{K_B}{K_{-1}}\right)C_B\right)}$		[21]
MDEA (B):	$k_{2,B}C_{CO_2}C_B$		
Blended amines MEA (B) +MDEA (C)	$\dfrac{K_{2,B}C_{CO_2}C_B}{1+\dfrac{1}{\left(\left(\dfrac{K_{H_2O}}{K_{-1}}\right)C_{H_2O}\right)}+\left(\left(\dfrac{K_B}{K_{-1}}\right)C_B\right)+\left(\left(\dfrac{K_C}{K_{-1}}\right)C_C\right)}$	$k_{2,C}C_{CO_2}C_C$	

Table 6: Kinetic parameters for pure of amine at T=298 K [17, 18, 19]

$k_{2,B}(m^3 mol^{-1} s^{-1})$	$\frac{k_{2,B} k_{H_2O}}{k_{-1}}(m^6 mol^{-2} s^{-1})$	$\frac{k_{2,B} k_B}{k_{-1}}(m^6 mol^{-2} s^{-1})$
	MEA (B)+H_2O	
6.358	$9.58*10^{-6}$	$1.58*10^{-3}$
	MDEA (B)+H_2O	
$5.21*10^{-3}$		

This figure shows that the outlet CO_2 concentration in the shell side increases with increasing gas flow rate. This is due to a decrease in the residence time of CO_2 in shell, which in turn reduces the removal percentage of CO_2 along the length of the module and further the reduction of CO_2 availability for the reactive absorption. The dimensionless CO_2 concentration along the length of the module for the different compositions of a blend of amines is presented in Figure 4. To evaluate the model, the results of this model were compared with the experimental data from references [17-18] and the results of COMSOL software. This comparison is illustrated in Figure 5.

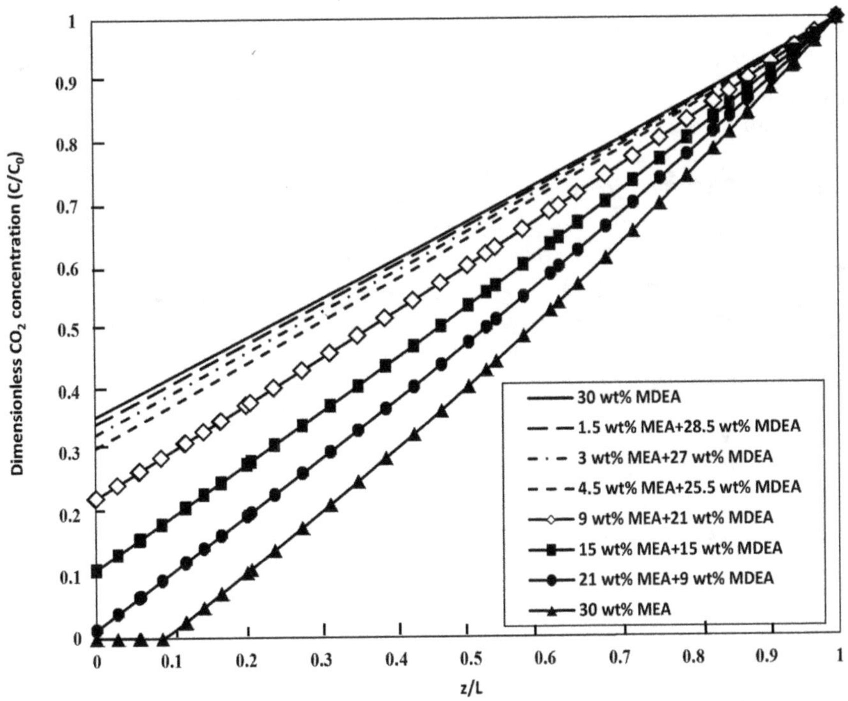

Figure 4: CO_2 concentration in the axial direction at different compositions of mixed amines; (n=3600, Q_{GAS}=500 mL/min, Q_{Liquid}=500 mL/min, $C_{o,gas}$=3.883 mol/m³, L=0.2286 m, ε/τ= 0.2, T = 313 K).

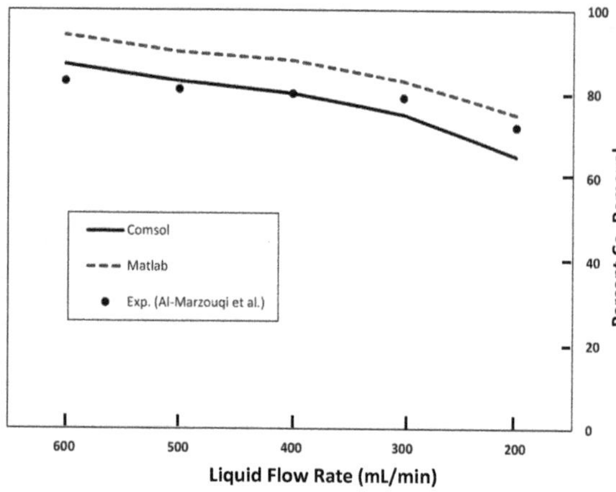

Figure 5: Comparison of results from this modeling with experimental data and COMSOL software results for MEA solution; n = 3600, Q_{Gas} = 500 mL/min, $C_{o,\ Gas}$ = 3.883 mol/m³, $C_{CO_2\text{-initial}}$ = 4 mol/m³, L = 0.2286 m, ε/τ = 0.2, T = 298 K)

The results show good agreement. The outlet CO_2 concentration at different liquid flow rates in the shell side is illustrated in Figure 6.

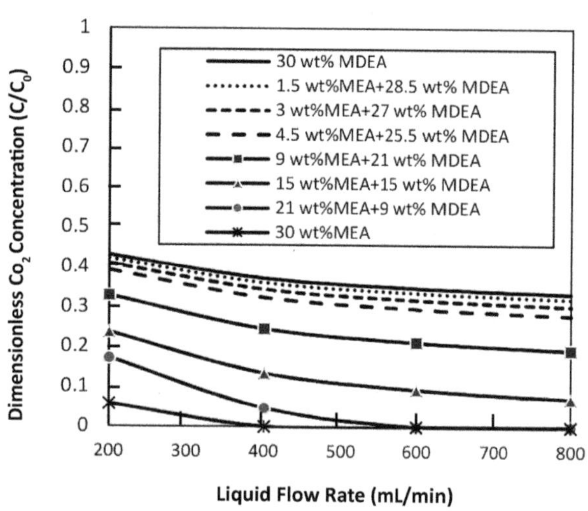

Figure 6: The effect of liquid flow rate on outlet CO_2 concentration in the membrane contactor for the absorption of CO_2. (n = 3600, Q_{Gas} = 200 mL/min, $C_{o,\ Gas}$ = 3.883mol/m³, L=0.2286m, ε/τ=0.2, T = 313 K).

As can be seen from this figure, the outlet CO_2 concentration drops with an increase in the liquid flow rate. This is due to an increase in the concentration gradients of CO_2 in the liquid phase; thus, the outlet CO_2 concentration in gas becomes less and the CO_2 removal percentage increases. As the absorbent moves faster, the CO_2 outlet concentration in gas phase rises, and the CO_2 removal rate augments. The outlet CO_2 concentration in the shell decreases with increasing the total number of fibers due to an increase in the interface of mass transfer contact. These results are presented in Figure 7. This figure shows that increasing the number of fibers in membrane modules will encourage the CO_2 removal percentage in the membrane contactor for the absorption of CO_2.

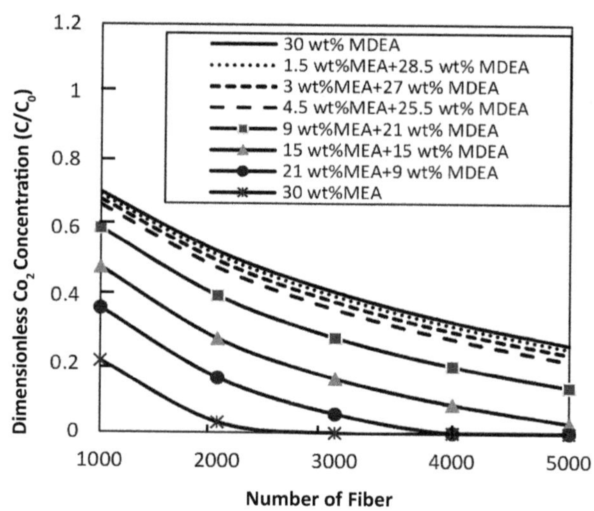

Figure 7: The effect of the number of fibers on outlet CO_2 concentration in the membrane contactor for the absorption of CO_2. (Q_{Liquid} = 500 mL/min, Q_{Gas} = 200 mL/min, $C_{o,\ Gas}$ = 3.883 mol/m³, L = 0.2286 m, Ɛ/τ = 0.2, T=313 K).

The effect of porosity-to-tortuosity ratio of the outlet CO_2 concentration is shown in Figure 8. As the porosity-to-tortuosity ratio increases, the outlet CO_2 concentration decreases because the diffusion coefficient of CO_2 in the membrane is a function of membrane porosity and tortuosity. By increasing the porosity-to-tortuosity ratio, the membrane mass transfer resistance decreases. Therefore, the total resistance to the mass transfer of CO_2 drops.

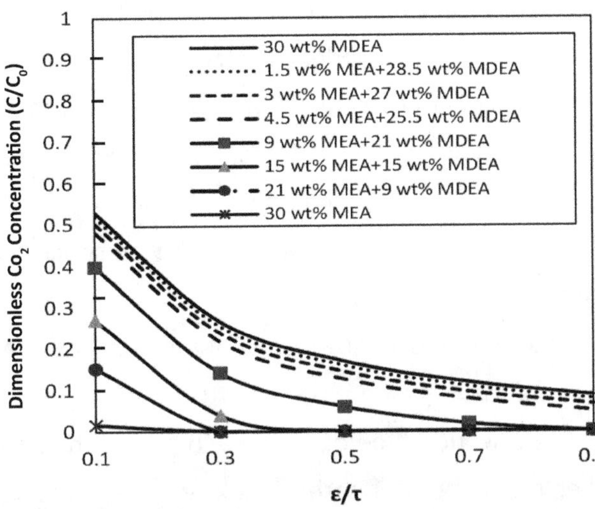

Figure 8: The effect of porosity-to-tortuosity ratio on outlet CO_2 concentration in the membrane contactor for the absorption of CO_2; (n = 3600, Q_{Liquid} = 500 mL/min, Q_{Gas} = 200 mL/min, $C_{o, Gas}$ = 3.883 mol/m³, L = 0.2286 m, T=313 K).

According to Equation 16, with increasing porosity-to-tortuosity ratio, which is provided by membrane manufacturer, the effective diffusion coefficient of CO_2 in the membrane increases.

$$D_{CO_2-membrane} = D_{CO_2-tube}\left(\frac{\varepsilon}{\tau}\right) \quad (16)$$

The outlet CO_2 concentration at different membrane lengths is demonstrated in Figure 9.

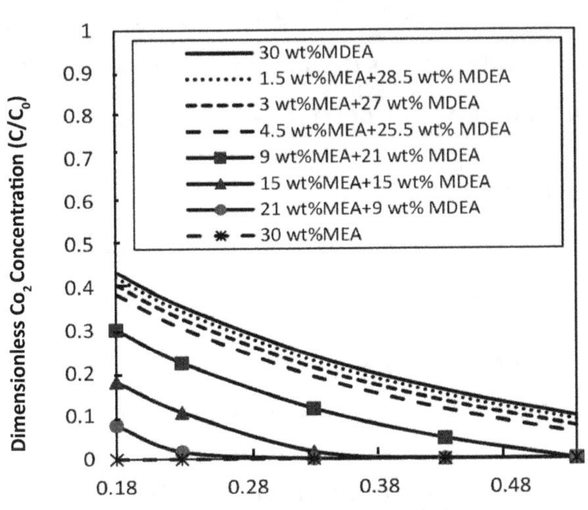

Figure 9: The effect of membrane length on outlet CO_2 concentration in the membrane contactor for the absorption of CO_2 (n = 3600, Q_{Liquid} = 500mL/min, Q_{Gas} = 200 mL/min, $C_{o, Gas}$ = 3.883 mol/m³, ε/τ = 0.2, T = 313 K).

As can be seen from this figure, the outlet CO_2 concentration decreases with increasing membrane length, which is because of an increase in the contact time of gas phase and liquid phase due to a longer membrane length. Furthermore, in this research the removal of CO_2 from gas mixture was modeled using a blend of amines. The results of modeling are shown in Figures 3 to 9. It can be inferred from these figures that the removal of CO_2 increases by increasing MEA concentration in the blend of MEA and MDEA. This is due to the fact that the absorption reaction rate of CO_2 in the presence of an MEA solution is much higher than that of CO_2 in the presence of an MDEA solution.

Also, the effect of liquid flow rate on the CO_2 removal percentage for various membranes such as PP, PVDF, and PTFE in the presence of MEA as the absorbent was studied; the results are shown in Figure 10.

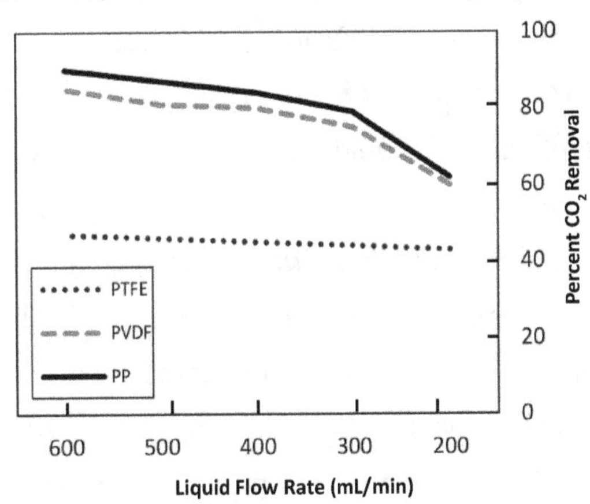

Figure 10: Effect of liquid flow rate on the CO_2 removal percentage for various hollow fiber membranes; ($C_{MEA-initial}$ = 10 mol/m³; $C_{CO_2-initial}$ = 4 mol/m³; Q_{Gas} = 500 mL/min).

The results show that the percentage of CO_2 removal of the polypropylene hollow fiber membrane is higher than that of the PVDF and PTFE hollow fiber membranes. Finally, the outlet CO_2 concentration at different solvent concentrations is demonstrated in Figure 11. This figure shows that by increasing the concentrations of the solvent, the outlet concentration of carbon dioxide drops.

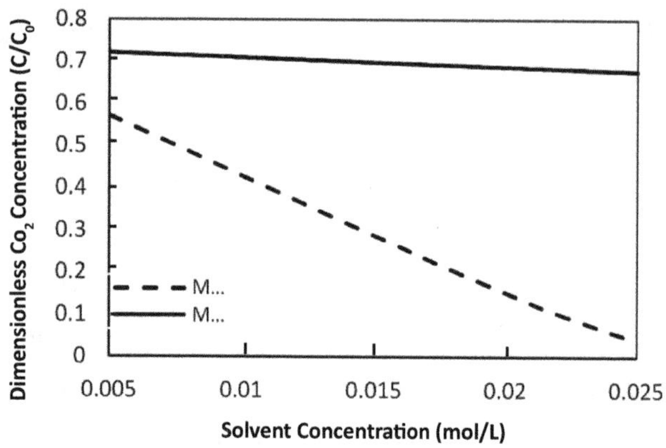

Figure 11: The effect of solvent concentration on the concentration of the outlet CO_2.

CONCLUSIONS

In this work, mathematical modeling was performed in axial and radial directions for the removal of CO_2 in the hollow fiber membrane contactors in the presence of MEA, MDEA, and a blend of them in a fully-wet condition for a countercurrent gas–liquid flow arrangement. Both of axial and radial diffusions have been considered in three segments of shell, membrane, and tube. The effects of liquid flow rate, gas flow rate, membrane length, porosity-to-tortuosity ratio, and total number of fibers were studied. The results of modeling indicated that the pure aqueous solution of MEA is the most suitable absorbent for the absorption of CO_2 among a mixture of MEA and MDEA solution. This modeling showed that the removal of CO_2 increases by adding MEA into MDEA aqueous solution. The reason is that the absorption reaction rate of CO_2 in the presence of an MEA solution is much higher than that of CO_2 in the presence of an MDEA solution. Also, the results of modeling showed that the removal of CO_2 is increased by decreasing gas flow rate, by increasing liquid flow rate, by extending membrane length, by increasing porosity-to-tortuosity ratio, and by raising the total number of fibers in the membrane contactor. The experimental data agreed well with the mathematical model developed in this work. Furthermore, the results show that the percentage of CO_2 removal by the polypropylene hollow fiber membrane is higher than that of the PVDF and PTFE hollow fiber membranes in the presence of MEA as the absorbent.

Moreover, the following results of this study can be used for future works:

1- The selection type of amine and type of hollow fiber membrane for carbon dioxide separation;

2- The determination of gas and liquid flow rate range for carbon dioxide separation process in hollow fiber membranes.

Furthermore, the following items can be studied for further research in the future:

1- Ionic liquids can be used instead of the amines;

2- A comparison can be drawn between the physical and chemical solvents;

3- A comparison can be made between the co-current and counter-current operations;

4- The separation percentage of CO_2 in a hollow fiber membrane contactor and in the absorption tower with amines may be compared;

5- Removing carbon dioxide and hydrogen sulfide simultaneously can be studied.

ACKNOWLEDGEMENTS

The authors are grateful to the Shiraz University of Technology for supporting this work.

NOMENCLATURES

$C_{CO_2\text{-tube}}$	: Concentration of CO_2 in the tube ($mol.m^{-3}$)
$C_{CO_2\text{-membrane}}$	: Concentration of CO_2 in the membrane ($mol.m^{-3}$)
$C_{CO_2\text{-shell}}$	: Concentration of CO_2 in the shell ($mol.m^{-3}$)

$C_{solvent-tube}$	: Concentration of solvent in the tube (mol.m^{-3})
$C_{solvent-membrane}$	: Diffusion coefficient of CO_2 in membrane (m^2.s^{-1})
$C_{CO_2-shell,initial}$	: Diffusion coefficient of CO_2 in shell (m^2 .s^{-1})
D_{CO_2-tube}	: Diffusion coefficient of CO_2 in tube (m^2.s^{-1})
$D_{CO_2-membrane}$	: Diffusion coefficient of CO_2 in membrane (m^2.s^{-1})
$D_{CO_2-shell}$	: Diffusion coefficient of CO_2 in shell (m^2.s^{-1})
$D_{solvent-tube}$	: Diffusion coefficient of solvent in tube (m^2.s^{-1})
$D_{solvent-membrane}$	: Diffusion coefficient of solvent in membrane (m^2.s^{-1})
$\overline{u}$	: Average velocity (m.s^{-1})
ε	: Porosity (-)
L	: Length of fiber (m)
M	: Physical solubility (-)
n	: Total number of fibers (-)
Q_{gas}	: Gas flow rate (L.min^{-1})
Q_{liq}	: Liquid flow rate (L.min^{-1})
r_1	: inner tube radius (m)
r_2	: Outer tube radius (m)
r_3	: inner shell radius (m)
R_{CO_2}	: Overall reaction rate of CO_2 (mol.s^{-1})
$R_{solvent}$	: Overall reaction rate of solvent (mol.s^{-1})
$V_{z-shell}$	: Velocity in shell (m.s^{-1})
V_{z-tube}	: Velocity in tube (m.s^{-1})
τ	: Tortuosity (-)
Subscripts and Superscripts	
1	: inner tube
2	: outer tube
3	: inner shell

REFERENCES

1. Zhang J., Webley P. A. and Xiao P., "Effect of Process Parameters on Power Requirements of Vacuum Swing Adsorption Technology for CO_2 Capture from Flue Gas," *Energy Conversion and Management*, **2008**, *49*(2), 346-350.
2. Powell C. E. and Qiao G. G., "Polymeric CO_2/N_2 Gas Separation Membranes for the Capture of Carbon Dioxide from Power Plant Flue Gases," *J. Member Sci.*, **2006**, *279*, 1-10.
3. Kohl A. L. and Nielsen R., "Gas Purification," Houston Texas: Gulf Publishing Company, **1997**.
4. Wang R., Zhang H. Y., Feron P. H. M., and Liang D. T., "Influence of membrane wetting on CO_2 capture in microporous hollow fiber membrane contactors," *Separation and Purification Technology*, **2005**, *46*, 33-40.
5. Singh P. and Versteeg G. F., "Structure and Activity Relationships for CO_2 Regeneration from Aqueous Amine-based Absorbents," *Process Safety and Environmental Protection*, **2008**, *86*, 347–359.
6. Mavroudi M., Kaldis S. P., and Sakellaropoulos G. P., "A Study of Mass Transfer Resistance in Membrane Gas-liquid Contacting Process," *Journal of Membrane Science*, **2006**, *272*, 103-115.
7. Wang R., Li D. F., Zhou C., Liu M., and et al., "Impact of DEA Solutions with and without CO_2 Loading on Porous Polypropylene Membranes intended for Use as Contactors," *Journal of Membrane Science*, **2004**, *229*, 147-157.
8. Lv Y. X., Yu X. H., Tu S. T., Yan, J.Y., and et al., "Wetting of Polypropylene Hollow Fiber Membrane Contactors," *Journal of Membrane Science*, **2010**, *362*, 444-452.
9. Lv Y. X., Yu X. H., Tu S. T., Yan J. Y., and et al., "Experimental Studies on Simultaneous Removal of CO_2 and SO_2 in a Polypropylene Hollow Fiber Membrane Contactor," *International Conference on Applied Energy*, ICAE 2011, Italy, **2011**.
10. Qi Z. and Cussler E. L., "Microporous Hollow Fibers for Gas Absorption," *J. Member Sci.*, **1985**, *23*(3), 321-330.

11. Karoor S., and Sirkar K. K., "Gas Absorption Studies in Microporous Hollow Fiber Membrane Contactors," *J. Member Sci.*, **2001**, *194*(4), 57-64.
12. Dindore V. Y., Brilman D. W. F., and Versteeg G. F., "Modeling of Cross-flow Membrane Contactors: Mass Transfer with Chemical Reactions," *J. Member Sci.*, **2005**, *255*, 275-285.
13. Wang R., Li D., and Liang D., "Modeling of CO_2 Capture by Three Typical Amine Solution in Hollow Fiber Membrane Contactors," *Chem. Eng. Proc.*, **2004**, *43*, 849-860.
14. Al-Marzougi M., El-Nass M., Marzouk S., Abdulkarimi M., and et al., "Modeling of CO_2 absorption in membrane contactors," *Sep. Purification Technology*, **2008**, *59*(1), 286-292.
15. Happel J., "Viscous Flow Relative to Arrays of Cylinders," *AIChE J.*, **1959**, *5*(1), 174-182.
16. Bird R. B., Stewart W. E., and Lightfoot E. N., "Transport Phenomena," New York: Wiley, **1960**.
17. Paul S., Ghoshal A. K., and Mandal B., "Removal of CO_2 by single and Blended Aqueous Alkanol amine Solvents in Hollow-Fiber Membrane Contactor: Modeling and Simulation," *Ind. Eng. Chem. Res.*, **2007**, *46*, 2576-2585.
18. Al-Marzougi M., El-Nass M., Marzouk S., and Abdullatif N., "Modeling Chemical Absorption of CO_2 in Membrane Contactors," *Sep. Purifi. Technol.*, **2008**, *62*, 499-515.
19. Versteeg G. F. and Van Swaalj W., "Solubility and Diffusivity of Acid Gases (Carbon Dioxide, nitrous Oxide) in Aqueous Alkanol Amine Solutions," *Journal of Chemical and Engineering Data*, **1988**, *33*, 29-34.
20. Cussler E., "Diffusion: Mass Transfer in Fluid Systems, 1," *Cambridge University Press, Cambridge, Chapters*, **1984**, *1*, 9-21.
21. Hagewiesche D. P., Ashour S. S., Al-Ghawas H. A., and Sandall O. C., "Absorption of Carbon Dioxide into Aqueous Blends of Monoethanolamine and N-methyldiethanolamine," *Chemical Engineering Science*, **1995**, *50*, 1071-1079.

14

Implication of an Integrated Approach to the Determination of Water Saturation in a Carbonate Gas Reservoir Located in the Persian Gulf

Marzieh Neyzan Hosseini[1], Ezatallah Kazemzadeh[2*], Ehsan Sobhani[3], and Bita Arbab[4]

[1] Department of Geophysics, College of Basic Science, Kermanshah Branch, Islamic Azad University, Kermanshah, Iran
[2] Department of Petroleum Engineering, Islamic Azad University, Science and Research Branch, Tehran, Iran
[3] Department of Geophysics, College of Basic Science, Kermanshah Branch, Islamic Azad University, Kermanshah, Iran
[4] Department of Petrophysics, Iranian Offshore Oil Company, Tehran, Iran

ABSTRACT

Water saturation determination is one of the most important tasks in reservoir studies to predict oil and gas in place needed to be calculated with more accuracy. The estimation of this important reservoir parameter is commonly determined by various well logs data and by applying some correlations that may not be so accurate in some real practical cases, especially for carbonate reservoirs. Since laboratory core analysis data have a high accuracy, in this study, it is attempted to use core and geological core description data to present an improved method to determine an optimized cementation factor (*m*) and a saturation exponent (*n*) in order to evaluate water saturation within carbonate reservoirs compared to default values (*m=2, n=2, a=1*) in a carbonate gas reservoir located in the Persian Gulf. Based on integrating core petrography and velocity deviation log (VDL), core samples were classified based on the type of porosity and geology description, and then by employing log-log plots of formation resistivity factor (FRF) versus porosity and formation resistivity index (FRI) versus water saturation, saturation parameters (*m,n*) were determined for each classification. Utilizing default and optimized values of saturation parameters, water saturation logs were obtained through different conductivity models by employing Multi min algorithm. Then, optimized water saturation was compared to core data. Error analysis showed that water saturation data resulted in optimized saturation parameters having a lower average error of 0.08 compared to the default ones with an average error of 0.14, and based on cumulative histogram, optimized water saturation data are in good agreement with the trend of core water saturation.

Keywords: Water Saturation, Saturation Parameters, Petrography, Velocity Deviation Log

INTRODUCTION

One of the key petrophysical parameters for evaluating reserve estimation is water saturation. In order to calculate the hydrocarbon reserve, one needs to know the water saturation amount. The improper calculation of water saturation leads to great errors in reserve estimation. Traditional methods for obtaining water saturation were based on the assessment of core samples which were drilled using oil based mud. Nevertheless, due to the technical difficulties and costs, these methods could only measure saturation for few points.

***Corresponding author**
Ezatallah Kazemzadeh
Email: kazemzadehe@ripi.ir

In 1942, Archie proposed a sound physical and much easier method which relates water saturation to the formation porosity and resistivity. In a clean formation, Archie's equation is given by:

$$S_w{}^n = FRF\frac{R_w}{R_t} \quad (1)$$

where, S_w is water saturation, and n is an exponent of water saturation; R_w represents the formation brine resistivity, and R_t stands for the fluid resistivity of un-invaded zone (brine and hydrocarbon); *FRF* is the formation resistivity factor given by:

$$FRF = \frac{\alpha}{\phi^m} \quad (2)$$

where, α is tortuosity factor, and ϕ and m are porosity and an exponent depicting the cementation degree of a formation respectively.

Another useful representation of the Archie's equation is based on the formation resistivity index (FRI) defined as follows:

$$FRI = \frac{R_t}{R_o} = \frac{1}{S_w{}^n} \quad (3)$$

where, R_o is the resistivity of rock fully saturated with brine (formation water) [1].

There are a number of assumptions for Archie's equation, including (1) rock surface to be strongly water-wet and (2) pore system to be free of clay minerals. These assumptions are not always guaranteed, especially when one is dealing with carbonate rocks owing to complex pore geometry and a high degree of local heterogeneity.

Clay mineral in the rock matrix acts as an electrolyte conductor, so they introduce a new conduction path for ions similar to the formation water. Since 1950, the clay problem has been fully recognized and addressed. The clay models have been divided into two main groups [2]:

(1) Models based on the clay volume fraction such as Hossin, Simandoux, Bardon & Pied, Worthington, and Indonesia model. The disadvantage of these models is that they do not take account of the composition of constituent clays [2-6].

(2) Models based on the ionic double-layer phenomenon such as Waxman & Smith and Dual-water [7, 8].

The second group models tack account of the composition of constituent clays, but parameters such as cation exchange capacity (CEC) have directly been brought from core samples in laboratory, while for the first group clay volume fraction can indirectly be determined from well log data.

There have been efforts in the past for calculating water saturation exponent (*n*) from only well logs [9]. The water saturation exponent (*n*) has been reported to be equal to 2.0 for clean, consolidated, and water-wet sandstones and has been assumed equal to 2.0 in many petrophysical evaluation equations. In reality, however, *n* is not constant, but it varies as a function of different factors, including fluid distribution in pore spaces and rock wettability. The traditional method to determine *n* is plotting FRI versus water saturation on a logarithmic scale and calculating the slope of straight line fitting to the data points.

Many studies have noticed the proportionality of *m*, the cementation factor, to the formation porosity. Considering this, the number of correlations are established using porosity logs. However, in all of these works, the tortuosity of the pore network is neglected, and they concluded that the cementation factor depends on the pores geometry and the degree of their interconnectivity. Therefore, the variation of pore type is of significant importance with respect to *m*, and it is needed to group data to identify this variation. Uncertainty analysis has shown that among various parameters, the cementation factor has the highest impact. The common method to estimate *m* is the slope of the straight line fitting the formation resistivity factor versus porosity on a logarithmic scale. The best fit will achieve when the variation of the tortuosity is considered [10-16].

This study has been focused on the investigation of more accurate and optimum petrophysical parameters to calculate and determine water saturation profile by a suitable water saturation model for carbonate reservoirs of an offshore gas field in the Persian Gulf. Four wells have been drilled in this field and core analysis data are only available for one well, which is selected for our purpose.

EXPERIMENTAL PROCEDURES

Methodology

Geological core description and special core analysis data along with resistivity and porosity logs for one of the wells in the considered carbonate gas field were selected in order to determine water saturation. Figure 1 shows a part of petrography results; according to the petrography studies, carbonate reservoirs of the field mainly consist of limestone and dolomite. At many depths, pore spaces have been filled with anhydrite cement. Interparticle, moldic, and inter crystalline are porosity types in reservoirs, and in some limited depths vuggy porosity exists.

In this study, 97 core samples have been used for the core analysis. FRF and FRI measurements have been performed for 32 core samples. Relative permeability and irreducible water saturation data have been measured for 64 core samples. Since core water saturation data were not available, irreducible water saturation data in the reservoir section above oil-water contact and transition zone were used to compare with the water saturation obtained from petrophysical modeling.

Table 1 shows FRF and porosity measurements at a confining pressure of 6200 psi corresponding to the reservoir pressure.

		Lithology (%)							Bioclast (%)								Facies	
Plug No.	Depth (m)	Limestone (%)	Dolostone (%)	Anhydrite (%)	Sandstone (%)	Clay (%)	Matrix + Porosity (%)	Main Lithology	Fine skeletal debris	Large skeletal debris	Benthic forams	Planktonic forams	Peloid	ooid	Intraclast	oncoid	Facies code	Facies
01H	3457.25	100	0	0	0	0	100	l									4	mudstone
01V	3457.4	0	100	0	0	0	100	d									4	dolomudstone
02H	3457.5	0	95	0	0	0	100	d									4	dolomudstone
03H	3457.98	0	90	0	0	0	100	d									4	dolomudstone
04H	3458.13	70	25	5	0	0	100	dl		20			5	30			8	skeletal ooid grainstone
05H	3458.62	75	20	5	0	0	100	dl		10				70			8	ooid grainstone
06H	3458.84	92	5	3	0	0	100	l		40				30	5		6	ooid skeletal grainstone
02V	3458.94	98	0	2	0	0	100	l		15				75			8	skeletal ooid grainstone
07H	3459.14	72	1.6	6.4	0	0	100	l		10				80			8	skeletal ooid grainstone
08H	3459.45	93	0	0	0	0	100	l	10				20				5	peloid skeletal wackestone
09H	3459.72	83.3	0	1.7	0	0	100	l		40				25			6	ooid skeletal grainstone
03V	3459.89	100	0	0	0	0	100	l	10					75			8	skeletal ooid grainstone

		Porosity(%)								Routin				
Plug No.	Depth (m)	Interparticle	intraparticle	moldic	intercrystalline	voggy	fracture	Fenestral	total porosity%	Porosity (%)	Permeability(mD)	Grain Density (g/cc)	Sedimentary Environment	
01H	3457.25								0				lagoon	
01V	3457.4								0				lagoon	sucrosic dolo micrite, neomorphism
02H	3457.5				5				5	6.684	0.605	2.76	lagoon	stylolite, sucrosic dolomite, planar dolomite
03H	3457.98				10				10	4.616	7.002	2.62	lagoon	stylolite, sucrosic dolomite,
04H	3458.13								0	3.037	0.032	2.74	lagoon	calcite cement, planar dolomite ,stylolite, micritized ooid, bivalve, blocky calcite cement,
05H	3458.62								0	9.443	16.490	2.82	ooid shoal	calcite cement, patchy anhydrite cement, bivalve, planar-e, non-planar, stylolite, isopach
06H	3458.84								0	6.917	0.399	2.83	shoal margin	calcite cement, planar-e, micritized ooid bivalve, isopachous cement, blocky cement
02V	3458.94								0	8.748	0.014	2.70	lagoon	gastropod, blocky cement, isopachous cement, replaced ooid, micrite envelope, bivalve d
07H	3459.14			20					20	22.818	0.743	2.69	ooid shoal	moldic porosity, blocky cement, compaction
08H	3459.45			2	5				7	7.394	1.223	2.42	lagoon	copmaction, bivalve, ostracod, heavily crystalized
09H	3459.72	2	5	5	3				15	10.323	0.304	2.70	shoal margin	calcite cement, styloli, bivalve debris
03V	3459.89								0				ooid shoal	microbial filament, isopachous cement, blocky cement, neomorphism, micrite envelope

Figure 1: A part of petrography studies for the core plugs of well.

Table 1: FRF and porosity measurements at the confining pressure (6200 psi) for core plugs.

No.	Core Depth (m)	Grain Density (g/mL)	Gas Permeability (mD)	Formation Resistivity FRF	Porosity by brine (v/v)
1	3460.50	2.71	1.87	173.02	0.19
2	3461.26	2.72	1.78	213.71	0.22
3	3464.43	2.74	0.06	355.11	0.09
4	3465.78	2.76	0.24	404.33	0.17
5	3469.82	2.88	0.39	657.81	0.05
6	3535.63	2.79	21.7	96.47	0.10
7	3536.16	2.74	19.2	50.56	0.15
8	3542.17	2.85	428	23.00	0.23
9	3542.64	2.86	241	18.26	0.25
10	3568.73	2.87	0.01	953.00	0.02
11	3569.87	2.88	7.93	65.74	0.13
12	3572.06	2.87	410	22.98	0.24
13	3575.11	2.75	0.35	230.76	0.11
14	3579.66	2.73	1.26	263.72	0.15
15	3581.13	2.69	0.01	596.40	0.16
16	3597.44	2.73	<0.01	187.37	0.11
17	3599.39	2.72	0.00	813.53	0.03
18	3601.76	2.71	<0.01	1842.63	0.00
19	3603.33	2.72	<0.01	1778.29	0.01
20	3625.27	2.88	2.71	510.74	0.04
21	3627.90	2.86	35.5	64.71	0.13
22	3634.95	2.88	<0.01	1068.85	0.04
23	3638.43	2.85	2.07	138.73	0.09
24	3645.49	2.87	0.72	59.44	0.14
25	3646.96	2.87	0.07	684.38	0.04
26	3654.25	2.85	24.4	20.30	0.22
27	3660.49	2.86	0.05	272.86	0.03
28	3685.37	2.86	30.1	465.51	0.07
29	3688.81	2.85	5.57	72.77	0.15
30	3690.37	2.84	8.21	42.80	0.19
31	3691.69	2.84	0.73	203.79	0.09
32	3695.40	2.88	0.48	167.26	0.17

Determination of Petrophysical Parameters Based on the Conventional Methods

Based on the Archie equation, log-log plots of FRF versus porosity (Figure 2) and FRI versus water saturation (Figure 3) were used to determine petrophysical parameters. It is worth mentioning that two data points were out of range and were, therefore, removed.

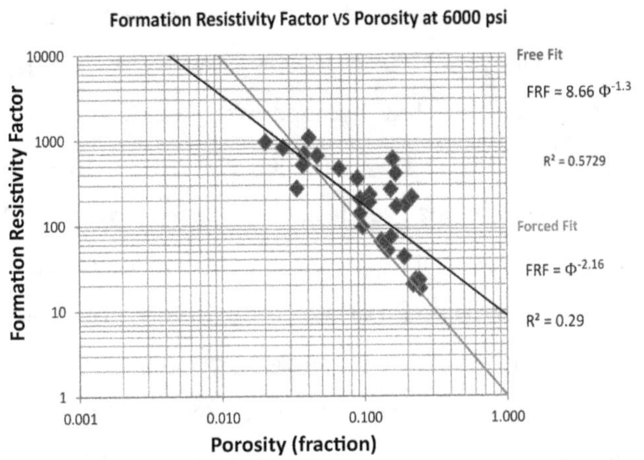

Figure 2: Determination of *m* based on log-log plot of FRF versus porosity using free and forced fit methods.

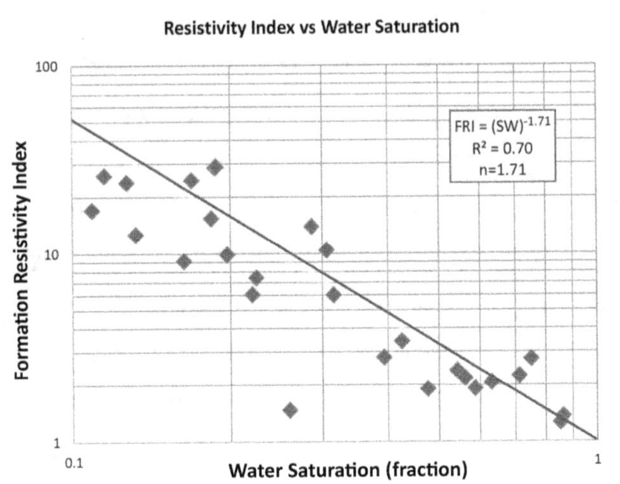

Figure 3: Determination of *n* based on the log-log plot of FRI versus water saturation.

According to Figure 2, in the forced fit method, the correlation coefficient of straight line (R^2= 0.29) is not good enough, so the obtained *m* is not acceptable. In addition, in the free fit method, the correlation coefficient (R^2=0.57) is not satisfying either, so the obtained *m* and *a* parameters will be uncertain.

An attempt was also made to determine *m* as a function of porosity (ϕ) and compare it with equations cited in the literature. Some of these commonly used equations are listed below:

- Borai established an equation for low porosity and tight carbonates as given below:
  $m=2.2-0.035/(\phi+0.042)$
- Shell recommended the equation below for evaluating low porosity carbonates:
  $m=1.87+0.019/\phi$
- Sethi used the below equation:
  $m=2.05+\phi$
- Rafiee et al. recommended a new equation for carbonate parts of the southwest Iranian oil fields as reads:
  $m=2.461-0.048/(\phi+0.031)$
- Asadollahi et al. presented a relation between cementation factor and porosity in Iranian carbonate formations for both low and high porosity reservoirs as defined below [17-19]:
  $m=1/(0.36-0.08\ln(\phi))$

Figure 4 illustrates the cementation factor values obtained by forced fitting method (*a=1*) versus porosity. The correlation coefficients (R^2) for both linear and power trend lines are not so suitable indicating that the *m* values belong to different rock types.

Figure 5 illustrates the graphs of Borai, Shell, Sethi, Asadollahi, and Rafiee relations, which show different trends with core data. Therefore, these formulas cannot be used for *m* values.

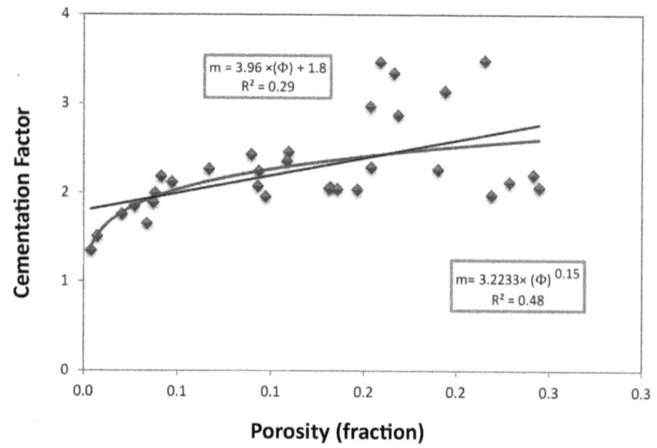

Figure 4: Relationship between cementation factor and porosity.

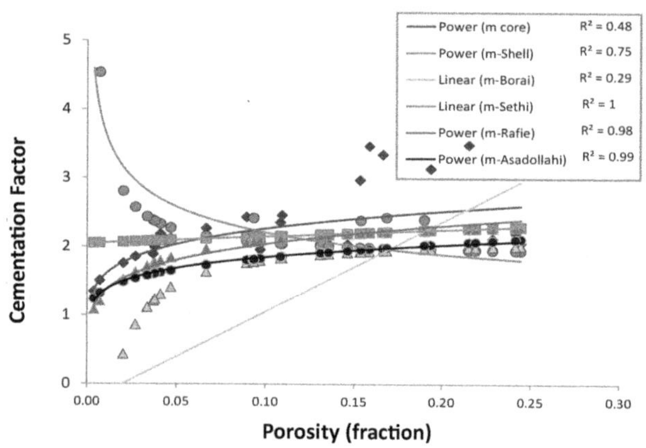

Figure 5: Cross plot of cementation factor versus porosity according to different empirical equations.

Determination of Petrophysical Parameters with Classifying Core Measurement Data Based on the Velocity Deviation Log (VDL)

The following equation is proposed to determine porosity from the transit time of compressional waves [20]:

$$\Delta T = (1-\Phi)\Delta T_{ma} + \Delta T_f \quad (3)$$

where, ΔT_{ma} is the transit time for the unit length of the rock matrix, and ΔT_f is the transit time for the saturating fluid; Φ is the porosity.

Most often, for two porous media with identical porosities there might be velocities which do not conform to the Wyllie equation owing to the dissimilarities in the type of porosity. The velocity deviation log is referred to as the difference between the velocities obtained from the actual sonic log and that obtained from a synthetic velocity log by means of porosity log data in the time average equation. This method is proposed by Anselmetti and Eberli to identify main porosity types in carbonate reservoirs. Generally, velocity is inversely proportional to porosity; however, the porosity types can potentially alter this relationship. In carbonate reservoirs, elastic properties are mostly governed by pore geometry [21, 22].

Anselmetti and Eberli studied the velocity deviation for five different types of porosity class corresponding to various samples to investigate the effect of pore geometry on velocity. They reported velocity deviations from positive to negative for various pore types, including intrafossil, moldic, interparticle, intercrystalline, and high micro-porosity. Pores with a low degree of connectivity within a densely cemented matrix usually give a positive deviation which corresponds to low permeability zones. For negative deviations, Anselmetti and Eberli indicated some factors other than lithology such as caving, borehole irregularities, and high content of free gas and/or fracture porosity [23].

According to the above explanations, velocity deviation logs can be obtained by the following equation:

$$VDL = V_{sonic} - V_{synthetic} \quad (4)$$

where, VDL is the velocity deviation log; V_{sonic} represents the velocity from sonic log, and $V_{synthetic}$ is the velocity calculated from one of the neutron, density, neutron–density logs, or core-plug porosity by using the Wyllie time average equation. Upon work done by Kazemzadeh et al., deviation logs derived from neutron–porosity logs showed better results for the determination of different pore types and petrofacies. In this study, the comparisons between different porosity logs also showed the same results. Therefore, neutron porosity log was used to obtain deviation velocity log [24]. Firstly, the environmental corrections were carried out on neutron-porosity log, and then, according to SGR log, it was corrected in terms of shale effect. Since the selected well had been drilled by water base mud, ΔT_f was considered 189.5 µs/ft. Having lithology log and transit time data for each mineral, ΔT_{ma} log was obtained by using Apparent Matrix Properties module in Geolog (version 7) software.

Considering the following relationships, real and synthetic velocity logs were calculated to obtain VDL log.

$$V_{sonic} = 304.8/\Delta T_{sonic} \quad (5)$$

$$V_{synthetic} = 304.8/\Delta T_{synthetic} \quad (6)$$

According to the obtained VDL log, as it is shown in Table 2, the values of velocity deviation log for all the cores samples were high and could not be classified with respect to pore types. Analyzing core petrography results, it was recognized that, in many of the samples, anhydrate (with a grain density higher than both dolomite and limestone) has filled pores in the form of patchy anhydrate cement, so it has increased real velocity and VDL.

Therefore, for these samples and those that have only moldic pore type, VDL will be high, and they cannot be classified in different groups. We can conclude that in situations like this VDL method will not be applicable to classifying core samples with respect to their pore types.

Table 2: VDL data based on shale corrected neutron porosity log at the core samples depth.

No.	Core Depth(m) (wire log depth matched)	VDL	Rock Category Code
			-500 <VDL<+500 : 1, +500 <VDL : 2, VDL<-500 : 3
1	3460.50	1021	2
2	3461.26	2718	2
3	3464.43	1310	2
4	3465.78	1751	2
5	3469.82	-1659	3
6	3535.63	4584	2
7	3536.16	2009	2
8	3542.17	2249	2
9	3542.64	2605	2
10	3568.73	1546	2
11	3569.87	1751	2
12	3572.06	1108	2
13	3575.11	2148	2
14	3579.66	2686	2
15	3581.13	1744	2
16	3597.44	1349	2
17	3599.39	1757	2
18	3601.76	577	2
19	3603.33	516	2
20	3625.27	2786	2
21	3627.90	3063	2
22	3634.95	2977	2
23	3638.43	3010	2
24	3645.49	1984	2
25	3646.96	2613	2
26	3654.25	2282	2
27	3660.49	2090	2
28	3685.37	2359	2
29	3688.81	3444	2
30	3690.37	3509	2
31	3691.69	2322	2
32	3695.40	3598	2

Determination of Petrophysical Parameters with Classifying Core Measurement Data Based on the Integration of Petrography and VDL Results

Knowing pore types and the geological description of the core samples based on the petrography data and considering VDL results, the core samples were classified in three groups as shown in Table 3. For each group, the log-log plots of FRF versus porosity are used for the determination of *m* in the forced fit, and FRI versus water saturation is employed to determine *n* (Figures 6-11).

Table 3: Classifying core samples according to pore types and geological description.

Rock Category	Geology Description	Data Frequency
1	Moldic pore type, without anhydrate	7
2	Interparticle and intercrystalline pore type, patchy anhydrate cement	15
3	Interparticle and Intercrystalline pore type, without anhydrate	10

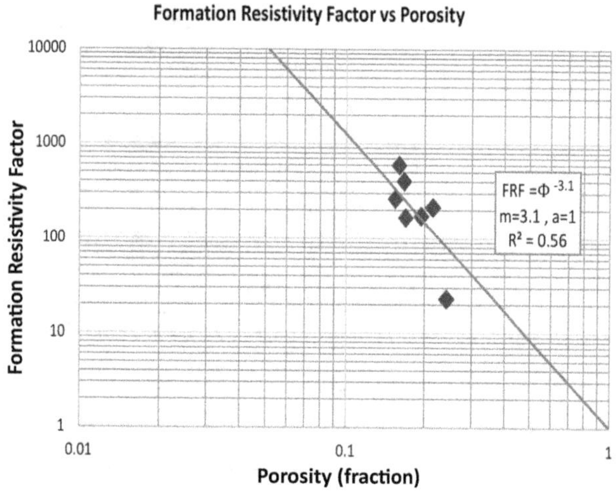

Figure 6: Relationship between FRF and porosity for rock category 1.

Table 4 summarizes the results of this classifying. As it is seen, the correlation coefficients (R^2) for all three groups (especially for group 2) are acceptable for the saturation parameters obtained.

Table 4: Summarized results of the determined petrophysical parameters for each category.

Category	*a*	*m*	R^2	*n*	R^2
1	1	3.1	0.56	1.93	0.62
2	1	2.18	0.90	1.57	0.90
3	1	1.67	0.73	2.27	0.67

In order to employ these petrophysical parameters as a continuous well log, it was required to obtain secondary (moldic) porosity log generated from the difference between neutron porosity log (total porosity) and sonic log (primary porosity). Therefore, based on the above explained classification, providing lithology and secondary porosity logs, the petrophysical parameters were obtained as a well log, and they were then applied to water saturation equations in petrophysical modeling (performed using Geolog commercial software).

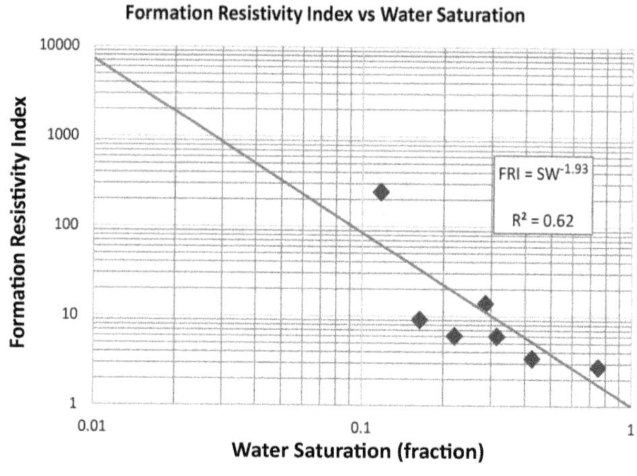

Figure 7: Relationship between FRI and water saturation for rock category 1.

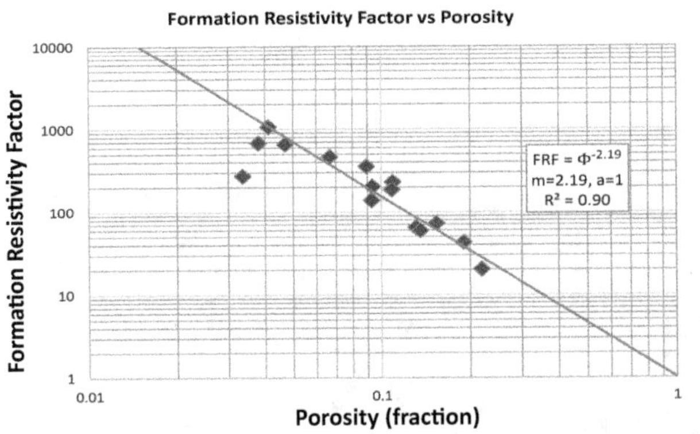

Figure 8: Relationship between FRF and porosity for rock category 2.

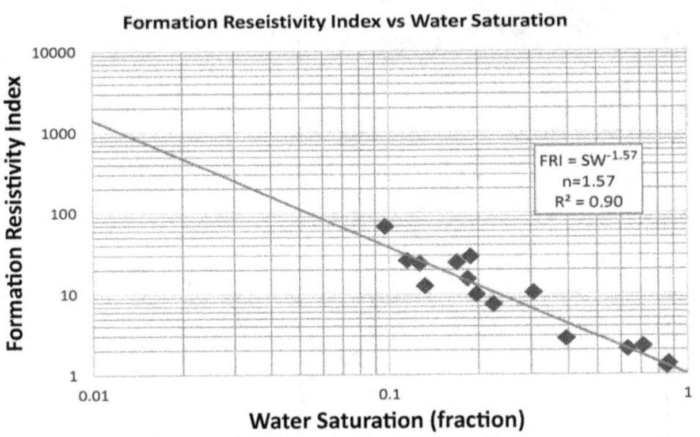

Figure 9: Relationship between FRI and water saturation for rock category 2.

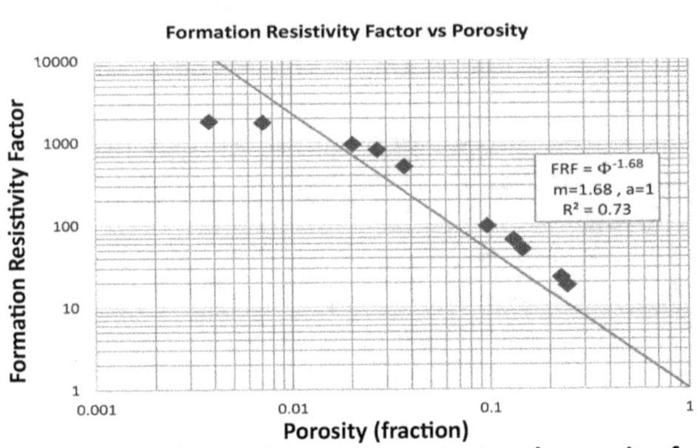

Figure 10: Relationship between FRF and porosity for rock category 3.

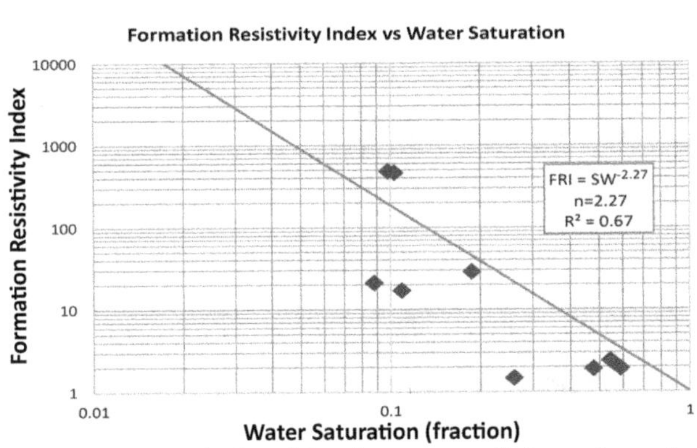

Figure 11: Relationship between FRI and water saturation for rock category 3.

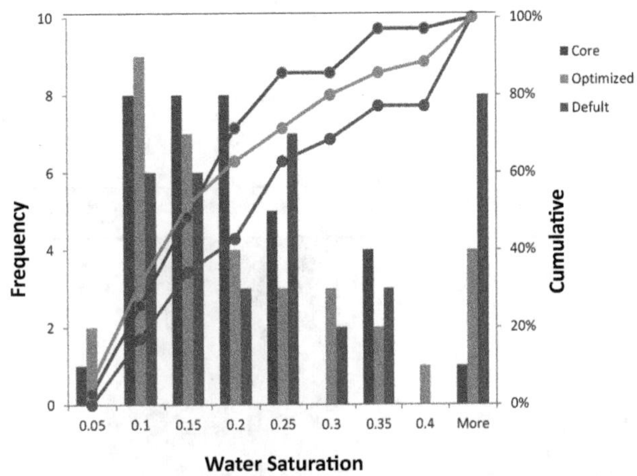

Figure 12: Cumulative histogram of the core and computed water saturation data.

Petrophysical Modeling and Analysis

As it was mentioned earlier, Geolog software, as a petrophysical analysis tool, was employed in this study to obtain water saturation log. To resolve a petrophysical problem, two different approaches are available. The first method is a deterministic method which is generally suitable for simple lithology problems. In this approach, porosity, shale fraction, and shale volumes (saturations) are computed sequentially. The second method is a multi-mineral approach referred to as Multimin. This approach is applied to complex lithology, where the minerals and rock pore fluids all affect the log responses. In this method, both log measurements and log response parameters are used in the log response equations, which will be solved simultaneously to compute the fraction volumes of formation minerals and fluids. Indonesia and other equations were selected here to compute fluid volumes and thus water saturation.

Figure 13 shows a layout of Multimin analysis results for the candidate well. The log tracks on the right are water saturation logs computed based on the Indonesia water saturation equation.

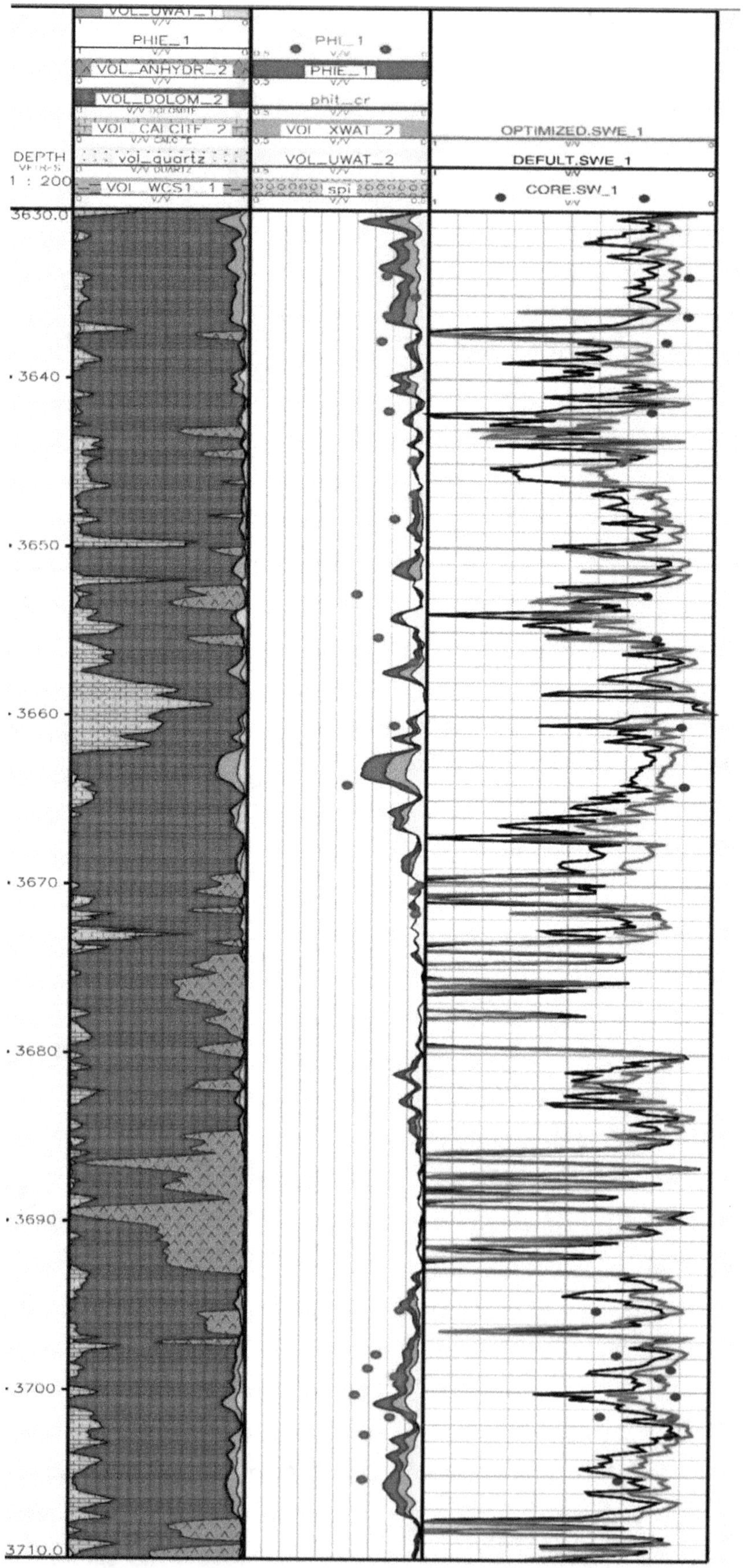

Figure 13: Computed water saturation (black and pink curves in the right track based on the default and the optimized petrophysical parameters) compared to the core data (blue dots).

The black curve refers to the water saturation computed based on default values (*m=2, n=2, a=1*), and the pink curve addresses the calculated water saturation based on the optimized parameters. Considering an average value of *n* instead of applying *n* as a variable resulted in a better match with the core data.

Several runs were performed according to different water saturation equations and different combinations of petrophysical parameters (i.e. as constant and/or variable values). Because of very low clay content in some parts of the reservoir, employing different water saturation equations did not make significant difference between the obtained water saturation logs. Moreover, other combinations of petrophysical parameters did not result in a better match with the core data.

In order to evaluate the accuracy of the optimized water saturation based on the default and the optimized petrophysical parameters compared to the core data, three error measures, including average error, standard deviation, and root mean square error (RMS) were used. The results of the error analysis are given in Table 5.

Table 5: Error analysis of the computed water saturation results based on the default and the optimized Archie's parameters.

Petrophysical Parameters	Average Error	Standard Deviation	Root Mean Square (RMS)
Default (m=2, n=2, a=1)	0.14	0.15	0.18
Optimized (*a*=1, *m*= variable, n= average value)	0.08	0.09	0.10

It is important to note that the error parameters for the optimized water saturation have considerably decreased compared to their default values, so using the optimized parameters for water saturation determination is recommended.

According to the cumulative histogram of water saturation data shown in Figure 12, both the optimized and the default data follow the trend of the core data. In addition, the optimized data are in good agreement with the trend of the core water saturation.

CONCIUSIONS

1. A considerable degree of uncertainty is observed when the cementation factor is determined according to the traditional method of calculating the slope of the forced and free straight line fitting to the data points on a log-log plot of FRI versus water saturation.
2. The cementation factor values determined according to the conventional methods of free and forced straight line fitting on the log-log plot of formation resistivity factor (FRF) versus porosity were uncertain.
3. According to cross plot of cementation factor values and porosity, a reliable equation to present cementation factor as a function of porosity was not achieved. Also, the empirical equations did not provide a good match with the core data points, and even Shell equation trend was in contrary to the trend of the core data.
4. Applying VDL approach to classify the core measurement data based on pore types is not applicable in cases that anhydrate exists in the core samples and fills pores as patchy cement.
5. Acceptable and reliable petrophysical parameters were determined by classifying the core measurement data based on core geological descriptions and VDL results.
6. The results of petrophysical modeling based on the multimineral approach through Multimin module of Geolog software showed that applying m as variable and n as an average value in Indonesia water saturation equation would give a better

match between the calculated water saturation log and the core data points compared to applying other combinations of petrophysical parameters.

7. In petrophysical modeling, because of a very low clay volume in some parts of reservoirs, employing different water saturation equations did not make a significant difference between the calculated water saturation logs.
8. Error analysis and cumulative histogram showed that the water saturation data resulted in the optimized petrophysical parameters compared to the default ones, had a lower error, and were in good agreement with the trend of the core water saturation.

ACKNOWLEDGMENTS

The authors warmly thank the Research and Technology Department of Iranian Offshore Oil Company (IOOC) for supporting this work.

REFERENCES

1. Archie G. E., "The Electrical Resistivity Log as an Aid in Determining Some Reservoir Characteristics," *Transactions of AIME,* **1941**, *146*, 54-62.
2. Worthington P. F., "The Evolution of Shaly-sand Concepts in Reservoir Evaluation," *The Log Analyst*, **1985**, *26*, 23-40.
3. Hossin A., "Calcul des Saturations en Eau par la Methode du Ciment Argileux," *Bulletin de l'Association Francaise des Techniciens du Petrole*, **1940**
4. Simandoux P., "Mesures Dielectriques en Milieu Poreux, Application a la Mesure des Saturations en Eau," *Etude du Comportement des Massifs ArgileuxRevue*, IFP Supplementary Issue, **1963**.
5. Bardon C. and Pied B., "Formation Water Saturation in Shaly Sands," Trans. SPWLA 10th Annual Logging Symposium, **1969**.
6. Poupon A. and Leveaux. J., "Evaluation of Water Saturation in Shaly Formations," SPWLA 12th Annual Logging Symposium, **1971**.
7. Waxman M. H. and Smiths L. J. M., "Electrical Conductivities in Oil-bearing Shaly Sands," *Society of Petroleum Engineers Journal*, **1968**, *8*, 107-122.
8. Clavier C., Coates G., and Dumanoir J., "The Theoretical and Experimental Bases for the 'Dual Water' Model for the Interpretation of Shaly Sands," *Paper SPE 6859*, **1977**.
9. Aguilera R., "Analysis of Naturally Fractured Reservoirs from Conventional Well Logs", *Journal of Petroleum Technology*, **1976**, *28*, 764-772.
10. Borai, A. M., "A New Correlation for the Cementation Factor in Low-porosity Carbonates," *SPE Formation Evaluation*, **1987**, *2*, 495-499.
11. Lucia F. J., "Carbonate Reservoir Characterization," *Springer*, **1999**, 226.
12. Ragland D. A., "Trends in Cementation Exponents (m) for Carbonate Pore Systems," *Petrophysics*, **2002**, *43*, 434-446.
13. Rezaee M. R., Motiei H., and Kazemzadeh E., "A New Method to Acquire m Exponent and Tortuosity Factor for Microscopically Heterogeneous Carbonates," *Journal of Petroleum Science and Engineering*, **2007**, *56*(4), 241-251.
14. Towel G., "An Analysis of the Formation Resistivity Factor-porosity Relationship of Some Assumed Pore Geometries," in 3rd Annual Logging Symposium Transactions, SPWLA, **1962**.
15. Focke J. W. and Munn D., "Cementation Exponent in Middle Eastern Carbonate Reservoirs," SPE Formation Evaluation, **1987**, 155-167.
16. Hamada G. M., Almajed A. A., and Okasha T. M., "Uncertainty Analysis of Archie's Parameters Determination Techniques in Carbonate Reservoirs," *Journal of Petroleum Exploration and Production Technology*, **2013**, *3*, 1-10.
17. Sethi D. K., "Some Considerations about the Formation Resistivity Factor-porosity Relations," 20th SPWLA Symposium, **1979.**
18. Rafiee S., Hashemi A., and Shahi M., "A New Cementation Factor Correlation in Carbonate Parts of Southwest Iranian Oil Fields," *Electronic Scientific Journal Oil and Gas Business*, **2013**, *2*, 227-251
19. Asadollah M., Keramati M., Bagheri A. M., and Haghighi M., "Investigation of the Effect of Cementation Factor on OOIP in an Iranian Carbonate Reservoir," *GEO 2006 Middle East Conference and Exhibition*, **2006.**
20. Wyllie M. R. J. and Gregory A. R., "Formation Factors of Unconsolidated Porous Media: Influence of Particle Shape and Effect of Cementation," *Journal of Petroleum Technology*,

1953, *5*, 103–110.

21. Anselmetti F. S. and Eberli G. P., "The Velocity-deviation Log: a Tool to Predict Pore Type and Permeability Trends in Carbonate Drill Holes from Sonic and Porosity or Density Logs," *AAPG Bull*, **1999**, *83*, 450–66.
22. Brie A, Johnson D. L. and Nurmi R. D., "Effect of Spherical Pores on Sonic and Resistivity Measurements," Society of Professional Well Log Analysts 26th Ann. Logging Symp, **1985**
23. Eberli G. P., Anselmetti F. S., and Incze M. L., "Factors Controlling Elastic Properties in Carbonate Sediments and Rocks," *The Leading Edge*, **2003**, *22*, 654–60
24. Kazemzadeh E., Nabi-Bidhendi M., Keramati Moezabad M., Rezaee M. R. and Saadat K., "A New Approach for the Determination of Cementation Exponent in Different Petrofacies with Velocity Deviation Logs and Petrographical Studies in the Carbonate Asmari Formation," *Journal of Geophysics and Engineering*, **2007**, *4*, 160-170.

15

Displacement Discontinuity Analysis of the Effects of Various Hydraulic Fracturing Parameters on the Crack Opening Displacement (COD)

Abolfazl Moradi[1*], Behzad Tokhmechi[1], Vamegh Rasouli[2], and Mohammad Fatehi Marji[3]

[1] Mining, Geophysics and Petroleum Engineering Depatment, Shahrood University of Technology, Shahrood, Iran

[2] Petroleum Engineering Department, University of North Dakota, USA

[3] Mining and Metallurgical Engineering Department, Yazd University, Yazd, Iran

ABSTRACT

The combination of horizontal drilling along with hydraulic fracturing has significantly improved the production of hydrocarbon reservoirs and made it possible to extract the relatively impermeable and uneconomical reservoirs. The production rate of oil and gas wells increases proportional to hydraulic fracture aperture or crack opening displacement (COD). This is an important parameter in fracture mechanics literature and hydraulic fracturing of hydrocarbon reservoirs. Despite the significance of COD there are a few analytical solutions for the estimation of COD under certain conditions. In this paper the effect of various parameters on COD is investigated semi-analytically. A higher order displacement discontinuity method is used to consider the effects of different parameters (Young's modulus, Poisson's ratio, internal pressure, maximum and minimum horizontal stresses, crack half-length and its inclination with maximum horizontal stress) on the COD in a hydraulic fracturing process under arbitrarily conditions. The coefficient of determination and standard error of the estimate were 94.35% and 4.37×10^{-4} respectively, showing a good agreement between the fitted equation and the numerical results. The effect of propagation and well radius on the maximum COD was also investigated. The results showed that COD increases almost linearly with the crack propagation and increase of well radius of hydraulic fractures (HFs). These effects are more significant when HFs are propagating in the direction of maximum horizontal stress. The proposed equation and the results from propagation of hydraulic fractures can be used in early stages of a hydraulic fracturing design.

Keywords: Crack opening displacement, hydraulic fracturing, higher order displacement discontinuity method, crack propagation.

***Corresponding author**

Abolfazl Moradi

Email: a.moradi@nioc.ir

INTRODUCTION

Application of horizontal drilling in conjunction with the hydraulic fracturing process has significantly improved the production of hydrocarbon reservoirs so that the extraction from otherwise uneconomical reservoirs is being possible today. Therefore, the propagation of hydraulic fractures has been the focus of many studies [1–5]. At least 6 important advantages will be achieved through horizontal drilling which includes:

- Reduction of water and gas coning.
- Increase of production rate
- Lower pressure drop around the wellbore
- Better sand production control
- Larger drainage coefficient pattern leading to higher recovery rate

Most studies in this field utilize linear elastic fracture mechanics (LEFM) principles to simulate the fracture propagation process. Stress intensity factor (SIF) and energy release rate form the foundation of LEFM theory. They are used to predict the fracture of a cracked brittle solid (such as rock) based on LEFM concept. In the theory of LEFM, the nonlinear effects are assumed to be minimal and negligible, and the cracked solid is treated as a linearly elastic medium [4]. However, the SIF, along with the energy release rate approach, would become futile under large-scale plastic deformations [3]. Several fracture parameters have been proposed to predict fracture of solids under nonlinear deformation conditions, such as the J integral which is a generalization of the energy release rate concept to nonlinear elastic materials by Rice [6], the crack opening displacement (COD) and the crack tip opening displacement (CTOD) by Wells [7,8], and the crack tip opening angle (CTOA). Wells in sequence of his two papers [7,8] proposed the use of COD as a fracture parameter. The COD criterion is equivalent to the effective stress intensity factor criterion under modest yielding conditions but can be extended to the large-scale yielding conditions when it is combined with the COD equation based on Dugdale model [8]. Aperture of a hydraulic fracture (HF) or COD plays an important role in the production of hydrocarbon reservoirs. The higher the value of COD is achieved in HFs; the higher production rate will be. However, most studies have been focused on HF propagation path and the parameters affecting the amount of COD have not been fully studied. While there are many analytically and numerically derived equations to estimate the SIFs and predict crack propagation paths and directions, there are not much of such equations in the case of COD prediction. The most well-known analytical solution for crack opening displacement proposed by Sneddon which only considers a horizontal crack under uniform internal pressure and far-field stress acting perpendicular to the crack faces. Once the crack rotates, or the far-field stress or stresses are not perpendicular to the crack faces, the Sneddon's analytical solution loses its applicability. Abdollahipour et al. [9] studied the COD of hydraulic fractures using discrete element method (DEM). They proposed an equation to predict COD in HFs in direction of one of the principal stresses. In the current study, a huge amount of numerical modellings has been carried out to study the effects of various parameters (Young's modulus E, Poisson's ratio ν, internal pressure P, far-field stresses σ_H and σ_h, crack inclination angle, θ and crack half-length, a) on COD. A multi-parameter regression is performed on theses numerical results in order to propose a relation between COD and the mentioned parameters. The effect of crack propagation and well radius on the maximum COD of HFs have been studied too.

EXPERIMENTAL PROCEDURES

Higher Order Displacement Discontinuity Method

Quadratic element displacement discontinuity which is based on analytical integration of cubic collocation shape functions over collinear or straight-line displacement discontinuity elements [10] is used to achieve higher accuracy. Many studies have shown a significant increase in accuracy of the method when higher order elements have been used [11–14]. The far field in-situ stresses are changed to local normal and shear stresses on each element using appropriate stress transition equations. Then the resulted normal or shear local stresses are summed up with the corresponding boundary conditions (if there is any stress boundary) to produce the actual stress state on each element.

Quadratic Displacement Discontinuity Element

The quadratic element displacement discontinuity is based on analytical integration of quadratic collocation shape functions over collinear, straight-line displacement discontinuity elements. Figure 1 shows the quadratic displacement discontinuity distribution, which can be written in a general form as:

$$D_i(\varpi) = N_1(\varpi)D_i^1 + N_2(\varpi)D_i^2 + N_3(\varpi)D_i^3,\ i = x, y \qquad (1)$$

Where, D_i^1, D_i^2, and D_i^3 are the quadratic nodal displacement discontinuities, and

$N_1(\varpi) = \varpi(\varpi - a_2 - a_3)/[(a_1 + a_2)(a_1 + 2a_2 + a_3)]$ and
$N_2(\varpi) = -(\varpi + a_1 + a_2)(\varpi - a_2 - a_3)/[(a_1 + a_2)(a_2 + a_3)]$
and

$$N_3(\varpi) = \varpi(\varpi + a_1 + a_2) / [(a_2 + a_3)(a_1 + 2a_2 + a_3)] \qquad (2)$$

are the quadratic collocation shape functions. It should be noted that a quadratic element has 3 nodes, which are the centers of the three elements within the batch element.

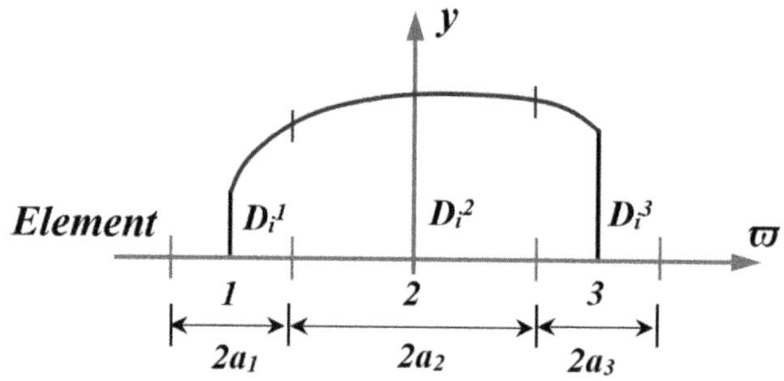

Figure 1: Quadratic collocation for the quadratic element displacement discontinuity.

Crouch has expressed the displacements and stresses for a line crack in an infinite body along the plane y= 0 (x-axis), in terms of single harmonic functions g(x,y) and f(x,y) [15], in which the displacements are

$$u_x = [2(1-\nu)f_{,y} - yf_{,xx}] + [-(1-2\nu)g_{,x} - yg_{,xy}] \text{ and}$$
$$u_y = [(1-2\nu)f_{,x} - yf_{,xy}] + [2(1-\nu)g_{,y} - yg_{,yy}] \qquad (3)$$

and the stresses are:

$$\sigma_{xx} = 2G_s[2f_{,xy} + yf_{,xyy}] + 2G_s[g_{,yy} + yg_{,yyy}] \text{ and}$$
$$\sigma_{yy} = 2G_s[-yf_{,xyy}] + 2G_s[g_{,yy} - yg_{,yyy}] \text{ and}$$
$$\sigma_{xy} = 2G_s[2f_{,yy} + yf_{,yyy}] + 2G_s[-yg_{,xyy}] \qquad (4)$$

Where $f_{,x}$, $g_{,x}$, $f_{,y}$, $g_{,y}$, etc. are the partial derivatives of the single harmonic functions f(x,y) and g(x,y) with respect to x and y, in which these potential functions for the quadratic element case can be found from:

$$f(x,y) = \frac{-1}{4\pi(1-\nu)} \sum_{j=1}^{3} D_x^j F_j(I_0, I_1, I_2) \text{ and}$$

$$g(x,y) = \frac{-1}{4\pi(1-\nu)} \sum_{j=1}^{3} D_y^j F_j(I_0, I_1, I_2) \qquad (5)$$

The common function F_j, is defined as:

$$F_j(I_0,I_1,I_2)= \int N_j(\varpi)\ln\left[(x-\varpi)+y^2\right]^{\frac{1}{2}}d\varpi, \quad j=1, \text{to } 3 \qquad (6)$$

Where the integrals I_0, and I_1 and I_2 are expressed as follows:

$$I_0(x,y)=\int_{-a}^{a}\ln\left[(x-\varpi)^2+y^2\right]^{\frac{1}{2}}d\varpi$$
$$=y(\varphi_1-\varphi_2)-(x-a)\ln(r_1)+(x+a)\ln(r_2)-2a$$

and

$$I_1(x,y)=\int_{-a}^{a}\varpi\ln\left[(x-\varpi)^2+y^2\right]^{\frac{1}{2}}d\varpi$$
$$=xy(\varphi_1-\varphi_2)+0.5\left(y^2-x^2+a^2\right)\ln\frac{r_1}{r_2}-ax$$

and

$$I_2(x,y)=\int_{-a}^{a}\varpi^2\ln\left[(x-\varpi)^2+y^2\right]^{\frac{1}{2}}d\varpi=$$
$$\frac{y}{3}(3x^2-y^2)(\varphi_1-\varphi_2)+\frac{1}{3}(3xy^2-x^3+a^3)\times$$
$$\ln(r_1)-\frac{1}{3}(3xy^2-x^3-a^3)\ln(r_2)-\frac{2a}{3}(x^2-y^2+\frac{a^2}{3}) \qquad (7)$$

The terms φ_1, φ_2, r_1 and r_2 in this equation are defined as:

$$\varphi_1=\arctan(\frac{y}{x-a}),\quad \varphi_2=\arctan(\frac{y}{x+a}),$$
$$r_1=\left[(x-a)^2+y^2\right]^{\frac{1}{2}}, and\ r_2=\left[(x+a)^2+y^2\right]^{\frac{1}{2}} \qquad (8)$$

Verification of Higher Order Displacement Discontinuity

The accuracy and applicability of the higher order displacement discontinuity method (using quadratic displacement discontinuity elements) is tested against two well-known problems (i.e. the problems of a central slant crack in an infinite body and a suddenly pressurized crack). The element length has been kept constant and equal to 5cm for all numerical analyses unless told otherwise.

Central Slant Crack in av Infinite Body

Central slant crack problem (see Figure 2) is used to verify the accuracy of the proposed quadratic formulation of the DDM. The slant angle, γ = 60° and a half crack length d =0.5 m are assumed in the analysis. A crack tip element length to half crack length ratio, L/d= 0.1 has been used for the numerical solution of the problem (L is crack tip element length). The analytical solution of the first and second mode stress intensity factors (SIFs), K_I and K_{II}, for the center slant crack problem are given as [16,17].

$$K_I=\sigma\sqrt{\pi d}\sin^2\gamma \text{ and}$$
$$K_{II}=\sigma\sqrt{\pi d}\sin\gamma\cos\gamma \qquad (9)$$

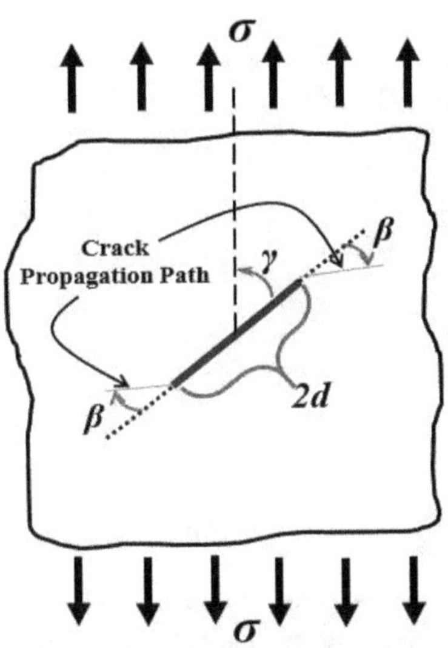

Figure 2: Central slant crack in an infinite plate.

Figure 3 compares the normalized analytical and numerical SIFs ($K_I/\sigma\sqrt{\pi a}$ and $K_{II}/\sigma\sqrt{\pi a}$) for various number of elements. The best results are achieved for 30 quadratic displacement discontinuity elements (i.e. element length of almost 3 Cm). For the higher number of elements, the crack tip element may loses its significance resulting in lower accuracy of SIF prediction [4]. Using 30 elements the error in estimating SIFs is less than 1%, showing the high accuracy of the proposed method.

A Suddenly Pressurized Crack

A suddenly pressurized crack in an infinite body is used to verify the proposed quadratic element formulation of the displacement discontinuity method for COD estimation.

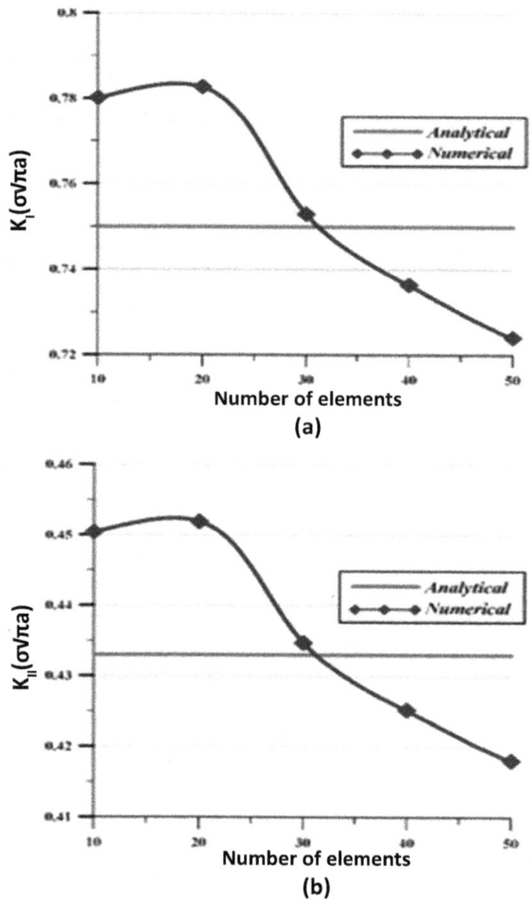

Figure 3: Comparison of numerical and analytical results of stress intensity factors a) Normalized K_I b) Normalized K_{II}.

Figure 4 shows the problem of a suddenly pressurized crack. A thin crack with length of 2a is subjected to constant internal pressure P. Sneddon calculated the exact amount of relative normal displacement of the crack surfaces (i.e. crack opening displacement or COD) in an elastic medium [18].

$$COD = \frac{4P\left(1-\nu^2\right)}{E}\sqrt{a^2-x^2} \qquad (10)$$

where $-a \leq x \leq a$.

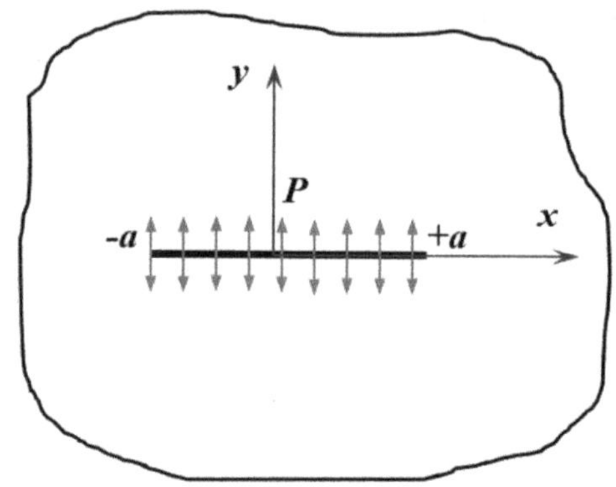

Figure 4: A suddenly pressurized crack problem.

Figure 5 compares the analytical and numerical predictions of COD. As it can be seen the numerical results are in good agreement with the analytical values. The error in COD estimation is less than 1% therefore, the proposed numerical model predicts COD with a significant accuracy.

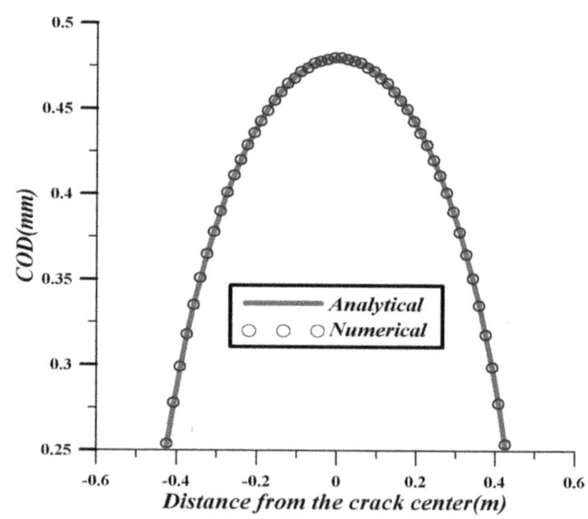

Figure 5: Comparison of the analytical values and numerical prediction of COD for a pressurized crack problem.

Crack Propagation Prediction

The pre-cracked rock-like cylindrical specimens with 60 mm diameters and 120 mm lengths are specially prepared by mixing the Portland Pozzolana cement (PPC), fine sands and water. The mechanical properties of the un-cracked rock-like specimens are obtained by using laboratory tests following ISRM standards. The mechanical properties used in the present analysis are: Compressive strength, σ_c=28 MPa; Young's modulus, E=15 GPa, Brazilian tensile strength, σ_t=3.81 MPa; Poisson's ratio, ν =0.21.

Some uniaxial compression tests are conducted on the rock-like specimens containing three cracks 1, 2 and 3. These cracks are created by inserting three thin steel shims with 10 mm width and 1 mm thickness in molds (before casting the specimens). The uniaxial compressive stress, σ was uniformly applied and the loading rate was kept at 0.5 MPa/s during the tests.

Figure 6 illustrates the geometry and loading condition of a pre-cracked specimen with three cracks. Cracks 1 and 2 (Iso-path cracks) have constant orientations of α= 60°. The third crack is oriented at different angles with respect to the direction of cracks 1 and 2 i.e. at angles β=0°, 45° and 90° (in a counterclockwise direction) (see Figure 6). Figure 7 demonstrates the experimental and numerical results of the specimens. As it can be seen the numerical results in Figure 7(b) have successfully captured the behavior of the experimental specimens in Figure 7(a) showing the accuracy and applicability of the numerical method in prediction of crack propagation.

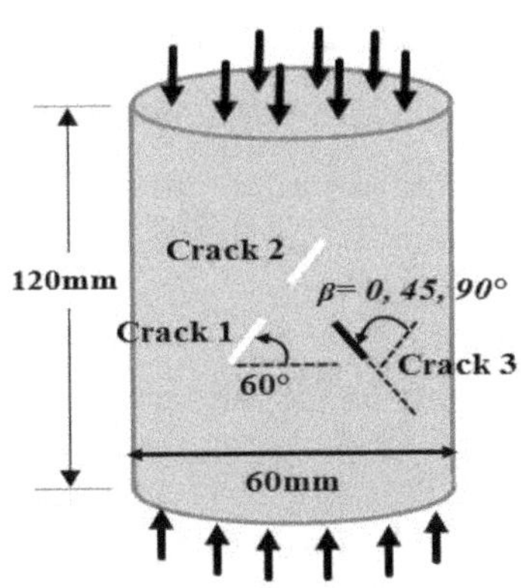

Figure 6: Geometry of the two cracks in a pre-cracked rock-like specimen under uniaxial compression.

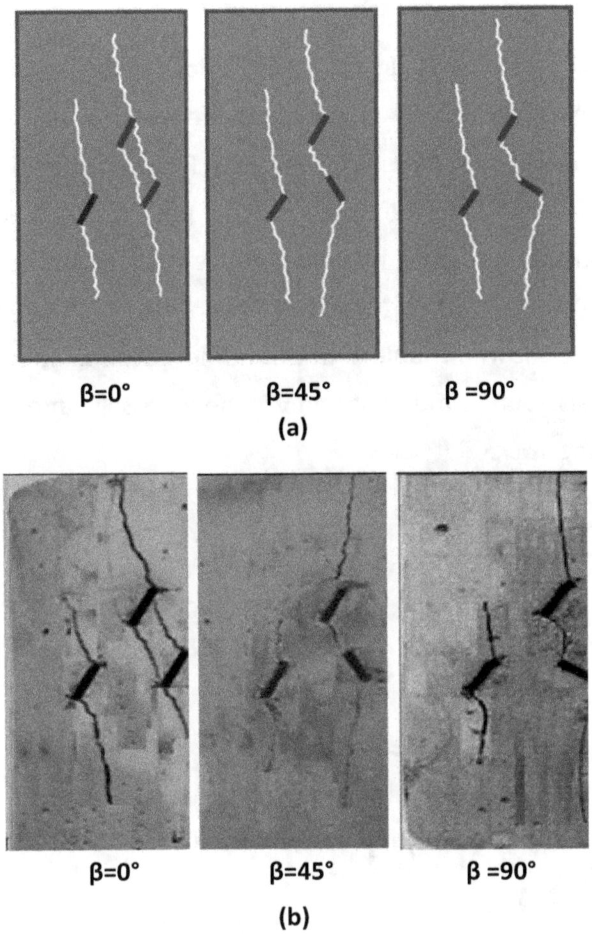

Figure 7: Comparison of crack propagation between experimental (a) and numerical(b) crack propagation for specimens (with three cracks) under uniaxial compression.

RESEARCH METHODOLOGY

A higher order displacement discontinuity method using quadratic element formulation is utilized to model the hydraulic fracture propagation. In this analyses, the rock mass is assumed to be impermeable and a uniform pressure of P is considered in the hydraulic fracture. For the case of far-field stresses, the maximum horizontal stress σ_H is applied in x-axis direction and the minimum horizontal stress σ_h is applied in y-axis direction (Figure 8). A fracture toughness of K_{IC}= $3MPa\sqrt{m}$ is assumed based on the average value of this parameter in many studies of reservoir conditions [19–24].

NUMERICAL MODELING

Almost 1500 different conditions have been modeled to fully capture the effect and behavior of each parameter on the COD. Figure 6 shows a schematic view of the model and parameters used in the numerical modeling. Angle θ changes from 0° to 75° with 15° intervals. The other parameters are shown in Table 1. In this numerical simulation, E is Young's modulus and v is the Poisson's ratio of the rock.

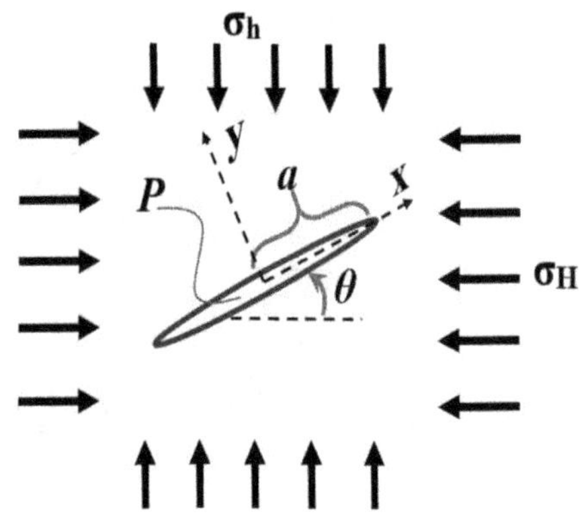

Figure 8: Schematic of models.

Table 1: Various values of parameters used in the numerical modeling.

P/σ_H	σ_h/σ_H	Crack half length(m)	E(GPa)	ν
1	0	0.25	30	0.20
1.5	0.35	0.50	40	0.25
2	0.70	0.75	50	0.30

Multivariate Regression

In order to propose a relation to predict COD by taking into account the effect of various parameters, a multivariate non-linear regression analysis is used. One of the classical problems in this statistical analysis is to find a suitable relationship between a response variable (in this case COD) and a set of regression variables (in this case, the parameters; E, v, P/σ_H, σ_h/σ_H, a, x, and θ) under suitable assumptions on the errors. The usual assumption is that the errors are independent, identically distributed random variables. In many situations the response function is a nonlinear function in both regression variables and the parameters (which is the case in the current model).

A merit function (which measures the agreement between the input data and a regression model) was chosen as follows:

$$\chi^2 = \sum_{i=1}^{N} \left\{ \frac{y_i - y(x_i)}{\sigma_i} \right\}^2 \qquad (11)$$

where σ_i is the standard deviation of the ith data point. y_i is the actual (numerical) value of COD, $y(x_i)$ is the predicted COD, x_i are independent variables. The Levenberg-Marquardt method is used for the initial estimation. Eq. (12) shows the best fit on the data. Coefficient of determination, R^2 for this equation is 94.35%. The standard error of the estimate is 4.37×10^{-4} further showing the goodness of the performed fit.

$$COD = 0.214 \frac{(1-\nu^2)(P-\sigma_h)(1+\cos^2\theta)}{2E\sigma_H}\left(\sqrt{a^2 - r^2}\right) \qquad (12)$$

where $r = \sqrt{x^2 + y^2}$ is the length along the crack direction, x and y are in crack local coordinates shown in Figure 8. It should be noted that for a horizontal crack (θ=0, y=0) the proposed equation simplifies to an equation very similar to the analytical solution proposed by Sneddon in Eq. (10).

As derived in an experimental study by Gruesbeck and Collins a crack width to average particle diameter of 3 will handle any concentration of sand (proppant) [25]. Equation 12 may be used in early stages of hydraulic fracturing treatment design to make sure the minimum desired width of hydraulic fracture has been obtained for safe passage of proppant.

A fracture propagation pressure or so called "fracture gradient" of order of 0.7 psi/ft is necessary to initiate the fracture propagation in hydraulic fractures [26].

One of the direct applications of the proposed equation in the paper is determination of dimensionless conductivity of a hydraulic fracturing treatment, leading to FOI, the expected folds of increase of productivity due to a fracture job[26].

RESULTS AND DISCUSSION

Effect of Crack Propagation on Initial Crack Length COD

The effect of crack propagation on COD of the initial crack is also investigated (for the same models of the previous section). The numerical results showed that propagation of HFs increases the maximum COD of the initial crack length. The relation is almost linear in all cases. As shown in Figure 9 for three different conditions, a linear equation is fitted with a high coefficient of determination (almost equal to 1). The x-axis stands for the ratio of new crack length to initial crack length and y-axis represents the maximum COD in the initial crack length. In the present analyses, for set (a) the parameters are $\sigma_h/\sigma_H=0.7$, $P/\sigma_H=1.5$, θ=0°, E=30GPa, ν=0.20, and a=0.50 m, for set (b) $\sigma_h/\sigma_H=0.7$, $P/\sigma_H=1.0$, θ=30°, E=50GPa, ν=0.25, and a=0.75 m and for set (c) $\sigma_h/\sigma_H=0.35$, $P/\sigma_H=1.0$, θ=60°, E=40GPa, ν=0.30, and a=0.25 m. As it can be seen, the value of COD has increased almost linearly for every increment of crack propagation regardless of the mechanical or geometrical properties of the model. It should be also noted that as the angle, θ increases, the effect of crack propagation on COD decreases (comparing the coefficient of x in the fitted equations shown in Figure 9). The maximum effect of crack propagation on COD is when the HF is in direction of maximum horizontal stress (i.e. θ=0°). This result is in agreement of many experimental and field studies showing perforation and crack propagation in direction of the maximum horizontal stress is the most efficient form of hydraulic fracturing [3,27-29].

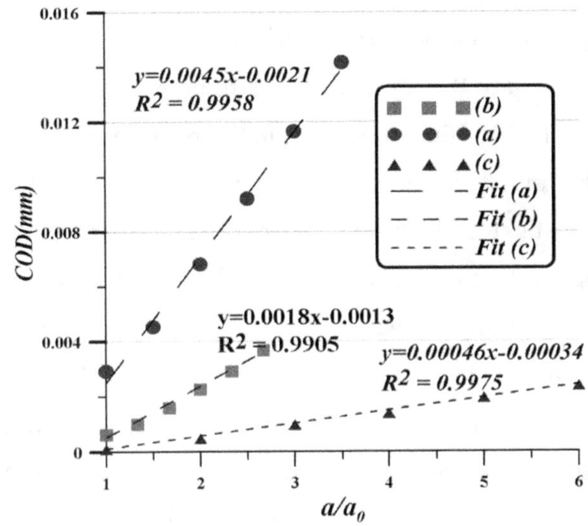

Figure 9: Increase of COD of initial crack with propagation.

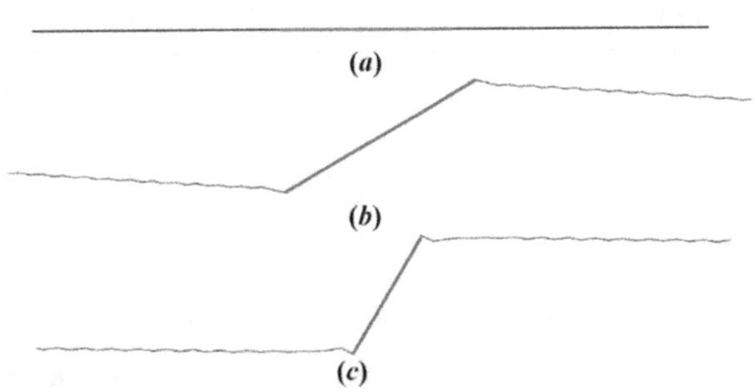

Figure 10: Crack propagation for data sets of Figure 9.

The increase of COD improves the production rate by increasing the volume of hydrocarbon entering the crack and finally reaching the well. Figure 10 shows the propagation of data set (c) in Figure 10. As expected, these results show that the crack propagation is in the direction of maximum horizontal stress. It should be mentioned that angle of initial crack in relation to stress field does affect crack propagation in early steps of the numerical simulation. However, this near crack tip effect is not clearly visible in Fig 10 since all cracks reorient themselves in direction of maximum stress as dictated by the theory of LEFM.

Effect of Well Radius on COD

A hydraulic fracture in a well stimulation process starts from the well wall and propagates through the hydrocarbon reservoir. It is interesting to investigate the effect of well size on the maximum obtainable COD. Therefore, in the present analyses the effect of well radius on the COD is also investigated. Five different radiuses are selected. Figure 11 shows a typical model used for analysis of well radius effects on COD. Radius R is changed from 0.1 m to 0.5 m with 0.1 m intervals. The internal pressure has been kept constant and equal to 60 MPa in all models. The maximum COD occurs at the contact of hydraulic fracture and the oil or gas well. The results show that the size of well has the same trend (but as will be explained in following, not the same significant effect) as the previous part. As it is shown in Figure 12 for various inclination angles, the maximum COD changes almost linearly with the well radius. As for the previous section comparing the coefficient of x in the fitted equations (Figure 12), the maximum effect of well radius is when HFs have the least angle with the maximum horizontal stress (i.e. for θ=30° in Figure 12).

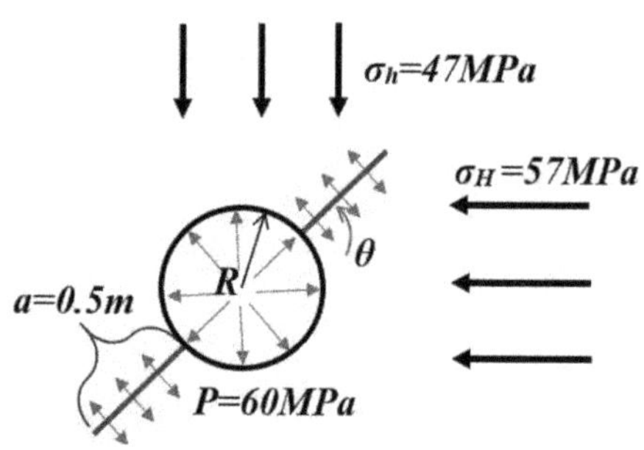

Figure 11: Schematic of the model for analysis of effect of well radius on maximum COD.

As seen in Figure 12 the value of COD has not changed substantially with well radius. This shows that well radius has marginal effect on COD.

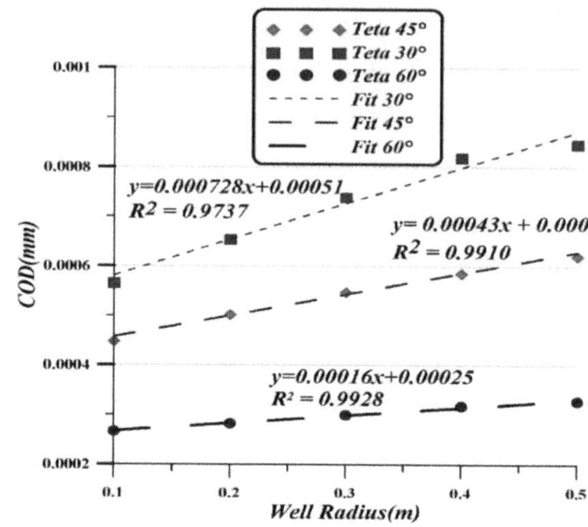

Figure 12: Effect of well radius on maximum COD of hydraulic fractures.

CONCLUSIONS

Crack opening displacement (COD) is an important parameter in fracture mechanics literature andhydraulic fracturing of hydrocarbon reservoirs. The existing analytical solution for COD considers a very simple problem limiting its applicability in the estimation of COD under a certain condition. In this paper, the effect of various parameters on COD was investigated. A higher order displacement discontinuity method was modified to analyze the different conditions. The applicability and accuracy of the model was tested against the two existing analytical solutions. The error in COD was less than 1%, showing the high accuracy of the proposed numerical model. Nearly 1500 models were built and run to capture the effect of all parameters.
An equation was fitted on the resulted predictions of COD of an HF in any arbitrarily condition. The coefficient of determination and standard error of the estimated COD were 94.35% and 4.37×10^{-4} respectively, showing a good agreement between

he fitted equation and numerical results. Since the numerical model was able to predict COD of the analytical problem with less than 1% error, the resulted equation is expected to predict COD of various conditions with an accuracy of the same order (there is a marginal difference between numerical and the proposed equation results which may negligibly increase the error of the proposed equation more than the numerical results). It was also observed that in case of crack propagation and increase of well radius, the COD increases almost linearly regardless of the other parameters in the model.

NOMENCLATURES

COD	: Crack Opening Displacement
HFs	: Hydraulic Fractures
SIFs	: Stress Intensity Factors

REFERENCES

1. Zhou D., Zheng P., He P., and Peng J., "Hydraulic Fracture Propagation Direction during Volume Fracturing in Unconventional Reservoirs," *J. Pet. Sci. Eng.*, **2016,** *141*, 82-89.
2. Bunger A. P., Gordeliy E., and Detournay E., "Comparison between Laboratory Experiments and Coupled Simulations of Saucer-shaped Hydraulic Fractures in Homogeneous Brittle-Elastic Solids," *J. Mech. Phys. Solids.,* **2013**, *61*, 1636-1654.
3. Abdollahipour A., Fatehi Marji M., Yarahmadi-Bafghi A., and Gholamnejad J., "Simulating the Propagation of Hydraulic Fractures from a Circular Wellbore Using the Displacement Discontinuity Method," *Int. J. Rock. Mech. Min. Sci.,* **2015**, *80*, 281–291.
4. Abdollahipour A., "Crack Propagation Mechanism in Hydraulic Fracturing Procedure in Oil Reservoirs," University of Yazd, **2015**.
5. Abdollahipour A., Fatehi Marji M., Yarahmadi-Bafghi A., and Gholamnejad J., "Numerical Investigation on the Effect of Crack Geometrical Parameters in Hydraulic Fracturing Process of Hydrocarbon Reservoirs," *J. Min. Environ.*, **2016,** *7*, 205-214.
6. Rice J. R., "A Path Independent Integral and the Approximate Analysis of Strain Concentration by Notches and Cracks," *ASME. J. Appl. Mech.*, **1968**, *35*, 379-386.
7. Wells A. A., "Unstable Crack Propagation in Metals: Cleavage and Fast Fracture," *Crack Propag. Symp.*, Cracnfield, UK, **1961**.
8. Wells A. A., "Application of Fracture Mechanics at and beyond General Yielding," *Br. Weld. J.*, **1963**, *11*, 563–570.
9. Abdollahipour A., Fatehi Marji M., Yarahmadi-Bafghi A. R., and Gholamnejad J., "DEM Simulation of Confining Pressure Effects on Crack Opening Displacement (COD) in Hydraulic Fracturing," *Int. J. Min. Sci. Technol.,* (In Press), **2016.**
10. Shou K. J. and Crouch S. L., "A Higher Order Displacement Discontinuity Method for Analysis of Crack Problems," *Int. J. Rock. Mech. Min. Sci. Geomech. Abstr.*, **1995**, *32*, 49-55.
11. Haeri H., Shahriar K., Fatehi Marji M., Moarefvand P., "A Coupled Numerical-experimental Study of the Breakage Process of Brittle Substances," *Arab. J. Geosci.*, **2013.**
12. Haeri H., Shahriar K., Fatehi Marji M., and Moarefvand P., "Investigating the Fracturing Process of Rock-like Brazilian Discs Containing Three Parallel Cracks under Compressive Line

Loading," *Strength Mater,* **2014**, *46*, 404-416.

13. Behnia M., Goshtasbi K., Fatehi Marji M., and Golshani A., "Numerical Simulation of Crack Propagation in Layered Formations," *Arab. J. Geosci.*, **2013**, *7*, 2729-2737.
14. Haeri H., Shahriar K., Fatehi Marji M., Moarefvand P., "A Boundary Element Analysis of the Crack Propagation Mechanism of Random Micro Cracks in Rock-like Specimens under Uniform Normal Tension," *J. Min. Environ.*, **2014**, *6*, 73-93
15. Crouch S. L., "Analysis of Stresses and Displacements around Underground Excavations an Application of the DDM," Minneapolis, **1976**.
16. Whittaker B. N., Singh R. N., and Sun G., "Rock Fracture Mechanics, Principles Design and Applications," Amsterdam: Elsevier, **1992**.
17. Tada H., Paris Paul C., and Irwin G. R. "The Stress Analysis of Cracks Handbook," 3rd ed., New York: ASME, **2000.**
18. Sneddon IN., "Fourier transforms," New York: McGraw-Hill Book Company, **1951.**
19. Abdollahipour A., Fateh-Marji M., Yarahmadi-Bafghi A., and Gholamnejad J., "Time-Dependent Crack Propagation in a Poroelastic Medium Using a Fully Coupled Hydromechanical Displacement Discontinuity Method," *Int. J. Fract.*, **2016**, *199*, 71–87.
20. Abdollahipour A., Marji M. F., Bafghi A. Y., and Gholamnejad J., "On the Accuracy of Higher Order Displacement Discontinuity Method (HODDM) in the Solution of Linear Elastic Fracture Mechanics Problems," *J. Cent. South University.*, **2016**, *23*, 2941–2950.
21. Hossain M. D. M. and Rahman M. K., "Numerical Simulation of Complex Fracture Growth during Tight Reservoir Stimulation by Hydraulic Fracturing," *J. Pet. Sci. Eng.*, **2008**, *60*, 86–104.
22. Suarez Y. a., Chen Z., Rahman M. K., Rahman S. S., and et al., "Unsuccessful Hydraulic Fracturing Cases in Australia: Investigation into Causes of Failures and their Remedies," *J. Pet. Sci. Eng.*, **2007**, *57*, 70–81.
23. Rahman M. M., Rahman M. K., and Rahman S. S., "An Integrated Model for Multiobjective Design Optimization of Hydraulic Fracturing," *Pet. Sci. Eng.*, **2001**, *31*, 41–62.
24. Rahman M. M., Hossain M. M., Crosby D. G., and Rahman M. K., "Analytical, Numerical and Experimental Investigations of Transverse Fracture Propagation from Horizontal Wells," *J. Pet. Sci. Eng.*, **2002**, *35*, 127–150.
25. Gruesbeck C. and Collins R. E. "Particle Transport through Perforations," *SPE J.*, **1978,** *22*, 857–865.
26. Smith M. B. and Montgomery C. T., "Hydraulic Fracturing," London, **2015**.
27. Yew C. H., Mear M. E., Chang C. C., and Zhang X. C., "On Perforating and Fracturing of Deviated Cased Wellbore," *SPE Annu. Tech. Conf. Exhib.*, Houston, TX: Society of Petroleum Engineers, **1993**.
28. Craig D. P., Brown T. D., and Ely J. W., "Field Investigation of Heat Transfer in Hydraulic Fractures and the Effect of Heat Transfer on Fracturing Fluid Design," *Annu. Tech. Conf. Exhib. SPE,* Denver: Society of Petroleum Engineers, **1996.**
29. Pearson C. M., Bond A. J., Eck M. E., and Schmidt J. H., "Results of Stress-oriented and Aligned Perforating in Fracturing Deviated Wells," *J. Pet. Technol.*, **1992**, *44*, 10-18.

Comparison of the Separation Characteristics of Polyvinyl Alcohol Membrane in Dehydration of Benzene and Toluene Using Pervaporation

Ghasem Khatinzadeh*, Ahmad Roozbahani, and Ali Mehdizadeh

Chemical, Polymeric & Petrochemical Technology Development Research Division, Research Institute of Petroleum Industry (RIPI), Tehran, Iran

ABSTRACT

Pervaporation (PV) is a well-established membrane technique used in separation; especially dehydration of organic solvents. Along with other parameters such as permeate pressure and feed temperature; feed composition is an important parameter, which affects separation characteristics and membrane performance in pervaporation. In this paper, the separation characteristics of a hydrophilic polyvinyl alcohol (PVA) membrane in the dehydration of benzene and toluene by pervaporation in the temperature range of 30-60 °C and permeate pressure of 7 mmHg have been compared. The results show that the difference in the molecular sizes of benzene and toluene and their mutual solubilities with water cause differences in separation parameters, so that for example at 50 °C, in dehydration of benzene water flux and water concentration in the permeate are 0.53 (g/hm^2) and 2.03 (wt.%) respectively, while the corresponding values in toluene dehydration are 0.61 (g/hm^2) and 3.19 (wt.%) respectively. Therefore these amounts result in higher pervaporation separation index (PSI) for toluene compared to benzene.

Keywords: Pervaporation, Benzene, Toluene, Dehydration, Poly (vinyl alcohol).

INTRODUCTION

In order to avoid deactivation of catalyst and/or corrosion of equipment, it is often necessary to remove water from solvents during their application in chemical processing industries. The traditional processes for dehydration of solvents such as distillation have operational difficulties and consume large amounts of energy [1]. Pervaporation is an efficient membrane-based process used in the economical separation of liquid mixtures given some of its advantages including highly effective separation, no pollution, less installation space and energy saving [2]. This method has attracted the attention of specialists in chemical and related fields such as biochemical and petrochemical industries. Separation of liquid mixtures by partial vaporization through a membrane (non-porous or porous) is the separation principle in pervaporation. The driving force for the pervaporation process is the difference in chemical potential, corresponding to the concentration gradient between phases on the opposite sides of the interfacial barrier. The

***Corresponding author**
Ghasem Khatinzadeh
Email: khatinzadehgh@ripi.ir

separation in PV occurs because of the different rates of sorption and diffusion of the feed components through the membrane [3]. The main industrial application of pervaporation even today is the dehydration of organic liquids; especially low molecular weight alcohols such as ethanol and isopropanol [4-6]. However, water removal from other solvents has also been considered in recent years [7-8].

Among aromatic compounds, dehydration of benzene using pervaporation has been focused on. Yu Lixin et al. have used PVA hydrophilic membrane in benzene dehydration at a temperature of 70° C and pressure of 800 (Pa) [9]. In addition, Gutch et al. have used aromatic polyamide membranes for the same purpose [10]. Li et al. have studied benzene dehydration in laboratory and pilot scales using modified hydrophilic membranes in 2002 [11]. Khatinzadeh et al. have studied benzene dehydration at low permeate pressures [12].

In dehydration of organic solvents by PV, the membrane should have sufficient affinity for water molecules so that water can be preferentially sorbed and transported at a rapid rate affecting an efficient separation. Due to high hydrophilicity, PVA has a greater tendency for polar water molecules than less polar organic molecules [13].

Although the solubility and permeability of water molecules in the membrane play fundamental and determining roles in the performance of hydrophilic membranes, the nature of organic solvent molecules can considerably affect these parameters. For example, in the dehydration of tetrahydrofuran and 1,4-dioxane using several hydrophilic membranes, Vijaya Kumar Naidu et al. found out that water flux values are higher for water+THF mixtures than water+1,4-dioxane mixtures. They concluded that the flux for water increases steadily for water+THF mixtures probably due to a higher preferential interaction of water with the membrane in the presence of THF compared with 1,4-dioxane [14]. In other words, the differences between the properties of 1,4- dioxane and THF in the presence of water lead to varying interactions of water molecules with the membrane, resulting in the difference in the selectivity of the membrane to water molecules. This effect has been considered in water removal from benzene and toluene by pervaporation using polyvinyl alcohol membrane in this work. Although benzene and toluene possess very similar chemical properties, the difference in their molecular sizes and water solubility may affect membrane selectivity to water molecules in the presence of these solvents. The roles of the latter parameters have not been studied much in previous works on dehydration of aromatic solvents. In addition, the effect of temperature, a very effective parameter on membrane performance, has been studied in this work.

EXPERIMENTAL PROCEDURES

Figure 1 shows the experimental set-up used in pervaporation experiments.

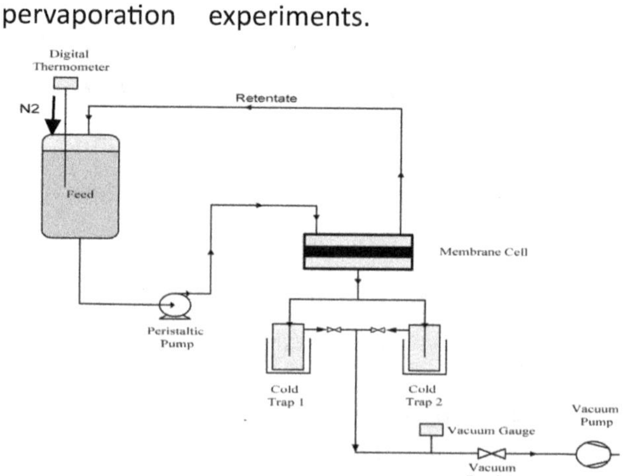

Figure 1: Schematic diagram of the experimental apparatus.

The membrane module used in this study was of a plate-frame type made of stainless steel, and the membrane was supported on a porous disc with an effective membrane area of 177 cm^2. In all experiments, a commercially available membrane from Sulzer Chemtech (PERVAP 2201) was used. The membrane was a composite one with a hydrophilic top layer of PVA (thickness=2 μm) on a polyacrylonitrile (PAN) porous support (thickness=80 μm). Before starting the PV experiments, the test membrane was equilibrated with the feed solutions for about 3 hrs. In all pervaporation experiments, the feed was circulated over the membrane with a flow rate of 1.1 L/min using a peristaltic pump. A temperature probe attached to the feed tank was used to monitor the temperature during each run. The feed tank was maintained in the 30-60 °C temperature range by using a temperature controller. The permeate side of the membrane was maintained at the desired pressure using a vacuum pump. The permeate samples were condensed and collected in cold traps immersed in a liquid nitrogen bath. The water content of the samples including feeds permeates and retentates was determined by Karl Fischer method (ASTM D1533) using a Metrohom-831 KF Coulometer. Permeate fluxes (g/hm^2) were gravimetrically obtained using Eq. (1):

$$J=M/At \tag{1}$$

where J is the permeate flux, M is the permeate weight (g), A (m^2) is the membrane area and t (hr) is operation time. Separation factor (α) was evaluated using Eq. (2):

$$\alpha = \frac{Y_W / Y_B}{X_W / X_B} \tag{2}$$

where α is the separation factor and X and Y are the mass fractions of the feed and the permeate, respectively. Subscripts W and B stand for water and benzene or toluene, respectively.

Benzene and toluene used in these experiments were provided by Isfahan Petrochemical Complex and contained 150 ppm of water, as determined by Karl Fischer method. All experiments were carried out in N_2 atmosphere in order to prevent moisture from entering the system.

RESULTS AND DISCUSSION

Figure 2 shows the performance of the membranes in the dehydration of benzene and toluene in terms of water concentration in the permeate at different feed temperatures and constant permeate pressure (7 mmHg). As observed, the water concentration in the permeate is markedly higher than that in the feed. Therefore, the membrane is more selective toward water than benzene and toluene molecules.

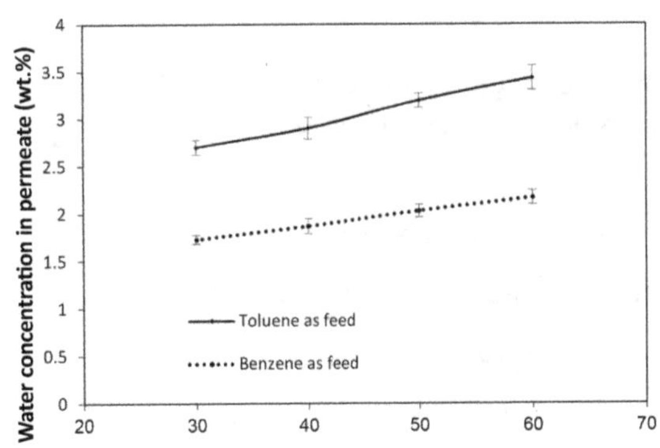

Figure 2: Water concentration in the permeate as a function of feed temperature (permeate pressure: 7 mmHg).

It is well known that hydrophilic membranes such as PVA can develop hydrogen bond interactions with water leading to preferential sorption and diffusion of water through the barrier membrane compared to organic solvents [15]. Furthermore,

Figure 2 shows well that water concentration in the permeate in toluene dehydration is greater than that of benzene dehydration. In other words, the membrane exhibits higher selectivity to water molecules in the presence of toluene compared with benzene. The investigation of water flux values also verifies this.

The water fluxes in the dehydration of toluene and the benzene as a function of temperature are shown in Figure 3. Water flux values are higher for toluene compared with benzene.

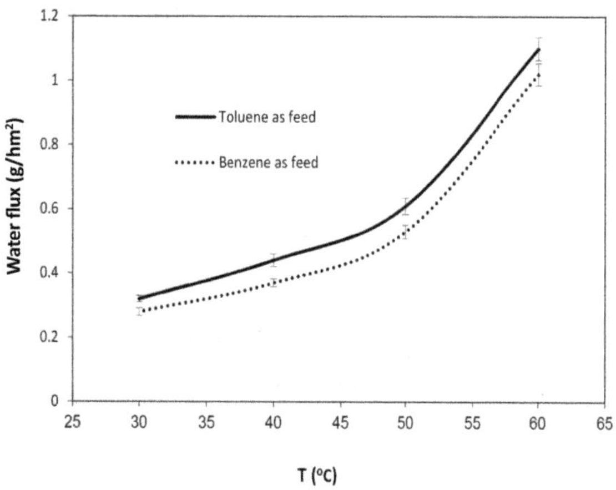

Figure 3: Water fluxes as a function of feed temperature (permeate pressure: 7 mmHg).

The transport of components of a mixture across the membrane in pervaporation is described by solution-diffusion model resulting from these processes in series: (1) solution of component molecules in the upstream surface of the membrane, (2) diffusion of the dissolved species across the membrane matrix and (3) desorption of the dissolved species in the downstream face of the membrane. These three fundamental processes also govern the mass transport across pervaporation membranes [16]. Therefore, according to the solution-diffusion mechanism, the permselective properties of pervaporation membranes are determined by solubility and diffusivity of the permeating components in the membrane [17].

$$\text{Permeability } (P) = \text{solubility } (S) \times \text{diffusivity } (D) \quad (3)$$

The solubility of a compound is the amount sorbedby the membrane under equilibrium conditions and is hence a thermodynamic parameter unlike diffusivity, which is a kinetic parameter quantifying the rate of permeation through the membrane. Solubility and diffusivity affect the selectivity as well as flux. Since both sorption and diffusion phenomena are dependent on the composition of the liquid mixture, the permeation properties are strongly influenced by the feed composition.

Therefore, the difference in the selectivity of the membrane to water molecules in the presence of benzene and toluene can be investigated from both solubility and diffusivity aspects. From the viewpoint of the solubility of water molecules in the membrane, it seems that the hydrophilic PVA membrane has a more effective interaction with water molecules in the presence of toluene, which may be due to the different effects of the molecules of these organic solvents on water molecules.

It is well established that the mutual solubilities of benzene-water are markedly larger than that of an alkane of similar size and an aromatic compound such as toluene or xylene. The mutual solubilities of some hydrocarbons with water are shown in Table 1. The relatively large mutual solubility of benzene with water compared with those of alkanes and aromatics is caused by the stronger solute-solvent Van der Waals interaction [18].

Table 1: Mutual solubilities of hydrocarbons and water at 25° C [19].

Compound	The solubility of hydrocarbon in water (ppm)	The solubility of water in hydrocarbon (ppm)
Benzene	1755.00	691
Toluene	573.00	543
o-Xylene	213.00	456
n-Hexane	12.40	90
n-Heptane	3.37	82

Therefore, the large mutual solubilities of water-benzene compared to water-toluene diminish the solubility of water molecules in the hydrophilic membrane, which results in lower water fluxes on the basis of the solution-diffusion mechanism. The increased water concentration in the permeate in both dehydration processes (Figure 2) indicates that water permeability (ratio of water to organic solvent flux) is increased by increasing feed temperature, as shown in Figure 4. In other words, the performance of the membrane is enhanced by increasing the feed temperature in water removal from both solvents. The closeness of the slopes of water permeability vs. temperature curves shows that water to solvent flux variations is close for both solvents at different temperatures.

Temperature is an important operating parameter in pervaporation because it affects both the sorption and diffusion rates. By increasing the temperature, the thermal motion of the polymer chains of the membrane increases, which increases the interspaces between the polymer chains. This can result in the enlargement of free volume inside the membrane, leading to easy permeation of the constituents through the membrane. The vapor pressures of all the components in the feed mixture increase with increasing the feed temperature, but the vapor pressure at the permeate side is not affected. As a result, the driving force increases with increased feed temperature and hence the permeation flux increases as well [20].

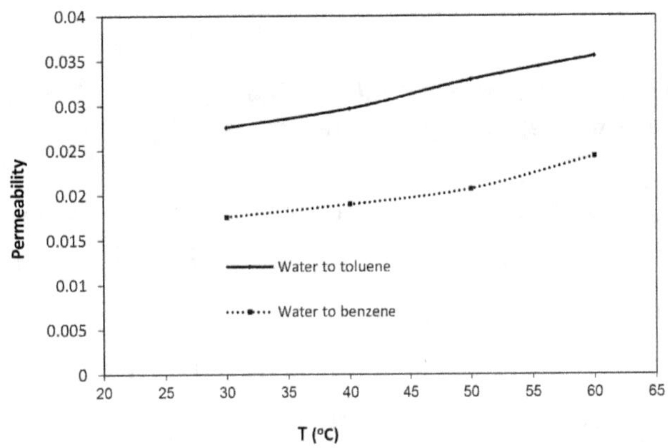

Figure 4: Water permeability in dehydration of benzene and toluene as a function of feed temperature (permeate pressure: 7 mmHg).

Increased water permeability at higher temperatures may be due to the increase in solubility, diffusivity or both. Furthermore, one-component permeability is dependent on the activation energy, according to Eq. (4):

$$J = J_0 exp(\frac{-E_p}{RT}) \quad (4)$$

where J_0 is permeability constant, E_p represents activation energy and R and T are the feed temperature and gas constant, respectively [21]. Therefore, water has a higher permeation at higher temperatures compared with toluene and benzene due to its larger activation energy.

Figure 5 shows benzene and toluene fluxes vs. temperature. Both benzene and toluene fluxes increase with an increase in temperature.

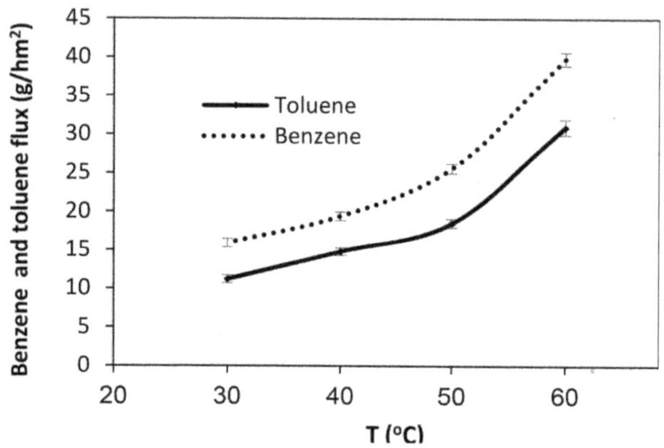

Figure 5: Benzene and toluene fluxes as a function of feed temperature (permeate pressure: 7 mmHg).

The difference between benzene and toluene fluxes is another reason for the different performances and selectivity of the membrane in the dehydration of these solvents. However, considering the hydrophilic character of the membrane and the small difference between the polarities and solubilities of toluene and benzene in the membrane, it seems that diffusion is the dominant factor in the flux differences of these two non-polar solvents. Diffusivity depends on the geometry of the membrane as well as its glassy rubber state. A small molecule will diffuse through a membrane more easily than a large one. Furthermore, diffusivity depends on the geometry of the penetrant. In other words, as the molecular size increases, the diffusion coefficient decreases. Dependency on size is determined by Stokes-Einstein equation presented in Eq. (5), which is the relation between frictional resistance and the radius of the diffusing component [22]:

$$f = 6\pi\eta r \qquad (5)$$

The diffusion coefficient is inversely proportional to the frictional resistance as:

$$D = \frac{KT}{f} \qquad (6)$$

According to Eqs. (5) and (6), the smaller size of the diffusing component causes frictional resistance and thus increases permeability coefficient.

The difference in the molecular size of benzene (0.59 nm) and toluene (0.68 nm) favors the permeability of the smaller molecule. In fact, the smaller benzene molecules can penetrate more easily than toluene molecules.

High benzene flux compared with toluene reduces membrane selectivity to water molecules, according to Eq. (2). Membrane performance is a function of water flux and selectivity to water molecules. The pervaporation separation index (PSI) was used to evaluate the overall PV performance [23]. This was calculated using Eq. (7):

$$PSI = J\,(\alpha-1) \qquad (7)$$

where J and α are flux and selectivity, respectively.

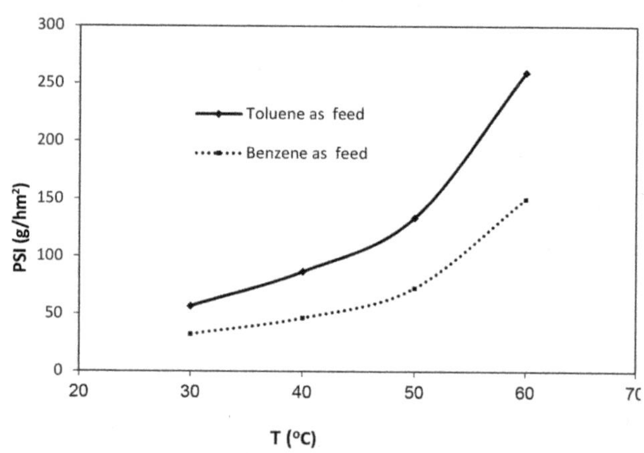

Figure 6: PSI for dehydration of toluene and benzene as a function of feed temperature (permeate pressure: 7 mmHg).

PSI values in Figure 6 show that the hydrophilic PVA membrane performs better in toluene dehydration compared to benzene in the temperature range of 30-60 °C. Thus, the differences in molecular size and mutual solubilities of these solvents with water lead to different separation characteristics, which cause different performances of polyvinyl alcohol hydrophilic membrane in the dehydration of benzene and toluene.

CONCLUSIONS

The higher concentration of water in the permeate in the dehydration of toluene by pervaporation indicates the more effective performance of PVA membrane in water removal from toluene compared to benzene, so that at 60 °C for toluene, PSI is 259.6 (g/hm^2) while the corresponding value for benzene is 150 (g/hm^2). In fact, the lower mutual solubility of toluene with water causes greater solubility of water molecules in the membrane, which results in more water permeability, based on the solution-diffusion mechanism. Furthermore, due to the higher diffusion coefficient, the smaller benzene molecules can diffuse more easily than toluene molecules, which causes higher benzene flux and lower concentration of water in the permeate. For example, at 60 °C in benzene dehydration, the water content in the permeate was 2.17 (wt. %) while the corresponding value in toluene dehydration was 3.43 (wt. %). However, the slopes of water permeability vs. temperature are close for both solvents at different temperatures.

On this basis, mutual solubility and molecular size considerably affect membrane performance. Moreover the data obtained in this work greatly help select the appropriate membrane for dehydration of aromatic compounds. However, the pervaporation process at low pressures is highly expensive. In addition obtaining supplementary data together with mass, energy and equipment cost calculation in order to optimize the separation conditions are necessary.

NOMENCLATURES

PV	: Pervaporation
PVA	: Polyvinyl Alcohol
PSI	: Pervaporation Separation Index

REFERENCES

1. Joshi S. and Fair J. R., "Adsorptive Drying of Hydrocarbon Liquids," *Ind. Eng. Chem. Res.*, **1991**, *30*(1), 177-185.
2. Neel J. and Huang R. Y. M., "Introduction to Pervaporation. In: Pervaporation Membrane Separation processes," Amsterdam, *the Netherlands: Elsevier Science*, **1991**, 36-37.
3. Bruggen B. V. and Luis P., "Pervaporation, in Progress in Filtration and Separation," R. J. Wakeman, Ed., Chapter 4, Elsevier Scientific, 1979, University of California, **2015.**
4. Ling L. Y., Chieh H. C., Yih L. J., and Ling L. Y., "Crosslinked Polybenzoxazine Based Membrane Exhibiting In-situ Self-promoted Separation Performance for Pervaporation Dehydration on Isopropanol Aqueous Solutions," *J. Membr. Sci.*, **2017**, *531*, 10-15.
5. Liu J. and Bernstein R., "High-flux Thin-film Composite Polyelectrolyte Hydrogel Membranes for Ethanol Dehydration by Pervaporation," *J. Membr. Sci.*, **2017**, *534*, 83-91.
6. Paramita D., Ray S.K Kuila S. B., Samanta H. S., and Singha N. R., "Systematic Choice of Crosslinker and Filler for Pervaporation Membrane: A Case Study with Dehydration of Isopropyl Alcohol–water Mixtures by Polyvinyl Alcohol Membranes," *Sep. Purif. Technol.*, **2011**, *81*(2), 159-173.
7. Chen M., Wu X., Soyekwo F., Zhang Q., and et al., "Toward Improved Hydrophilicity of Polymers of Intrinsic Microporosity for Pervaporation

Dehydration of Ethylene Glycol," *Sep. Purif. Technol.*, **2017**, *174*, 166-173.

8. Jen Han Y., Chiung Su W., Yih Lai J., and Ling Liu Y., "Hydrophilically Surface-modified and Crosslinked Polybenzimidazole Membranes for Pervaporation Dehydration on Tetrahydrofuran Aqueous Solutions," *J. Membr. Sci.*, **2015**, *475*, 496-503.
9. Lixin Y., Cuixian C., Liping Z., Jun Y., and Weijun J., "Study on the Dehydration of Organic Solvents by Pervaporation and the Mass Transfer Mechanism of Pervaporation," *Tsinghua Science and Technology*, **1996**, *1*, 351-356.
10. Gutch P. K., Pandey L. K., and Saxena C., "Dehydration of Benzene through Fluorine Containing Aromatic Polyamide Membrane by Pervaporatio," *J. Appl. Polym. Sci.*, **2008**, *110*, 203-209.
11. Li J., Chen C., Han B., and Peng, Y., "Laboratory and Pilot-scale Study on Dehydration of Benzene by Pervaporation," *J. Membr. Sci.*, **2002**, *203*, 127-136.
12. Khatinzadeh G., Mhadyarfar M., Mehdizadeh A., Esmailzadeh A., and et al., "Effect of Permeate Pressure and Feed Flow Rate on Benzene Dehydration by Pervaporation," *J. Pet. Sci. Technol.*, **2016**, *6*(2), 30-36
13. Immelman E., Sanderson R. D., Jacobs E. P., and Van Reenen A. J., "Poly(vinyl alcohol) Gel Sublayers for Reverse Osmosis Membranes. I. Insolubilization by Acid-catalyzed Dehydration," *J. Appl. Polym. Sci.*, **1993**, *50*, 1013-1034.
14. Vijaya K. N. B., Krishna R. K.S. V., and Aminabhavi T. M., "Pervaporation Separation of Water + 1,4-dioxane and Water + Tetrahydrofuran Mixtures Using Sodium Alginate and its Blend Membranes with Hydroxyethylcellulose-A Comparative Study," *J. Membr. Sci.*, **2005,** *260,* 131–141.
15. Kuan Y. H., Ren J., Hua M. X., and Liang X. Z., "Dehydration of Ethyl Acetate Aqueous Solution by Pervaporation Using PVA/PAN Hollow Fiber Composite Membrane," *Desalination*, **2011**, *280*, 252–258.
16. Binning R. C., Lee R. J., Jenning J. F., and Martin E. C., "Separation of Liquid Mixtures by Permeation," *Ind. Eng. Chem.*, **1961**, *53*, 6-50.
17. Wijmans J. G. and Baker R. W., "The Solution-Diffusion Model: A Review," *J. Membr. Sci.*, **1995**, *107*, 1-27.
18. Graziano G., "Benzene Solubility in Water: A Reassessment," *Chem. Phys. Lett.*, **2006**, *429,* 114-118.
19. Polak j. and Benjamin C. Y. L., "Mutual Solubilities of Hydrocarbons and Water at 0 and 25 °C," *Can. J. Chem.*, **1973**, *51*, 4018-4023.
20. Svang-Ariyaskul A., Huang R. Y. M., and Douglas P. L., "Blended Chitosan and Polyvinyl Alcohol Membranes for the Pervaporation Dehydration of Isopropanol," *J. Membr. Sci.*, **2006**, *280*, 815-823.
21. Cabasso I., Grodzinski J. J., and Vofsi D., "Polymeric Alloys of Polyphosphonates and Acetyl Cellulose: Sorption and diffusion of benzene and cyclohexane," *J. Appl. Polym. Sci.*, **1974**, *18*, 2117-2136.
22. Jyoti G., Keshav A., and Anandkumar J., "Review on Pervaporation: Theory, Membrane Performance, and Application to Intensification of Esterification Reaction," *Journal of Engineering*, **2015**, 1-24.
23. Nawawi M. G. and Huang R.Y.M., "Pervaporation Dehydration of Isopropanol with Chitosan Membranes," *J. Membr. Sci.*, **1997**, *124*, 53-62.

17

An Elastomer Additive Improving Elastic Properties of Heavy Weight Oil Well Cement: A Laboratory Study

Hamid Soltanian[1]*, Reza Khalokakaie[1], Mohammad Ataei[1], and Ezatallah Kazemzadeh[2]

[1] School of Mining, Petroleum and Geophysics Engineering, University of Shahrood, Iran
[2] Division of Petroleum Engineering, Faculty of Upstream Petroleum Industry, Research Institute of Petroleum Industry (RIPI), Tehran, Iran

ABSTRACT

The use of elastomer additives to solve the problems in oil well cementing has been investigated in recent years by several research groups in the petroleum industry. This study includes the laboratory examination of the effect of elastomer additives on the physical properties of heavy-weight oil well cement. In the research process, a candidate well is selected and the properties of the cement slurry used in a problematic section of the well are tested in the laboratory. Then, elastomer additives are added as an elastic agent and the improvements in the cement slurry and stone properties are studied. This article discusses the problems associated with the conventional heavy-weight oil well cement used in the candidate well and reports the detail of the improvements in cement properties obtained by adding an elastomer additive to the cement slurry formulation as an elastic agent. These properties include cement slurry rheological properties, free water, fluid loss, thickening time, cement stone elasticity properties, and compressive strength. The elastomer additive increases the yield point and plastic viscosity, but it decreases the free water and fluid loss of cement slurry. In addition, the cement stone compressive strength decreases; however, there is an optimum concentration of the elastomer additive at which the maximum compressive strength is reached. Moreover, the elasticity properties of the cement stone are improved and a lower value for the Young's modulus and a higher value for the Poisson's ratio are achieved. The theories supporting the results are discussed in the discussion section. The results of this study can be used to optimize the cement slurry design in any given set of conditions.

Keywords: Elastomer Additive, Oil Well Cement, Compressive Strength, Young's Modulus, Poisson's Ratio

INTRODUCTION

Primary cementing in oilfield applications includes the placement of cement slurry between the drilled formations and the well casing. The main goals of well cementing are casing protection and zonal isolation [1-4]. Zonal isolation must be maintained throughout the lifetime of the well. Correct cement design and cement placement are the key factors in achieving the primary cementing goals. The required properties for the cement stone and slurry are determined by the well conditions. These properties include cement slurry weight, cement slurry rheology, cement

***Corresponding author**
Hamid Soltanian
Email: soltanianh@ripi.ir

slurry thickening time, cement stone compressive and tensile strength, cement stone elasticity properties, and cement stone durability. The set cement sheath should withstand the stresses induced by the well events and maintain integrity during the life of the well [5-8]. Mechanical forces from the adjacent formations may damage the cement sheath and casing string. It has been found that in order to prevent mechanical damage to the cement, the best type of cement in terms of mechanical durability is the one with higher elasticity properties than the adjacent formation. The uniform distribution of the in situ stress acting upon the outside of the cement sheath can be achieved by a softer and more ductile cement stone, which means lower values of Young's modulus and higher value of Poisson's ratio. This will also enhance the collapse resistance of the casing string [9-11]. Successful application of high-density elastic cements to solve HP/HT challenges in the south Texas is reported in the literature [11- 12]. The elastic properties of the cement stone can be improved by adding an elastomer additive in low volumetric fractions to the cement formulation. Similar works are carried out by other researchers to improve the elastic properties of medium weight oil well cement by adding latex and fiber additives. In studies conducted by Khalaf (1985), the effects of styrene-butadiene latex and fiber polymers on the elastic behavior of the cement stone are verified. Five light- to medium-weight cement slurry samples with different formulations are prepared and the mechanical properties of the cement stones are determined. Figure 1 shows the mechanical behavior of the five cement stone samples, cement stone without additives, standard cement stone, cement stone containing fiber, cement stone containing latex, and cement stone containing fiber and latex. From Figure 1, cement stones including fiber and latex show higher tensile strength and more elastic behavior comparing with other cement stones [13]. In studies conducted by Le Roy-Delage (2000), in order to design proper cement slurries for long term zonal isolation purposes, common oil well cement additives with elastic additives and expanding additives are used. Table 1 shows the Young's modules and compressive strength of 16 ppg cement stones cured at 114°C. From the Table 1, in cement samples containing an elastic additive Young's modules and compressive strength are decreased, but the cement stone elastic behavior is increased [6]. Heavy weight oil well cements with proper mechanical properties are used by Barry Wray in 2009 in the south Texas HP/HT oil wells to increase the oil well life time. Nitrogen gas as an impact damper to increase the mechanical strength and fiber additives to increase the tensile strength of the cement stones are used in the cement slurry formulations. The formulations of the cement slurries are not reported in the article and just the cementing operation results are analyzed by strain analysis software [12]. None of the above examples is suitable for Iran's HP/HT oil and gas well conditions because of low cement slurry weight. In this study, for the first time, a new different elastomer additive is used in heavy weight oil well cement. Maroon oil field has been known in the oil industry for its HP/HT Gachsaran formation. Drilling wells to safely and effectively extract the contents of these reservoirs often proves to be a formidable exercise. In addition to the difficulties of dealing with high pressure and temperature, low-pressure zones can also be present. In recent years, well designs have also become more complex as fields mature and operators search for new sources of oil and gas. Cementing the casing and liner strings in place in these wells is a critical phase in the well construction process. The cement sheath is vital since it serves to protect and support the casing and deliver the necessary zonal isolation for the well. Once the well is completed, a whole new set of challenges that can stress an initially competent cement sheath are encountered. For example, some Maroon field HP/HT wells require high-pressure fracturing operations to unlock their full production potential, while, in other wells, the heavy weight drilling fluid, with which the well is drilled, is swapped out to lightweight brine at some point during the well completion process. In both cases, the loads imposed on the cement sheath can be extreme, potentially leading to debonding and even the catastrophic failure of a conventional cement sheath. When this occurs, the situation generally cannot be effectively remediated, and a well operator might find that they have lost a wellbore which is costly to construct.

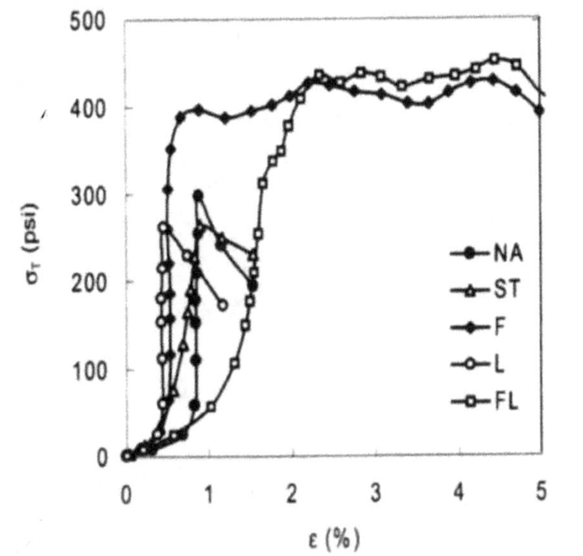

Figure 1: Tensile strength (σ_T) versus strain (ε) curve [13].

Table 1: Compressive strength and Young's modulus for 16 lbm/gal designs at 114°C (238°F) [6].

Formulation	Flexible Additives	Young's Module (MPa)	Compressive Strength (MPa)
G	Without	9041	39.2
H	With	6242	35.5
I	With	2594	21.1
J	With	1023	10.9

EXPERIMENTAL PROCEDURES

Materials

The materials used in this study include API well cement class E from Tehran Cement Company, tap water, Hidense as weighting agent, fluid loss control (Hallad 413), friction reducer (CFR-3) from Halliberton Company, boric acid as a retarder and salt (NaCl) from Petrokavir Company, antifoam (D047) from Dowell Company, and an elastomer additive. Figure 2 shows the shape and size distribution of the elastomer additive. The elastomer is a kind of modified terpolymer rubber containing 0.1 to 1 percent carboxylated monomer to enhance the adhesion of rubber to cement and improve the interaction at their interface. Rubbers have got the highest Poisson's ratio between other materials and are the best choice for increasing the Poisson's ratio of cement and decreasing its elastic modulus.

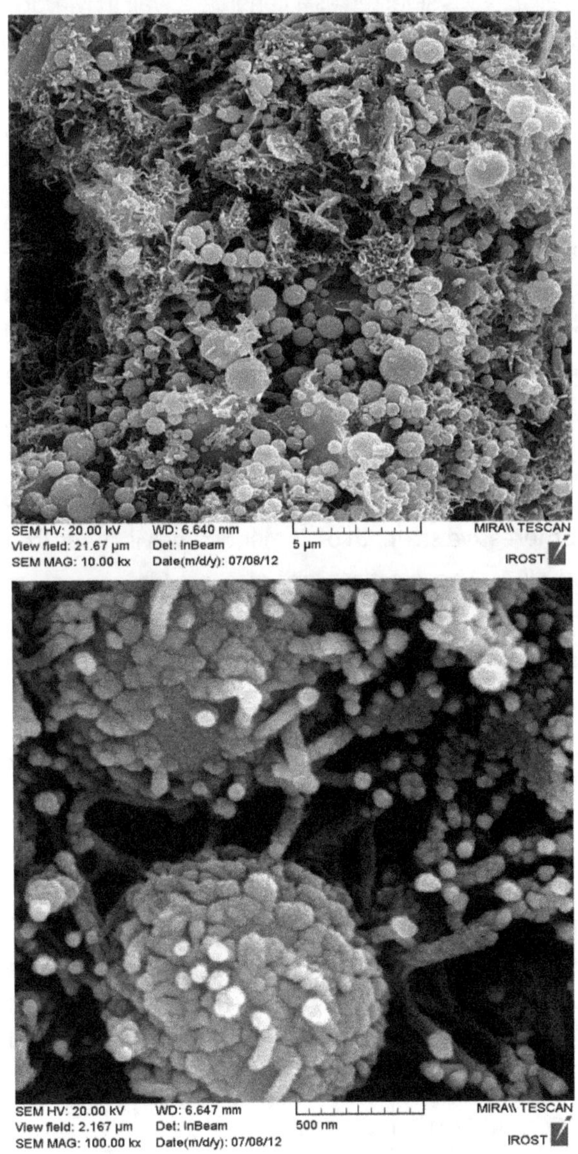

Figure 2: SEM micrograph of the elastomer additive.

The elastomer particles adsorb and damp the stress and stop the crack growth. The specifications of the elastomer additive are presented in Table 2.

Table 2: Physical and chemical properties the elastomer additive.

Size (micron)	Specific gravity	Humidity (%)	Color	Phase
75	1.1	4-5	Dark brown	solid

Study Method

This study is based on the laboratory testing of the improvements in the cement slurry and stone by using an elastomer additive at five concentrations. Well # 179 is selected as a candidate well in Maroon oilfield in the southwest of Iran. Recently, many casing collapse problems have occurred in this field in 9 5/8" casing string, which covers Gachsaran formation. Gachsaran formation is made of gypsum-anhydrite rocks and exhibits a plastic behavior. This imparts a horizontal stress to the casing string. More elastic cement sheath can enhance the protection of casing. The conventional and modified cement slurries are mixed in the laboratory and their properties including weight, rheology, free water, fluid loss, and thickening time are measured according to API RP 10B-2. The compressive strength of cement stones is measured by UCA. The elastic properties are measured on cylindrical samples by an uniaxial compressive strength tester equipped with axial and radial strain gauges (Figure 3).

The elastic properties of cement stone is determined based on the ISRM and ASTM rock mechanic standards. The cylindrical cement stone samples are prepared by curing the cement slurry for 48 hours at 210°F and 3000 psi in a rubber mold with a diameter of 68 mm and a length of 210 mm in a curing chamber. After the ends of the sample are cut flat, they are cooled down and kept saturated by merging in the saturated water; the loading rate is 4000 psi/min which is the maximum loading rate of the available equipment. Table 3 shows the conventional cement recipe in well # 179 and four other recipes modified by different amounts of the elastomer additive. Table 4 shows the elasticity properties measured for five samples.

Figure 3: Uniaxial compressive strength tester equipped with axial and radial strain gauges.

Table 3: Cement slurry compositions.

Component	Unit	Slurry 1	Slurry 2	Slurry 3	Slurry 4	Slurry 5
Cement Class E	lb	110	110	110	110	110
Elastomer Additive	lb/SK	-	3.3	6.6	9.9	13.2
Salt	lb/ SK	15.51	15.51	15.51	15.51	15.51
Hidense	lb/ SK	60.5	60.5	60.5	60.5	60.5
Boric Acid	lb/ SK	0.33	0.33	0.33	0.33	0.33
Fluid Loss Control	lb/ SK	-	0.33	0.33	-	-
Cement Friction Reducer	lb/ SK	-	0.55	0.55	0.11	1.32
Antifoam	Gal/ SK	0.02	0.02	0.02	0.02	0.02
Water	ft^3	0.6696	0.6696	0.6696	0.6696	0.6696

Table 4: Cement stones elasticity properties.

Sample	Elasticity Properties	
	Young's Modulus (GPa)	Poisson's Ratio
1	2.86	0.04
2	2.76	0.06
3	2.48	0.16
4	1.94	0.25
5	1.76	0.31

RESULTS AND DISCUSSION

Rheological Properties

According to the measured rheological properties of five slurry samples (Table 6), the increase of the elastomer additive will increase the plastic viscosity and yield point of the cement slurry. To have proper pumpable slurries, cement friction reducer additive is added in different amounts to the samples and the small differences in the rheological properties of 5 slurry samples are due to their different compositions. Sample number 1 had the settling problem of the Hidense and samples number 2 to 5 needed the cement friction reducer. Thicker slurry has more suspending capability and leads to a more homogenized cement sheath and constant cement stone properties along the casing string. The settling of the Hidense on the real well will result in a cement sheath with nonuniform properties along the well. At the bottom, high concentration of the Hidense will lead to heavy cement with a low compressive strength, and at the top, there will be light, thin, and two-phase slurries.

Free Water and Fluid Loss

Free water and fluid loss of cement slurries are determined according to the recommended method of API-RP-10B **[14].** Free water and fluid loss decrease by increasing the elastomer additive content (Table 5). It could also be deduced that the effectiveness of the elastomer additive in decreasing the free water and fluid loss decreases by increasing the concentration. Similar results are found in other studies [15-17]. High free water of the cement slurry causes problems in zonal isolation especially in horizontal wells and creates heterogenic cement stone properties along the cement sheath. High free water means thicker cake and weaker bonding between cement sheath and the formation. Loss of the water content of the cement slurry to the formation causes the cement slurry dehydration and increases the formation damage. It will also reduce the possibility of annular bridging by dehydrated cement [18-19]. Fluid loss control is also important in controlling the water-sensitive formation and gas migration problem.

Table 5: Cement slurries properties.

Slurry Number	Slurry Weight (PCF)	Free Water (cc)	Fluid Loss(mL) (T=180°F, P=1100 psi)	Thickening Time(min) (T=180°F, P=9000 psi)	Compressive Strength 24 hr(Psi) (T=210°F, P=3000 psi)
Slurry 1	149	1.5	30	145	3650
Slurry 2	147	1	27	160	3100
Slurry 3	145	1	30	180	2770
Slurry 4	143	0.5	31.5	193	2150
Slurry 5	141	0	33	203	1900

Thickening Time and Compressive Strengths

The elastomer additive in the cement slurry increases the thickening time, which will decrease the amount of the retarder additive needed and the total cost of the cement slurry (Table 5). The compressive strength of five cement samples is measured by ultrasonic cement analyzer (UCA). Figures 4-8 show the compressive strength of the 5 samples. It can be seen from these figures that the compressive strengths of the samples containing the elastomer additive are lower. According to Figures 4-8, by increasing the concentration of the elastomer additives up to 13.2 lb/sk in slurries, compressive strength decreases.

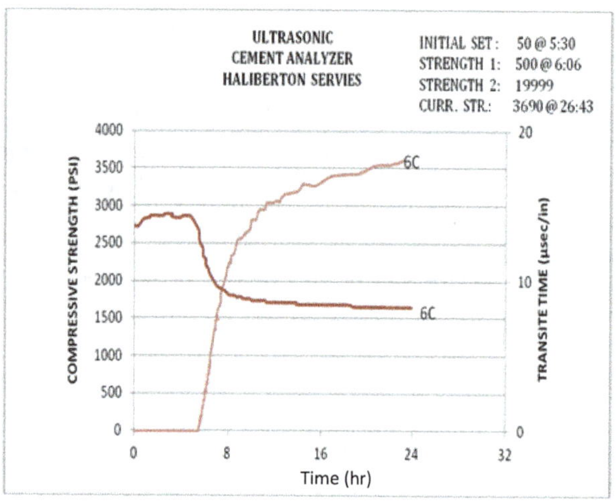

Figure 4: Compressive strength for cement slurry number 1.

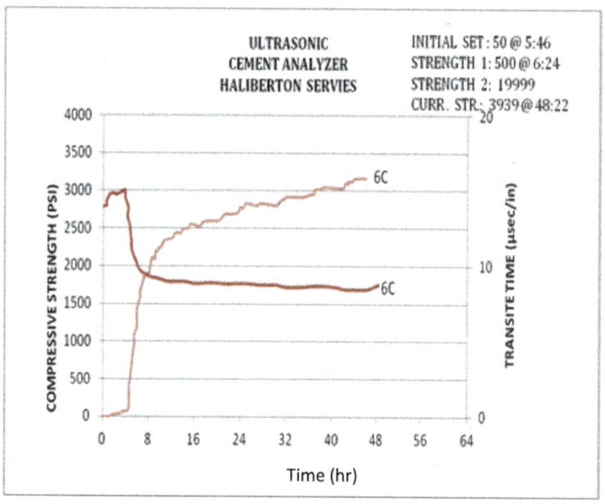

Figure 5: Compressive strength for cement slurry number 2.

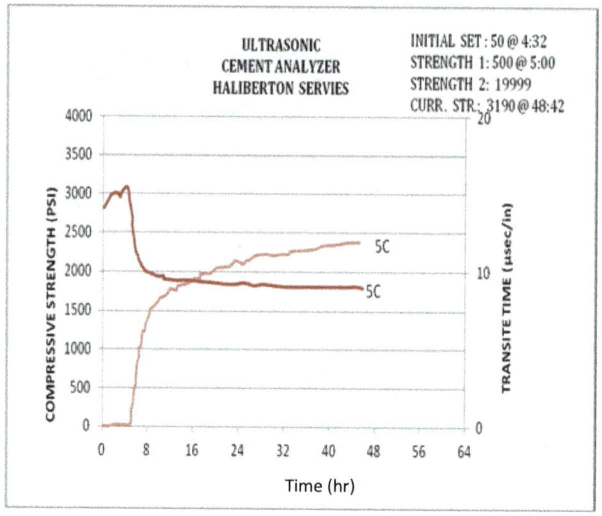

Figure 6: Compressive strength for cement slurry number 3.

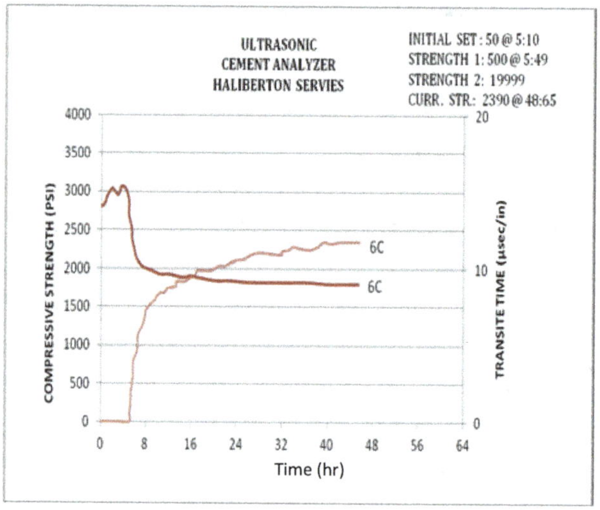

Figure 7: Compressive strength for cement slurry number 4.

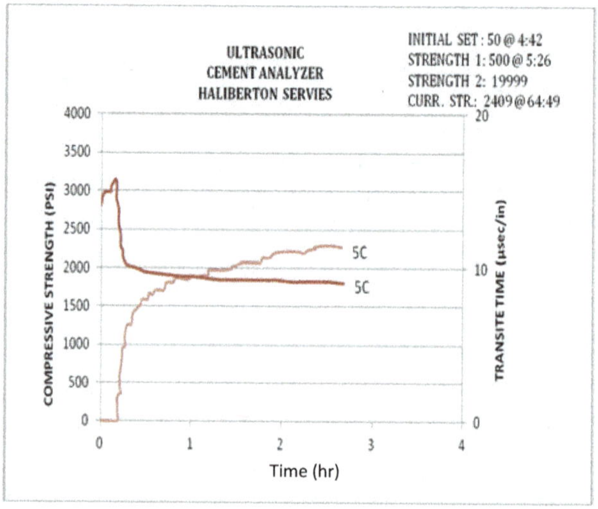

Figure 8: Compressive strength for cement slurry number 5.

Improving the Mechanical Properties of Cements

As could be expected, the mechanical properties of cement are strongly dependent on the particular additives incorporated to the slurry. Even when cements prepared with no admixtures or with a standard slurry design presenting high compressive strengths, they show a fragile behavior characterized by a low tensile strength (σ_T) and toughness (T). On the other hand, its high Young's modulus (E) and low Poisson's ratio (v) reduce its deformation capacity. The use of the elastomer additive improves the elastic behavior of cement as it reduces its Young's modulus and increases the Poisson's ratio. The elastic properties of the cement stone for the modified cement compositions with 4 concentrations of the elastomer additives are also measured in this study (Table 3). Figures 9-13 show the axial and radial stress-strain curve of five cement stone samples. The Young's modulus and Poisson's ratio are determined from the curve. Young's modulus is the slope of the best straight line to the axial strain-stress curve and the ratio of diametrical strain and axial strain is defined as Poisson's ratio. Similar to the previously mentioned properties, an increase in the Poisson's ratio and a decrease in Young's modulus can be observed (Table 4). Elastic cement systems showed an improvement in the collapse resistance of the casings. Casing collapse resistance is very sensitive to cement Poisson's ratio and cement Young's modulus [11]. A higher Young's modulus means that cement stone will have temporary and reversible deformations under a certain stress and will withstand higher stresses without breaking. A higher Poisson's ratio can distribute stresses and prevent the creation of high-stress points. The forces will be distributed along the cement sheath and the whole cement sheath will show an elastic behavior avoiding breakage by elastic deformation. In other studies, it has been shown that the compressive and flexural strengths of cement mortars are enhanced with adding latex [7].

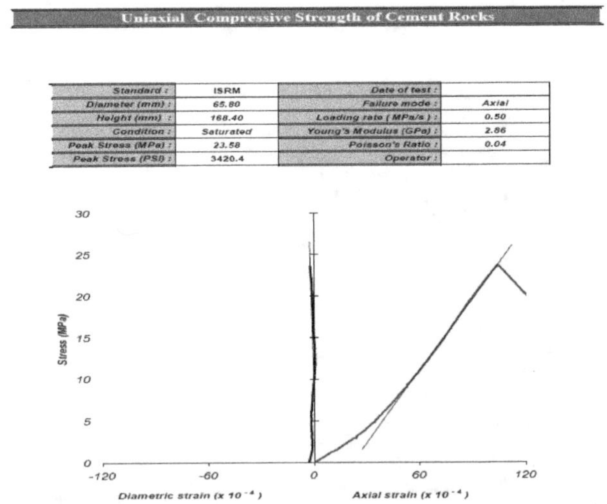

Figure 9: Axial and radial stress strain curve for cement slurry number1.

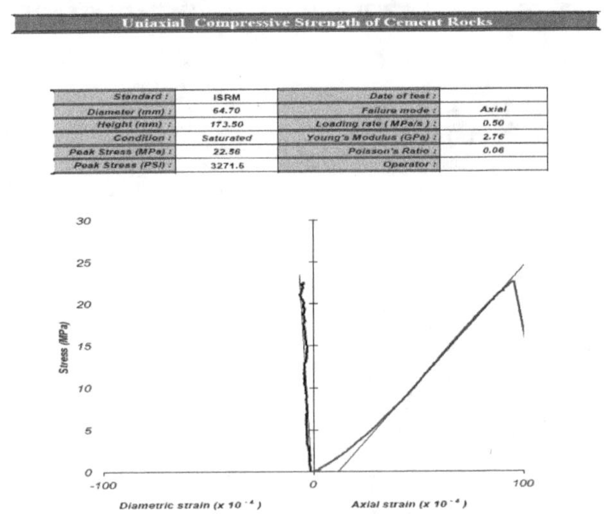

Figure 10: Axial and radial stress strain curve for cement slurry number 2.

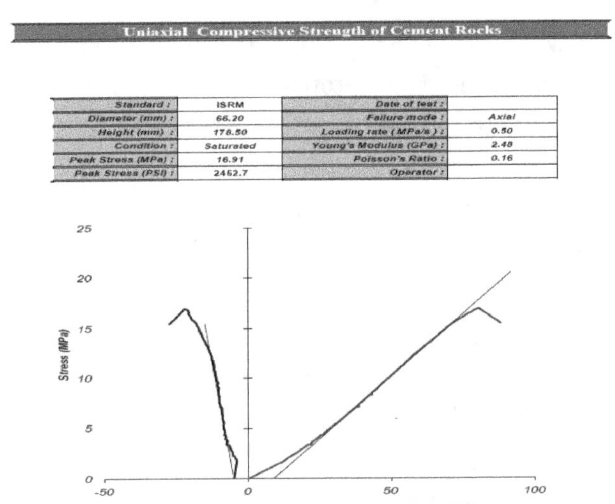

Figure 11: Axial and radial stress strain curve for cement slurry number 3.

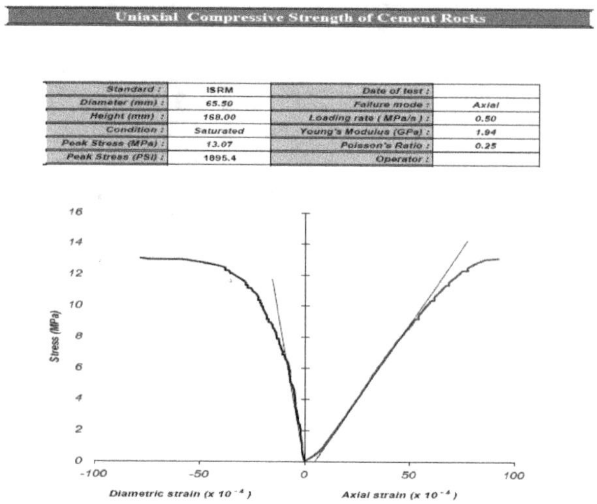

Figure 12: Axial and radial stress strain curve for cement slurry number 4.

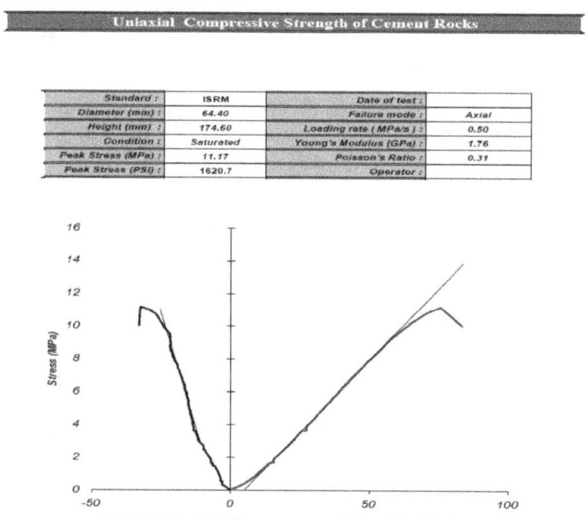

Figure 13: Axial and radial stress strain curve for cement slurry number 5.

In Maroun oil field, Gachsaran formation with plastic properties transmits the overburden pressure to the casing and leads to casing collapse [6-7]. The conventional cement slurry used in cementing this string has very poor properties and cannot resist against the stresses. Improving the elastic properties of the cement sheath covering this string can help solve the problem. Casing collapse resistance is improved by lowering the Young's modulus and increasing the Poisson's ratio [11-12]. More elastic cement can resist higher stresses [8] and the deformable cements can distribute the stress evenly [7]. The properties of the heavy-weight oil well cement used in this section of the well are improved by using the elastomer additives and the following results are obtained:

1. Elastomer additives increase the viscosity of cement slurry. High surface area of the elastomer additives in the cement slurry, when the water content is kept constant, will lead to more water adsorption by the solid content, and thereby decreasing the free water. The elastomer additives fill the voids between cement grains decreasing the volume of pores between them, which finally results in less free water. Therefore, there is a higher internal friction between solid particles, and viscosity rises.

2. The elastomer additives decrease the free water, fluid loss but increase the thickening time of the samples.

3. Adding the elastomer additives decreases the compressive strength but increases the elastic properties of the cement stone. Slurry samples numbers 2, 3, 4, and 5 have a lower compressive strength than the base cement. By increasing the elastomer additive content the compressive strength drops. Adding an elastic phase to the cement composition can improve the elastic properties of the cement stone, which means achieving a decrease in Young's modulus and an increase in Poisson's ratio. Cement stone matrix consists of two elastic and rigid phases. Under the exposed tensions, rigid phase resists and the elastic phase deforms. By increasing the elastic phase the whole system tends to increase the overall elasticity, which results in the elastic phase playing the major role.

CONCLUSIONS

This study confirms the results obtained in the previous studies and provides new experimental results on the improvement of the elastic properties of cement stone by elastomer additives. Elastomers improved the properties of the cement slurry and stone. From SEM micrographs (Figure 14) analysis, the elastomer additive acts as bridges between cracks and pores and guarantees the force transfer under stress. Elastomers keep their granular nature under downhole conditions even when they are soft and melted. They are like rubber particles or liquid droplets in the matrix of cement stone, and they do not form films on hydrated cement particles. They act as stress release spots, which improves the mechanical properties of the cement stone. Adding the elastomer additives decreases the compressive strength and Young's modulus, but it increases the Poisson's ratio of the cement stone samples. Furthermore, the elastomer additive in the cement slurry increased the viscosity and improved the suspending ability of the slurry. Free water and fluid loss also decreased by increasing the concentration of the elastomer additive. The effectiveness of the elastomer additive in decreasing the free water and fluid loss is higher at lower concentrations. It was also observed that the thickening time increased by increasing the concentration of the elastomer additives. The elastomer additive replaces the cement particles and decreases the hydration rate and retarder additive consumption, which consequently cuts the overall costs. More tests at higher concentrations of the elastomer additive have not been carried out since the purpose of the current study is to design a heavy weight oil well cement (heavier than 140 PCF) with the minimum required compressive strength, which, in this case, is 1600 psi; higher concentrations of the elastomer additive will lead to a further decrease in the slurry weight and reduces the compressive strength of the cement stone to the extent that it would be out of our acceptable range.

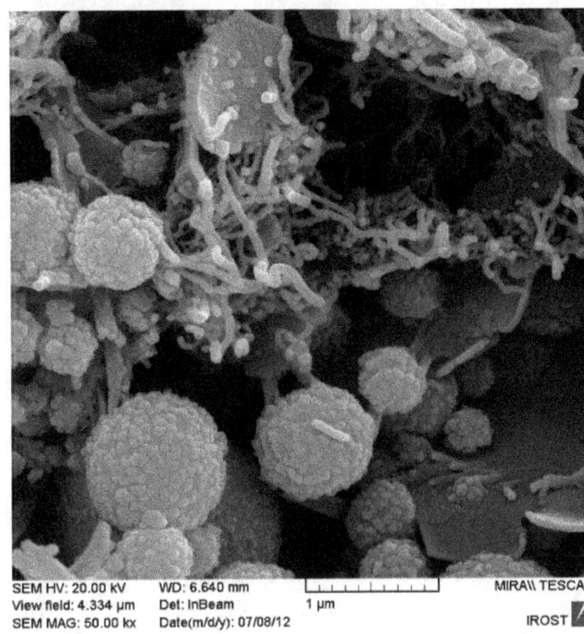

Figure 14: SEM micrograph of the cement stone including the elastomer additive.

Table 6: Cement slurries rheological properties.

Slurry number	(θ_{600})	(θ_{300})	(θ_{200})	(θ_{100})	(θ_6)	(θ_3)	PV (cP)	YP (lb/100ft^2)
Slurry 1	>300	273	190	107	15	13	249	24
Slurry 2	>300	250	177	98	20	16	228	22
Slurry 3	>300	286	181	100	19	15	249	17
Slurry 4	>300	276	165	107	17	13	253.5	22.5
Slurry 5	>300	248	177	98	29	19	225	23

REFERENCES

[1] Boukhelifa L., Moroni N., James S. G., Le Roy-Delage S., et al., "Evaluation of Cement Systems for Oil and Gas Well Zonal Isolation in a Full Scale Annular Geometry," SPE 87195-MS, Dallas, Texas, **2004**.

[2] Bannister C.E., "Evaluation of Cement Fluid Loss Behavior under Dynamic Conditions," SPE 7592-MS, Houston, Texas, **1978**.

[3] Destrade M., Gilchrist M. D., Motherway J., and Murphy J. G., "Slight Compressibility and Sensitivity to Changes in Poisson's Ratio,"

International Journal for Numerical Methods in Engineering, *90*, 403-411, **2011**.

[4] Fleckenstein W. W., Eustes A. W., Rodriguez W. J., Berger A., et al., "Cemented Casing: the True Stress Picture," AADE-05-NTCE-14, National Technical Conference and Exhibition, Houston, Texas, **2005**.

[5] El-Hassan H., Sultan M., Saeed A. A., Johnson C., et al., 'Using a Flexible, Expandable Sealant System to Prevent Microannulus Formation in a Gas Well: A Case History," SPE 92361, SPE Middle East Oil & Gas Conference, Bahrain, 12–15 March, **2005**.

[6] Le Roy-Delage S., Baumgarte C., Thiercelin M., and Vidick B., "New Cement Systems for Durable Zonal Isolation," SPE 59132-MS, IADC/SPE Drilling Conference, New Orleans, Louisiana, USA, 23-25 February, **2000**.

[7] Mack J. and Dillenbeck L. R., "Cement: How Tough is Tough enough? A Laboratory and Field Study," SPE 78712-MS, SPE Eastern Regional Meeting, Lexington, Kentucky, USA., 23-26 October, **2002**.

[8] Morris W., Criado M. A., Robles J., and Bianchi G., "Design of High Toughness Cement for Effective Long Lasting Well Isolations," SPE 81001-MS, SPE Latin American and Caribbean Petroleum Engineering Conference, Port-of-Spain, Trinidad and Tobago, 27-30 April, **2003**.

[9] Mueller D. T., "Producing Stress-resistant High-temperature/High-pressure Cement Formulations through Microstructural Optimization," SPE 84562-MS, SPE Annual Technical Conference and Exhibition, Denver, Colorado, USA, 5-8 October, **2003**.

[10] Ravi K., Bosma M., and Gastebled O., "Safe and Economic Gas Wells through Cement Design for Life of the Well," SPE 75700-MS, SPE Gas Technology Symposium, Calgary, Alberta, Canada, 30 April-2 May, **2002**.

[11] Yuan Z., Schubert J., Teodoriu C., and Gardoni P., "HPHT Gas Well Cementing Complications and its Effect on Casing Collapse Resistance," SPE 153986-MS, 28-30 March, Mumbai, India, **2012**.

[12] Wray B., Bedford D., Leotaud L., and Hunter B., "The Application of High-Density Elastic Cements to Solve HP/HT Challenges in South Texas: The Success Story," SPE 122762-MS, SPE Annual Technical Conference and Exhibition, New Orleans, Louisiana, USA, 4-7 October, **2009**.

[13] Khalaf F. and Cairo U., "Increasing Casing Collapse Resistance against Salt-induced Loads," SPE 13712-MS, 11-14 March, Bahrain, **1985**.

[14] "API Recommended Practice for Testing Well Cements," RP 10B API, Washington, **2010**.

[15] Nelson E., Guillot D., *Well Cementing*, 2nd Ed., Sugar Land, Tex.: Schlumberger, **2006**.

[16] Feder J., "Casing and Cementing," 3rd Ed., Austin, Texas: The University of Texas at Austin-Petroleum Extension Service, **2001.**

[17] Feder J., *Well Cementing (Oil and Gas Production)*, Austin, Texas: The University of Texas at Austin-Petroleum Extension services, **1983**.

[18] Rahman S. S., and Chilinggar G. V., "Casing Design: Theory and Practice," Amsterdam: Elsevier, **1995**.

[19] Ravi K., Bosma M., and Hunter L., "Optimizing the Cement Sheath Design in HPHT Shearwater Field," SPE 79905-MS, 19-21 February, Amsterdam, Netherlands, **2003**.

18

Thermal-oxidative Degradation of PGA, PLLA, and Random Binary PLLA-PGA Copolymers

Reza Asadi-Malekshah[1], Rasool Mohammadikhah[2] and Jamshid Mohammadi-Rovshandeh[3*]

[1]School of Chemical Engineering, College of Engineering, University of Tehran, Iran
[2]Process Development & Control Group, Process Development & Equipment Technology Division, Research Institute of Petroleum Industry, (RIPI), Tehran, Iran
[3]Caspian Faculty of Engineering, College of Engineering, University of Tehran, Rezvanshahr, Gilan, Iran

ABSTRACT

Dimerization process is essential for producing copolymers. The features of dimerization process like thermal-oxidative degradation should be well known to reach maximum efficiency and a superior reactor design. Also, the degradation mechanism of biodegradable polymers is important during sterilization processes. Thermal-oxidative degradation of PGA, PLLA, and their binary copolymers was investigated under isothermal heating as well as dynamic heating. All the samples were prepared by a polycondensation process and were characterized by TG, DTG, DSC, and HNMR analyses. Activation energy under dynamic heating was obtained by using Friedman plot. A new three stage mechanism, namely random, transition, and specific stages is proposed for dynamic heating degradation. Isothermal heating investigation is conducted under an inert atmosphere, and frequency factor and activation energy were achieved. It was found that the activation energy under isothermal heating is generally higher than that under dynamic heating. It was found that the rate of degradation increases significantly with an increase in temperature. The effects of pressure on the degradation rate were studied in different atmospheres with various oxygen partial pressures. Pressure effect was correlated by a second order polynomial in terms of total pressure. The obtained activation energies under isothermal heating were in good agreement with those reported by others. The complete kinetic scheme suitable for reactor design for the thermal-oxidative degradation of the samples was reported. Finally, the optimal operational conditions for the dimerization process were reported.

Keywords: Activation Energy, Dimerization, Frequency Factor, Reactor Design, Thermal-oxidative Degradation

INTRODUCTION

In the past three decades, biodegradable polymers such as polylactic acid (PLA), polyglycolic acid (PGA), and their binary copolymers have been in use for widespread applications [1,2]. The results of several histological studies on lactide, the dimer of PLLA, have verified that the human body can metabolize lactide into carbon dioxide and water without any undesirable response [3].

The majority of the commercially produced lactic acid (LA), the precursor of PLLA, is by the bacterial fermentation [4]. An extensively used method to prepare PLLA and PGA respectively from LA and glycolic acid (GA) was presented by Gilding and Reed [2]. More advanced methods for the

***Corresponding author**
Jamshid Mohammadi-Rovshandeh
Email: roshandeh@ut.ac.ir

pre paration of biodegradable polymers/copolymers can be found elsewhere [5]. Mohammadi-Rovshandeh and his co-workers synthesized and characterized novel biodegradable tri-block copolymers from L-lactide, glycolide, and polyethylene glycol/polypropylene glycol [6, 7]. Ajioka et al. reported a kind of one-step method, a direct solution condensation polymerization from LA and GA, to synthesize PLLA-PGA copolymers [8, 9]. Improved methods for synthesizing PLLA by direct dehydration polycondensation have been reported [10-12]. The process is much simpler than the ring-opening polymerization method demonstrated first by Carothers in 1932 [13]. Later on, there have been several attempts of condensation polymerization to gain more simplicity as one can find in reference [4] and references therein. The optimal synthetic conditions of PLLA and PLLA-PGA copolymers by means of melt copolycondensation directly from LA and/or GA with a feed molar ratio of 50/50 were presented [3, 14].

To meet different properties suitable for certain applications and specific requirements, random copolymers of PLLA-PGA can be made by binary combination of dimers with arbitrary weight/mole percentage of each one. It is crucial to determine characteristics of the polymer/copolymer for proper usage into vital environments after synthesizing. For example, nanoparticles of PLLA-PGA copolymers were used for encapsulating an efficient anti-cancer drug [15]. During sterilization, the poly mer/copolymer may unfavorably be decomposed due to an increase in temperature, so the thermal characteristics of polymer/copolymer should be known to hold it stable. Athanasiou et al. investigated sterilization, toxicity, biocompatibility, and clinical applications of PLIA-PGA copolymers [16]. On the other hand, it is useful to know the kinetics of thermal degradation to get a new sight of dimerization process and to design a proper reactor. Such dimer can produce high molecular weight polymer by the ring-opening polymerization method [16]. It was found that the purification of PLLA before heating increases the thermal stability by removing molecules with low molar mass and residual amounts of catalyst [17,18]. The thermal degradation of PLLA under nitrogen atmosphere was investigated by Nalbandi and found to be a single stage type with activation energy of 21-23 kcal.mol^{-1} [19]. The oxidative-thermal degradation of PLLA was investigated and it was found that the presence of oxygen/air aggravates polymer degradation with activation energy of 22-28 kcal.mol^{-1} [20, 21]. Wachsen et al. applied an interesting degradation-recombination scheme to modeling the thermal degradation of PLLA [22]. It was presented that the kinetic degradation constant (frequency factor) varies in the range of 107 s^{-1} and the energy of activation changes in the range of 28.7 kcal.mol^{-1} for crude PLLA which contains residual tin catalyst [23]. The kinetic parameters of the thermal degradation of PLLA in a blend with polyethylene (PE) were reported, from which it can generally be realized that adding PE increases the stability of PLLA [24]. Gupta and Deshmukh studied the thermal degradation of PLLA in the presence of air and found that the polym uyjer undergoes a thermal-oxidative degradation [20,21]. The thermal degradation of PLLA, PGA, and PLLA-PGA copolymers in a physical blend with poly (ε-caprolactone) (PCL) was studied under an inert atmosphere [25]. Regrettably, none of the above studies have reported the frequency factor and activation energy in their kinetic scheme for PLLA-PGA copolymers. This proposition makes the kinetic scheme unusable to establish a good reactor design. Moreover, pressure effects on the kinetic scheme should be known for designing reactor. The effects of pressure on the reaction rate for PLLA and PLLA-co-PGA have not been investigated so far. Hence there is an urgent need for new studies in this field. The main goal of this work is to find all unknown features affecting the reaction rate and kinetic parameters. The thermal degradation of PGA in the presence of vapor and at different pressures was investigated in the previous work, where it was concluded that the thermal degradation of pure PGA indeed conformed to a single-stage-type reaction at high temperatures [26].

There is only limited information on the mechanism of the thermal degradation of biodegradable polymers [25, 27-29]. It was reported that PGA is degraded by random chain scission through some changes in the degradation rate [30]. A new approach says that the mechanism of PGA thermal degradation changes from random chain scission to specific chain end scission when the temperature increases [25].

Apart from thermal degradation, hydrolytic/oxidative and in-vitro degradations can separately or simultaneously take place in the presence of relevant active agents like oxygen and vapor. In case of hydrolytic or in-vitro degradation, the rate of degradation is much lower than that of thermal and oxidative degradations [31-36]. Mixed degradation, e.g. thermal-hydrolytic or thermal-oxidative, has seldom been studied and only one report on the thermal-oxidative-hydrolytic degradation of PGA is available [26]; thus, it is of great importance to investigate mixed degradation in which the dimerization process is prevalently designed to be conducted under air atmosphere. The main objectives of our work are studying thermal-oxidative-hydrolytic degradation of crude samples and acquiring a reliable kinetic scheme suitable for reactor design. To our knowledge, no similar study has been accomplished so far. In this paper, the effects of temperature, polymer loading and water/oxygen partial pressure on the degradation rate are investigated. The frequency factor and activation energy of each crude sample are reported as well.

EXPERIMENTAL PROCEDURE

Raw Material

Glycolic acid (70% aqueous solution with a density of 1.26 gr/cm^3), L-Lactic acid (90% aqueous solution with a density of 1.21 gr/cm^3) and antimony (III) oxide (Sb_2O_3) as the catalyst were purchased from the Merck Company, Germany. A tiny amount of D-Lactic acid or D-L-Lactic acid (below 1%) was within L-Lactic acid, so PLA and its copolymers thorough this work are in the form of PLLA with high optical purity.

Synthesis Method

Condensation polymerization method was used to prepare all the samples due to simplicity and economical features. This method obtains samples with relatively low molecular weight. The preparation trend was more or less according to that of Gilding et al. [2], albeit with some innovations.

PGA

PGA preparation method can be found elsewhere [26]. The method is summarized in Table 1. The used setup for the preparation of samples is illustrated in Figure 1.

PLLA

At the beginning, 0.34 gr of antimony (III) oxide was added to 0.3 L L-lactic acid when mixing into a batch reactor equipped with an electrical mixer (40-60 rpm). The temperature of the homogenous solution was kept at 150°C for 2 hrs under a nitrogen atmosphere. In this stage, the un-bonded (extrinsic) water is vaporized. Then, the temperature was set on 175°C for 5 hrs in the presence of nitrogen (stage No. 2). The amount of water separated from the L-LA was frequently measured. The water is condensed into a double pipe heat exchanger linked to the top of the reactor (see Figure 1). The solution obtained from the previous stage is subjected to the temperature of 190°C for 90 min (stage No. 3). The end point of this stage is when the 90 percent of the total amount of theoretical water is extracted from the solution. In the second and third stages, a lot of bonded (intrinsic) water is chemically separated; nevertheless, there is still a little bit of water in the solution.

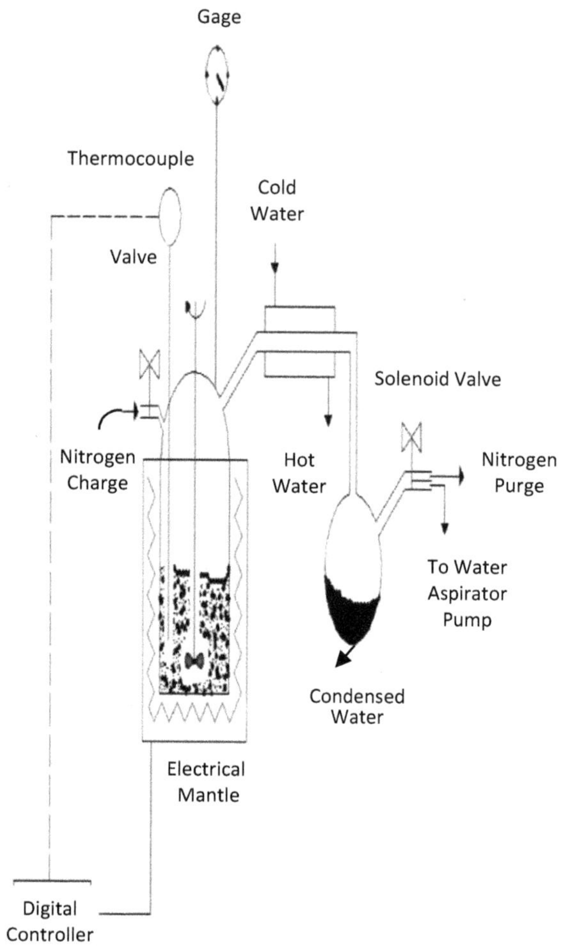

Figure 1: The setup used for polymer/copolymer preparation.

In the fourth stage, the temperature of the system decreased to 50°C and the nitrogen flow stopped while the vacuum pump (or water aspirator) was immediately allowed working to extract the rest of bonded water. This stage remained for 16 hrs and the temperature decreased to room temperature, afterwards. According to our experience, this method compared to other condensation polymerization styles obtains a good PLLA sample with relatively higher molecular weight and mechanical resistance.

Copolymers

Aqueous solutions of L-LA and GA with feeds ratios of 25/75, 50/50 and 75/25 (weight percent) were used to generate different PGA-PLLA copolymers, namely 25PGA-75PLLA, 50PGA-50PLLA and 75PGA-25PLLA. The method of synthesis resembles the one introduced in section 2.2.2 and can briefly be found in Table 1.

Material Balance

Table 2 shows the results of material balance for the preparation of crude samples. This is helpful to evaluate the yield of polymerization/copolymerization. The amount of water separated during preparation process determines progress and degree of completeness of polycondensation. The yield of dehydration is defined as:

$$\text{Yield} = \left(1 - \frac{|W_{total} - W_{sep}|}{W_{total}}\right) \times 100 \quad (1)$$

where W_{total} and W_{sep} are the weight of theoretical water and separated water respectively.

Table 1: Polymer/copolymer preparation method used in this study.

Polymer/ Copolymer	Stage No.							
	1		2		3		4 (vacuum)	
	t (min)	T (°C)	t (min)	T (°C)	t (min)	T (°C)	t (min)	T (°C)
PGA [26]	240	120	120	180	240	140	960	50
PLLA	60	150	300	175	90	190	720	150
25PGA-75PLLA	30	120	60	150	330	180	540	170
50PGA-50PLLA	60	120	300	180	30	190	540	170
75PGA-25PLLA	130	130	170	180	60	190	720	170

Table 2: Yield of dehydration for crude samples.

Polymer/ copolymer	W_{total}(gr)	W_{sep}(gr)	Yield (%)
PGA [26]	586.90	584	99.04
PLLA	101.64	99	97.40
25PGA-75PLLA	140.17	138	98.45
50PGA-50PLLA	171.44	168	97.99
75PGA-25PLLA	173.76	171	98.42

Characterization and Thermal Analysis

To characterize the prepared samples, we applied TGA, DTG, NMR, and DSC tests before the isothermal experiments get started. All the samples were not purified in all the tests, except for the HNMR test. All the copolymers were random type, as the HNMR test shows in Figure 4. The GPC tests for the evaluation of polymer/copolymer molecular weight were expensive and not applicable to PGA and to the copolymers containing up to 20 %wt. PGA because they become partially solved only in hexaflouroisopropanol. This solvent is very expensive and infrequent.

TG and DTG Measurements

A thermogravimetry analyzer (TGA Q50 V6.3) performed the dynamic heating degradation under an inert atmosphere of nitrogen (60 ml/min). The reference substance was platinum and the experiments were commenced from ambient temperature to 600°C. The sample weights were selected in the range of 25-35 mgr. The RAMP method (constant heating rate) was used for analyzing when the balance and sample gases were the same as argon at a flow rate of 40.0 ml/min and 60.0 ml/min respectively. From Figure 2 (TG and DTG plots obtained at a heating rate of 20°C/min), the global decomposition peaks for crude PGA, PLLA, 25PGA-75PLLA, 50PGA-50PLLA, and 75PGA-25PLLA are observed at 316.40, 373.09, 362.89, 369.56, and 369.8°C respectively.

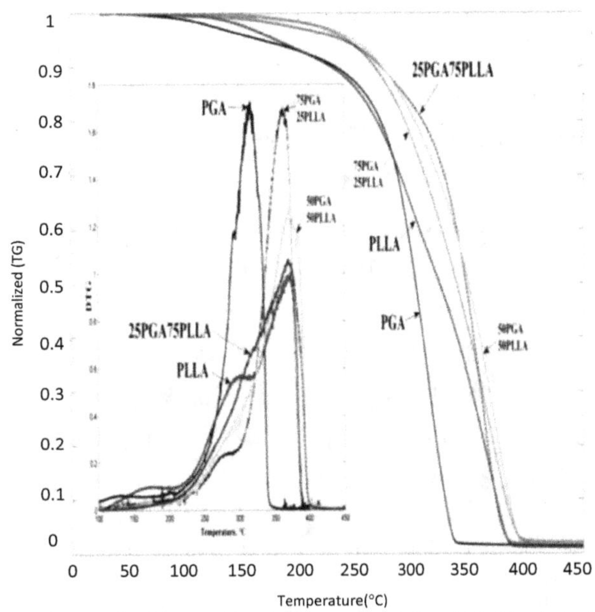

Figure 2: Normalized TG and original DTG plots for the crude samples.

Two paradoxically local maximum points are also observed for PLLA and 25PGA-75PLLA at 307 and 286°C. This treatment is, perhaps, due to impurities and various molecular weight chains. However, the obtained values for the first (local) decomposition peak temperature are in good agreement with those reported earlier [25]. The polymer/copolymer weight loss values are approximately 70%, 88%, 73%, 72%, and 78% respectively for PGA, PLLA, 25PGA-75PLLA, 50PGA-50PLLA, and 75PGA-25PLLA at the global decomposition peak. The weight loss values are 40% and 13% for PLLA and 25PGA-75PLLA at the first local peak respectively. Total polymer weight loss values at the end of degradation are determined as 99.57, 99.04, 98.53, 98.29, and 98.10% correspondingly taken place at 600, 600, 500, 500, and 500°C respectively for PGA, PLLA, 25PGA-75PLLA, 50PGA-50PLLA, and 75PGA-25PLLA. Indeed, Figure 2 shows that PLLA is more stable than PGA against thermal decomposition.

DSC Measurements

For better characterization, the glass transition and melting point of our samples were determined by RAMP method using DSC Q100 V9.0 analyzer. Between 1.5-3 mg of the crude samples was held in sealed aluminum crucibles when the heating rate was 20°C /min at an argon flow rate of 50 ml/min. The range from -90.14 to 397.24°C was the choice of beginning and ending temperatures respectively. Figure 3 reveals that glass temperatures (T_g) for 75PGA25PLLA and 50PGA50PLLA are 41.93 and 35.40°C respectively. The considerable crystallinity of PGA makes a dramatic endotherm in the DSC curve, showing that the melting point of crude PGA is 209.12°C. Surprisingly, there are three endotherms for PLLA: the first at 117.01°C (T_c), the second around 167°C, and the third at 187.1°C (the melting point).

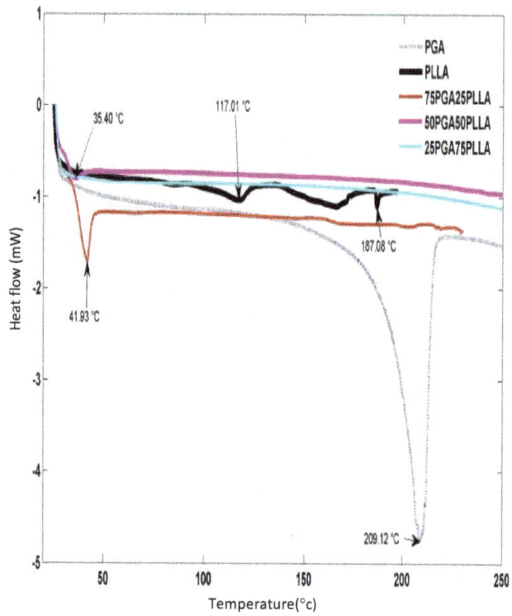

Figure 3: DSC analysis of the crude samples.

The first endotherm for PLLA is related to the temperature of crystallinity, as reported elsewhere [37]. As one can see in Figure 3, PLLA and 25PGA75PLLA do not have any glass temperature. The crystallinity of the synthesized samples can be extracted from the DSC analysis. The degree of crystallinity is in direct proportion to the area under the DSC curve. As clear, PLLA is partially crystalline, while PGA is fully crystalline. 75PGA25PLLA is relatively crystalline, whereas 50PGA50PLLA and 25PGA75PLLA are nearly amorphous. The degree of crystallinity and melting point of PLLA decrease with decreasing optical purity. The obtained PLLA has up to 99 percent L-unit; therefore the degree of crystallinity is between 45-70% [37]. The degree of crystallinity of PLLA in this work is calculated by mathematical relation and found to be 54.81%. PLLA's with optical purity less than 87.5 percent are amorphous. The crystallinity of 50PGA50PLLA and 25PGA75PLLA are 8.85 and 7.93 percent respectively.

NMR Measurements

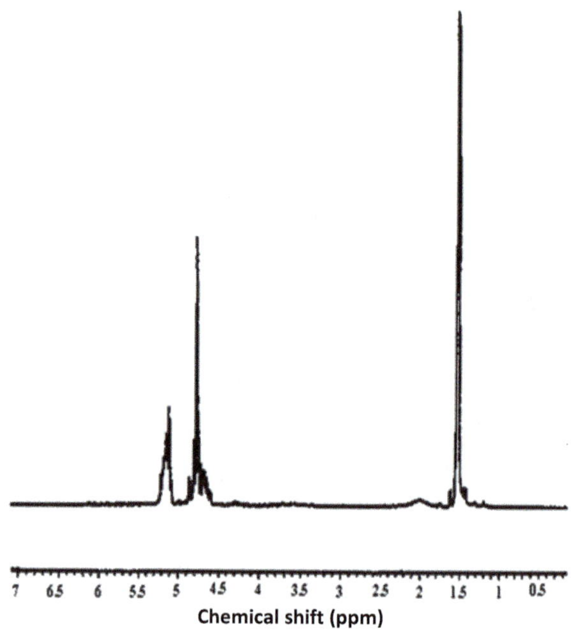

Figure 4: 1HNMR test for 25PGA75PLLA.

To find the structural details of copolymers (random or block), HNMR test was used for purified copolymers. Crude copolymers were purified by recrystallization in ethyl acetate (4 stages) to remove inferior peaks. As shown in Figure 4, the signals of 1HNMR spectra for 25PGA75PLLA split to distinct peaks due to the random distribution of monomers in the backbone structure. This treatment occurs for random copolymers.

Kinetic Scheme

Dynamic Heating

To search for the degradation mechanism of dynamic heating, the so-called Friedman plot shall be prepared from for example the data in Figure 2, where the variation of ln *(dα/dt)* against inverse temperature is depicted up to the decomposition peak temperature. Conversion, α, is defined by:

$$\alpha = 1 - \frac{m_A}{m_{A,0}} \tag{2}$$

All kinetic schemes assume that the isothermal rate of degradation is a linear function of a temperature-dependent rate constant, k, and a temperature-independent function of the mass concentration, m, in the below form:

$$-\frac{dm_A}{dt} = k_{th} f(m_A) \tag{3}$$

The general kinetic scheme may be made upon the reversible reaction defined as follows:

$$A \underset{k_2}{\overset{k_1}{\rightleftarrows}} \textit{Volatile + Dimer product + Solid residue}$$

The reaction is assumed to be irreversible, so k_1 becomes zero and k_2 changes to degradation rate constant. The temperature dependency of degradation rate constant under an inert medium can be expressed by an Arrhenius equation:

$$k_{th} = k_0 \exp(-E_a / RT) \tag{4}$$

where k_0, frequency factor, is temperature independent. The combination of Equation 3 and Equation 4 gives:

$$-\frac{dm_A}{dt} = f(m_A) k_0 \exp(-E_a / RT) \tag{5}$$

Biodegradable polymers are recognized to be degraded according to universal first order reaction rather than higher orders [29, 38,39].

$$f(m_A) = m_A^{\ n} \qquad n = 1 \tag{6}$$

Several empirical kinetic schemes were developed based on the fitting operation on the degradation data, but not based on well-known theories. As a merit, the first order reaction is the most used kinetic scheme, which has rigorous theoretical and experimental basics. For this reason, the first decomposition reaction is used in this work.

If the sample temperature is changed by a constant heating rate, $\beta = dT/dt$, the variation in the polymer weight can be analyzed as a function of temperature [39].

$$-\frac{dm_A}{m_A} = \frac{k_0}{\beta} \exp\left(-\frac{E_a}{RT}\right) dT \tag{7}$$

Friedman method is probably the most derivative technique for calculating activation energy [40-42]. This kinetic scheme would be valid to evaluate activation energy only under an inert atmosphere. Thus, an active atmosphere shall be investigated in the isothermal heating tests on the foundation of different kinetic schemes (See section 2.5.2.).

Isothermal Heating

The thermal-oxidative degradation of samples is investigated in isothermal heating because controlling oxygen/water partial pressure is simpler than dynamic heating. Once more, the relevant kinetic scheme may be made based on a first order reaction just as dynamic heating (Equation 3). The major difference is the definition of degradation rate constant so that it agrees with pressure variation and oxygen partial pressure. As previously reported [26], a kinetic model of Kelen, capable of inserting pressure effect into unimolecular decomposition rate, is used to estimate the degradation rate constant [43]. Kelen's basic approach to pseudo-unimolecular degradation rate constant is expressed in the following form:

$$k = k_{th}(T) F_c F_p \tag{8}$$

where, k_{th} is the true thermal degradation rate constant, which is evaluated under an inert atmosphere and it is only temperature dependent as Equation 4 dictates. The correction factors of F_c and F_p in Equation 8 depend on the polymer loading and total pressure. They are temperature-independent. The factor of F_p would compensate for the oxidative-hydrolytic contributions to the pseudo-unimolecular degradation rate constant. Commonly, functionality of F_p is determined based on the empirical correlations due to simplicity. Substituting Equation 6 into Equation 5 and making integration results in:

$$-\ln\frac{m_A}{m_{A,0}} = kt \tag{9}$$

The time variation of mass fraction retained during isothermal stage (Equation 9) can be simplified [25]:

$$\frac{m_A}{m_{A,0}} = 1 - kt \tag{10}$$

where

$$\alpha = kt \tag{11}$$

Equation 10 is just a deterministic approach toward Equation 9 (a truncated expansion for the exponential term); however, it gives reliable results for a wide range of degradation for a short period of time. Isothermal heating is investigated at different pressures associated with different oxygen/water partial pressures using Equation 10. The setup is founded to handle the planned air/water partial pressures [26].

RESULTS AND DISCUSSION

To estimate activation energy under dynamic heating, Friedman data points were prepared. Some of the doubtful data points were rejected. The conditions for dynamic heating can be found in section 2.4.1. The fitting process was accomplished by MATLAB 7.5.2007b Curve Fitting Toolbox. The linear least square optimization method of Trust-Region with a confidence level of 95% was applied. Variation of *ln(dα/dt)* versus *1000/T* is the Friedman plot, which is constructed from ascending portion of DTG curve until the main decomposition peak is appeared. The results indicate a change in the activation energy for PGA, PLLA, 25PGA-75PLLA, and 50PGA-50PLLA likely due to the change in the degradation mechanism from random chain scission at low temperatures to specific chain scission at high temperatures. For 75PGA-25PLLA any change in the slope is not noticeable. Lower activation energy for 25PGA-75PLLA is related to weight losses below 3%, whereas higher activation energy is obtained from the weight losses from 3% to 80%. Oddly enough, a transition zone during changing mechanism is evident for PGA and PLLA, while for copolymers two zones of random and specific scissions become merged. Thus as concluded in DSC section, the property of crystallinity can directly affect activation energy because crystalline structures need an extra energy to be activated. At low temperatures, PGA is degraded by ester interchange leading to the formation of larger olefinic chains with the liberation of smaller amount of glycolide (dimer) and methyl glycolate due to the decarboxylation of chain ends [44]. Conversely, at high temperatures, PGA is degraded by the decarboxylation of chain ends with glycolide and methyl glycolate as the main products [44]. Similar observations for PLLA were also reported [17]. Sivalingam et al. claimed that this procedure occurs for all aliphatic polyesters, while an obvious contradiction is here for 75PGA25PLLA, which does not show any changes in the activation energy and degradation mechanism [25].

The statistical parameters of fitting and the obtained activation energy are listed in Table 3, which shows a considerable decrease in the activation energy of the crude samples in comparison with that of the pure samples [25]. Fitted curves 1, 2, and 3 correspond to random, transition, and specific degradation mechanisms respectively. This mechanism (3 stages) is reported in the present work for the first time.

Table 3: Activation energy of dynamic heating for the crude samples.

Sample	Random E_a (kcal/mol)	Transition Ea (kcal/mol)	Specific E_a (kcal/mol)	R-squared of Fit 1	R-squared of Fit 2	R-squared of Fit 3	RMSE of Fit 1	RMSE of Fit 2	RMSE of Fit 3
PGA	7.27	21.40	13.72	0.956	0.992	0.962	0.0864	0.0627	0.0218
PLLA	5.04	13.30-17.58	9.85	0.922	0.988-0.943	0.993	0.0817	0.1593 - 0.0515	0.0133
25PGA75PLLA	5.47	14.51	27.38	0.822	0.980	0.998	0.1998	0.0619	0.0196
50PGA50PLLA	-	13.44	15.48	-	0.995	0.990	-	0.0223	0.0591
75PGA25PLLA	-	6.97	14.80	-	0.995	0.994	-	0.0108	0.0547

So far, the results are insufficient so as to complete kinetic scheme because frequency factor is still unknown. Besides, the effects of pressure and polymer loading on the kinetic parameters shall be investigated in an isothermal mode. Hence the kinetic scheme is improper for reactor design as the main goals of the present study. To study the effect of total pressure on the retained weight of samples, several experiments were carried out at different total pressures (absolute pressure) as well as initial oxygen/water partial pressures. The results are depicted in Figure 5 for PLLA, 25PGA-75PLLA, 50PGA-50PLLA, and 75PGA-25PLLA; similar results for PGA can be found elsewhere [26].

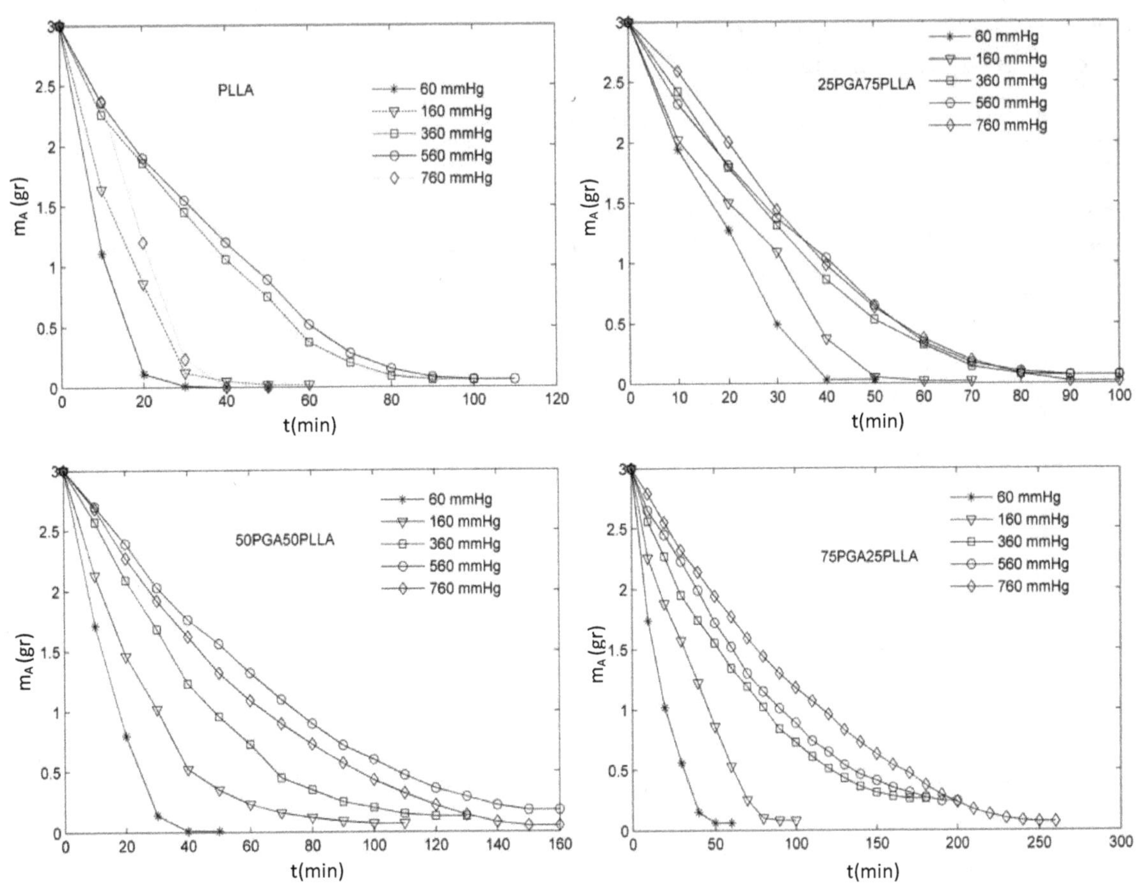

Figure 5: Variation of retained weight of samples with time at various total pressures; the temperature and polymer/copolymer loading are 260°C and 3 gr.

We expect the samples weight loss to be more observable at lower pressures (relative vacuum states) because of the Le Chatelier' principle. Relative vacuum state causes a decrease in the normal boiling point of the samples in favor of an increase in the rate of degradation. On the other hand, the presence of oxygen in the system (higher pressures) can increase the rate of degradation by oxidizing chain ends.

These two agents (low pressure against oxygen atmosphere) act oppositely in a way that for a relatively short while the first agent (vacuum) and for a long while the second agent (oxygen) are dominant, as it is clear from Figure 5. In a long period of time, the polymer chains become more fractured, and oxygen can consequently attack more carbon end atoms. Figure 5 shows that with an increase in the percentage of PLLA through copolymers, the second agent (oxidation) overcomes the first agent. Thus PGA is more resistance to oxidation than PLLA at any given time. Since the water aspirator pump was working at room temperature so as to induce a vacuum state, the minimum available pressure is identical to the water vapor pressure at 25°C, i.e. 23.78 mmHg [45]. Therefore, the total pressures of 60, 160, 360, 560, and 760 mmHg correspond respectively to 0.3963, 0.1486, 0.0661, 0.0425, and 0.0313 as water mole fractions and 0.127, 0.179, 0.196, 0.201, and 0.203 as oxygen mole fractions. All the mole fractions were calculated with the help of the science of phase equilibrium assuming that the partial pressure of volatile is negligible [46]. Equation 10 offers to make a linear fit plot on the data of $m_A/m_{A,0}$ versus time for obtaining degradation rate constant at different pressures. The slope of each linear plot gives us the relevant degradation rate constant. Table 4 summarizes the results of all the experiments. From this table, it is evident that the degradation rate at a pressure of 760 mmHg is surpassed by the one at 560 mmHg due to a greater oxygen mole fraction as demonstrated beforehand.

Table 4: Results of isothermal experiments.

Parameter in question	Test description	y_{water}	$y_{nitrogen}$	y_{oxygen}	P_{total} (mmHg)	L (gr)	T (°C)	k×10³ (1/min)	Adjusted R-squared	RMSE
Temperature	PLLA+ air+ water	0.023	0.772	0.205	760	10	180	0.178	0.998	0.011
							200	0.440	0.996	0.017
							220	1.335	0.993	0.020
							240	4.465	0.997	0.016
							260	13.205	0.996	0.021
							280	17.360	0.992	0.029
	25PGA75PLLA+ air + water	0.023	0.772	0.205	760	10	220	0.679	0.996	0.016
							240	2.722	0.998	0.009
							260	8.602	0.996	0.017
							280	14.862	0.992	0.030
	50PGA50PLLA+ air + water	0.023	0.772	0.205	760	10	220	0.541	0.999	0.021
							240	1.516	0.999	0.005
							260	6.897	0.991	0.017
							280	12.760	0.994	0.021

Parameter in question	Test description	y_{water}	$y_{nitrogen}$	y_{oxygen}	P_{total} (mmHg)	L (gr)	T (°C)	k×10³ (1/min)	Adjusted R-squared	RMSE
	75PGA25PLLA+ air + water	0.023	0.772	0.205	760	10	220	0.218	0.994	0.021
							240	1.012	0.999	0.005
							260	3.255	0.994	0.016
							280	10.601	0.987	0.026
Pressure	PLLA+ air+ water	0.291	0.560	0.149	60	3	260	48.170	0.939	0.121
		0.109	0.704	0.187	160			35.677	0.952	0.079
		0.049	0.752	0.199	360			13.941	0.979	0.044
		0.031	0.766	0.203	560			13.193	0.981	0.039
		0.023	0.772	0.205	760			31.600	0.981	0.056
	25PGA75PLLA+ air + water	0.291	0.560	0.149	60	3	260	27.331	0.986	0.042
		0.109	0.704	0.187	160			20.632	0.970	0.057
		0.049	0.752	0.199	360			16.680	0.987	0.037
		0.031	0.766	0.203	560			15.270	0.979	0.042
		0.023	0.772	0.205	760			16.428	0.993	0.026
	50PGA50PLLA+ air + water	0.291	0.560	0.149	60	3	260	36.672	0.980	0.052
		0.109	0.704	0.187	160			22.030	0.969	0.051
		0.049	0.752	0.199	360			14.773	0.999	0.006
		0.031	0.766	0.203	560			9.464	0.991	0.019
		0.023	0.772	0.205	760			11.362	0.996	0.012
	75PGA25PLLA+ air + water	0.291	0.560	0.149	60	3	260	26.810	0.928	0.095
		0.109	0.704	0.187	160			12.946	0.975	0.045
		0.049	0.752	0.199	360			8.964	0.974	0.032
		0.031	0.766	0.203	560			7.678	0.994	0.016
		0.023	0.772	0.205	760			6.544	0.995	0.013
Loading	PLLA+ air+ water	0.023	0.772	0.205	760	3	280	40.330	1.000	0.000
						5		35.801	0.986	0.042
						7.5		22.797	0.992	0.025
						10		17.360	0.992	0.029
						12		16.789	0.991	0.030
	25PGA75PLLA+ air + water	0.023	0.772	0.205	760	3	280	31.301	0.997	0.021
						5		24.883	0.993	0.033

Parameter in question	Test description	y_{water}	$y_{nitrogen}$	y_{oxygen}	P_{total} (mmHg)	L (gr)	T (°C)	k×10³ (1/min)	Adjusted R-squared	RMSE
						7.5		19.844	0.989	0.033
						10		14.862	0.992	0.030
						12		14.980	0.995	0.023
	50PGA50PLLA+ air + water	0.023	0.772	0.205	760	3	280	27.073	0.995	0.024
						5		19. 802	0.993	0.026
						7.5		15.763	0.987	0.028
						10		12.760	0.994	0.021
						12		11.572	0.993	0.021
	75PGA25PLLA+ air + water	0.023	0.772	0.205	760	3	280	18.634	0.990	0.029
						5		13.517	0.991	0.025
						7.5		12.345	0.991	0.022
						10		10.601	0.987	0.026
						12		9.726	0.987	0.024
Nitrogen atmosphere (flow rate of 377 cm³/min)	PLLA+ nitrogen	0	1	0	-	3	250	31.273	0.975	0.030
							270	56.536	0.986	0.022
	25PGA75PLLA+ nitrogen	0	1	0	-	3	250	9.683	0.980	0.035
							270	13.176	0.989	0.027
	50PGA50PLLA+ nitrogen	0	1	0	-	3	250	9.127	0.995	0.016
							270	14.699	0.985	0.029
	75PGA25PLLA+ nitrogen	0	1	0	-	3	250	3.342	0.988	0.015
							270	11.831	0.996	0.014

Among many tests done to investigate polymer/copolymer loading effect, we found that initial sample loading does not have any effect on the degradation rate constant as long as loading is higher than 10 gr. For loading below 10 gr, however, a difference in the degradation rate is observed likely due to the carbonization of the samples. It is discovered that at any given time, the rate of the carbonization of PLLA is higher than that of PGA. An extra methyl branch into the PLLA structure presumably increases the amount of carbonization. Loading (up to 10 gr) is an inert factor; thus the parameter of F_c in Equation 8 takes a unit value. Our conclusion for polymer loading effect is thoroughly consistent with that in Refs. [26, 47].

Several experiments were established at various temperatures at 760 mmHg. Sample loading for all the experiments in this section was 10 gr. The results are tabulated in Table 4, where it is clear that the degradation rate raises with an increase in temperature. Characteristic curves to evaluate the degradation rate constant for the samples at different temperatures are shown in Figure 6.

Statistical analyses such as adjusted R-squared and RMSE for all the fitted plots are reported in Table 4. The quantity of RMSE is a criterion for fitness quality and is defined as:

$$RMSE = \frac{\sqrt{\sum_{1}^{n}(A_{\exp} - A_{fitting})^2}}{n} \quad (12)$$

In fact, our basic measure of how closely a model fits some data is the Root Mean Squared Error (RMSE), which measures the average mismatch between each data point and the model (here is linear). The reported degradation rates in Table 4 were all obtained from Equation 10, which describes how this parameter varies with time.

We repeated the experiments with an adjusted R-squared lower than 0.90 to gain high accuracy.

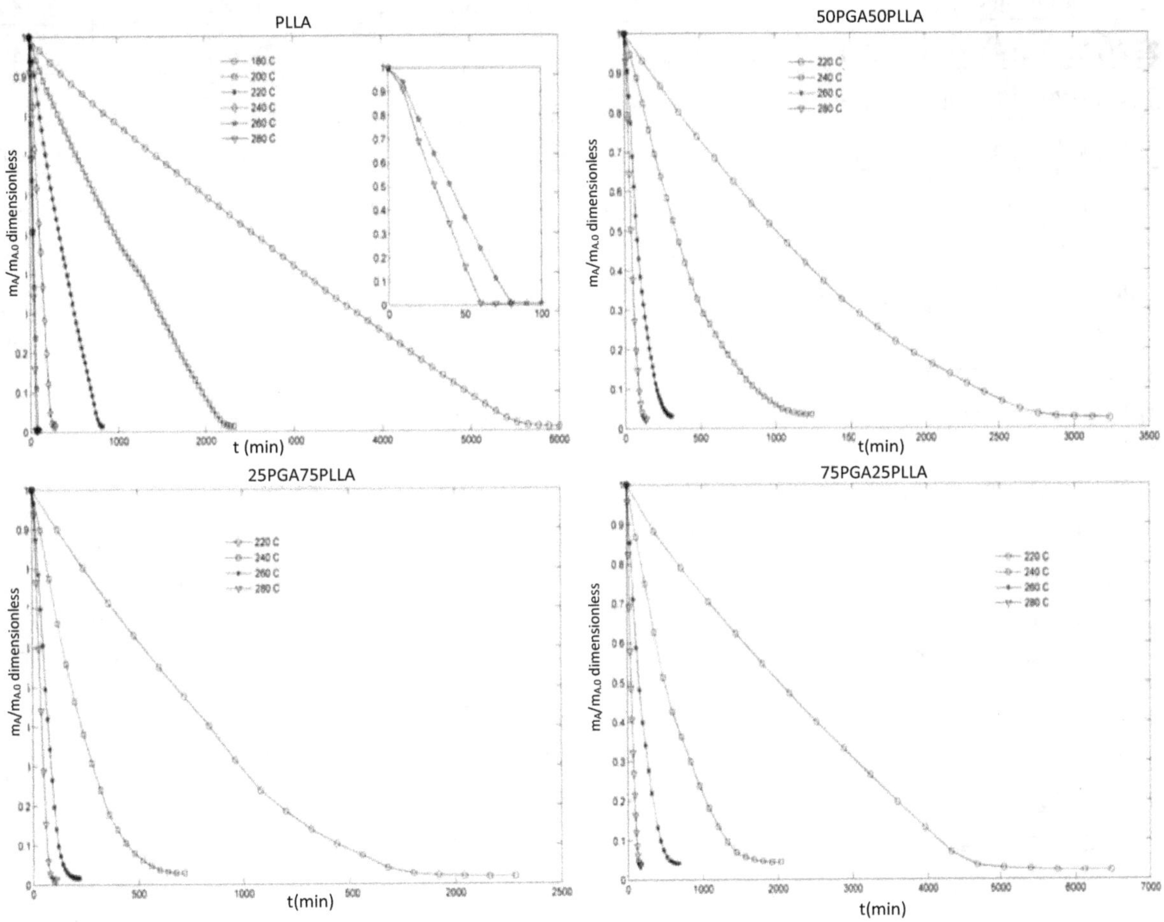

Figure 6: Variation of mass fraction retained with time at different temperatures at atmosphere pressure.

The true thermal degradation constant has to be acquired under an inert atmosphere and in the absence of oxygen and vapor. Before studying pressure dependency of degradation rate constant under thermal-oxidative-hydrolytic mode, this quantity must be specified. For all the samples, it is investigated at two temperatures as one can see the results in Table 4. Information is also available on true degradation constant of pure PGA [26]. Activation energy and frequency factor in different atmospheres were found considering Arrhenius plot. The results are incorporated into Table 5. Normal values (true values under an inert atmosphere) of activation energy and frequency factor are shaded in Table 5. It was found that activation energy in an oxygen/vapor atmosphere is higher than that in an inert atmosphere due to oxidative-hydrolytic-thermal degradation. In general, the activation energy obtained under isothermal heating is higher than the one obtained under dynamic heating. However, the obtained activation energies are in good agreement with those reported in literature, especially in Refs.[19-21, 25].

Table 5: Activation energy and frequency factor obtained in different atmospheres.

Test description	E (kcal/mol)	k_0 (1/min)	Adjusted R-squared	RMSE
PLLA+ nitrogen	16.707225	301341.6651	-	-
25PGA75PLLA+ nitrogen	8.6913318	41.58751492	-	-
50PGA50PLLA+ nitrogen	13.446213	3809.297241	-	-
75PGA25PLLA+ nitrogen	35.66856	2.71551×10^{12}	-	-
PLLA+ air+ water	24.481422	106536691	0.984	0.2345
25PGA75PLLA+ air + water	28.342206	2900067936	0.967	0.2483
50PGA50PLLA+ air + water	29.845608	9177310272	0.9727	0.2371
75PGA25PLLA+ air + water	34.772874	6.05306×10^{11}	0.9976	0.08072

The results summarized in Tables 4 and 5 are still incomplete without a correlation for F_p to be used for successful reactor design. The effect of pressure on the degradation rate constant (F_p) was investigated. When the normal frequency factor (k_0 or k_{th}) and F_c are given, F_p can be calculated by Equation 8. Table 6 includes the values of F_p acquired under different pressures. As demonstrated, F_p has a higher order of magnitude in the vacuum states. It first drops with pressure growth and then gradually rises since oxygen aggravates the degradation rate (Figure 7).

Table 6: True thermal degradation constant and F_p.

Test description	T (°C)	$k_{th}\times10^3$ (1/min)	P (mmHg)	F_p (dimensionless)
PLLA+ air+ water	260	42.304514	60	1.138649176
			160	0.8433379
			360	0.329539302
			560	0.311857973
			760	0.746965206
25PGA75PLLA+ air + water	260	11.326795	60	2.412950883
			160	1.821521445
			360	1.472614274
			560	1.348130694
			760	1.450366145
50PGA50PLLA+ air + water	260	11.633658	60	3.152232943
			160	1.893643427
			360	1.269849947
			560	0.813501652
			760	0.976648961
75PGA25PLLA+ air + water	260	6.3673	60	4.210575911
			160	2.033200886
			360	1.40781807
			560	1.205848633
			760	1.027751166

To design a reactor for conducting dimerization process, it is optimal to work away from the minimum point of F_p in every plot (Figure 7) because the rate of dimerization is the lowest there. The more the degradation rate is, the more desirable the dimer product is generated. In reactor engineering, it is important to maximize the rate of dimerization by setting operating conditions on the maximum value of F_p. The best option may be working under relative vacuum to avoid carbonization and oxidation and to have a greater degradation rate and consequently a higher dimerization rate. The functionality of F_p was found by fitting operation on the curves of F_p versus pressure, as can be seen in Table 7. It was found that a second order polynomial with respect to the pressure (mmHg) is capable of correlating the data very well.

In this stage, the kinetic scheme is ready for a comprehensive reactor design, in which the maximum rate of dimerization gives almost a pure dimer product. In our next work, we will centralize the reactor design for the dimerization process

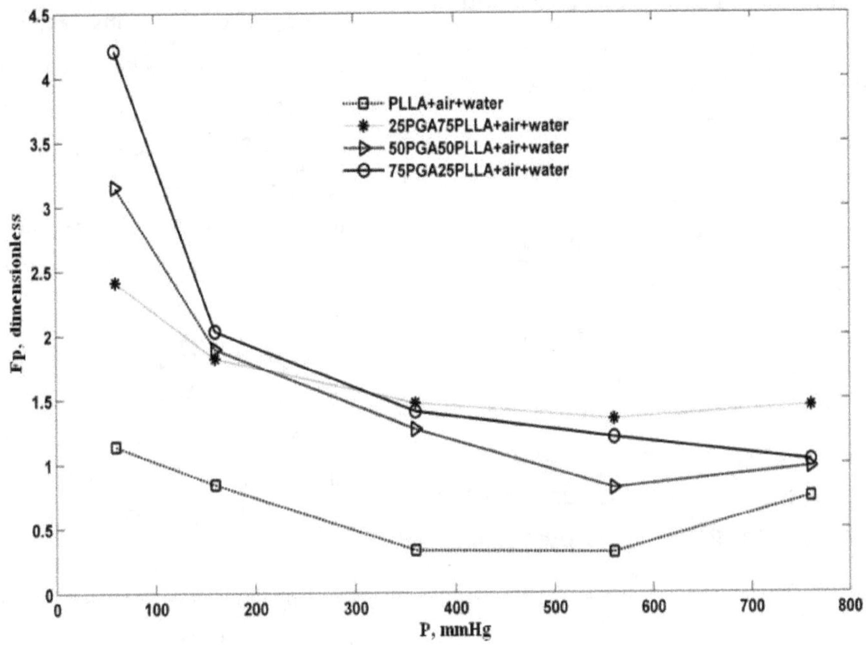

Figure 7: Variation of F_p versus total pressure.

Table 7: The correlations for F_p.

Sample	Correlation	Adjusted R-squared	RMSE
PLLA	$5.35050e - 6P^2 - 0.0050P + 1.4536$	0.9817	0.0479
25PGA-75PLLA	$4.2679e - 6P^2 - 0.0047P + 2.5962$	0.9265	0.1183
50PGA-50PLLA	$8.1082e - 6P^2 - 0.0094P + 3.5035$	0.9191	0.2701
75PGA-25PLLA	$1.0824e - 6P^2 - 0.0125P + 4.4546$	0.8726	0.6578

CONCLUSIONS

In this paper, the thermal-oxidative degradation of PLLA, PGA, and PLLA-PGA copolymers was studied at different water/oxygen partial pressures. The samples prepared by a new polycondensation method were characterized by TG, DTG, DSC, and HNMR analyses. Dynamic heating analysis showed that there may generally be three stages of degradation with different mechanisms as well as activation energies: random stage, transition stage, and specific stage. The obtained activation energies by dynamic treatment were in the expected range. Normal activation energy, frequency factor, and true degradation constant were evaluated in a nitrogen atmosphere and found to be in excellent agreement with those previously reported. It was found that the sample loading is inert, providing that the sample charge is up to 10 gr. It was found that the rate of degradation increases with a rise in temperature. It was also found that the presence of oxygen intensifies the rate of thermal degradation. Activation energies obtained in an oxygen atmosphere were higher than those obtained in an inert atmosphere. A second order polynomial in terms of total pressure was proposed to take pressure effect on the degradation rate constant into account. The complete kinetic scheme and optimal operational conditions were rendered for reactor engineering.

REFERENCES

[1] Dickers K. J., Huatan H., and Cameron R. E., "Polyglycolide-Based Blends for Drug Delivery: A Differential Scanning Calorimetry Study of the Melting Behavior," *Journal of Applied Polymer Science*, **2003**, *89*, 2937-2939.

[2] Doppalapudi S., Jain A., Khan W., and Domb A. J., "Biodegradable Polymers-an Overview," *Polymer Advanced Technology Journal*, **2014**, *25*, 427-435.

[3] Wang Z., Zhao Y., Wang F., and Wang J., "Syntheses of Poly (Lactic Acid-Co-Glycolic Acid) Serial Biodegradable Polymer Materials via Direct Melt Polycondensation and their Characterization," *Journal of Applied Polymer Science*, **2006**, *99*, 244-252.

[4] Gupta A. and Kumar V., "New Emerging Trends in Synthetic Biodegradable Polymers-Polylactide: A Critique," *European Polymer Journal*, **2007**, *43*, 4053-4074.

[5] Motta A. C. and Duek E. A. R, "Synthesis and Characterization of a Novel Terpolymer Based on L-Lactide, D, L-Lactide and Trimethylene Carbonate," *Materials Research Journal*, **2014**, *17*, 619-626.

[6] Farnia S. M., Mohammadi-Rovshandeh J., and Sarbolouki M. N., "Synthesis and Characterization of Novel ABA Triblock Copolymers from L-Lactide, Glycolide and PPG," *Journal of Applied Polymer Science*, **1999**, *73*, 633-637.

[7] Mohammadi-Rovshandeh J., Farnia S. M., and Sarbolouki M. N., "Synthesis and Characterization of Novel ABA Triblock Copolymers from L-Lactide, Glycolide and PEG," *Journal of Applied Polymer Science*, **1999**, *74*, 2004-2009.

[8] Ajioka M., Enomoto K., Suzuke K., and Yamaguchi A., "Basic Properties of Polylactic Acid Produced by The Direct Condensation Polymerization of Lactic Acid," *Bulletin of the Chemical Society of Japan*, **1995**, *68*, 2125-2131.

[9] Ajioka M., Suizo H., Higuci C., and Kashima T., "Aliphatic Polyester and their Copolymers Synthesized Through Direct Condensation Polymerization of Lactic Acid," *Polymer Degradation and Stability*, **1998**, *59*, 137-143.

[10] Takasu A., Narukawa Y., and Hirabayashi T., "Direct Dehydration Polycondensation of Lactic Acid Catalyzed by Water-Stable Lewis Acids," *Journal of Polymer Science*, **2006**, *44*, 5247-5253.

[11] Mohammadikhah R., Asadi-Malekshah R., and M. Rovshandeh J., "Preparation, Characterization and Thermal-Oxidative Degradation of Poly-L Lactic Acid (PLLA). 14th National Chemical Engineering Congress,"

Sharif University of Technology, **2012**, Tehran, Iran, 11215, 1-8.

[12] Nagarajan S., Tsibouklis J., and Reddy R. S., "Direct Condensation of D, L- And L-Lactic Acids: Effect of $Alcl_3$ Catalyst on Microstructural Arrangements of Homopolymers and Copolymers," *Iran Polymer Journal*, **2011**, *20*, 523-533.

[13] Carothers W., Dorough G., and Van Natta F., "Studies of Polymerization and Ring Formation," *Journal of the American Chemical Society*, **1932**, *54*, 761-772.

[14] Kim S. H. and Kim Y. H., "Direct Condensation Polymerization of Lactic Acid," *Macromolecular Symposia*, **1999**, 144, 277-287.

[15] Verderio P., Bonetti P., Colombo M., Pandolfi L., and Prosperi D., "Intracellular Drug Release from Curcumin-Loaded PLGA Nanoparticles Induces G2/M Block in Breast Cancer Cells," *Biomacromolecules Journal*, **2013**, *14*, 672–682.

[16] Athanasiou K. A., Niederauer G. G., and Agrawal C. M., "Sterilization, Toxicity, Biocompatibility and Clinical Applications of Polylactic Acid/Polyglycolic Acid Copolymers," *Biomaterials*, 1996, *17*, 93-102.

[17] Jamshidi K., Hyon S., and Ikada Y., "Thermal Characterization of Polylactides," *Polymer*, **1988**, *29*, 2229-2234.

[18] Zhang X., Wyss U., Pichora D., Goosen M., "An Investigation of the Synthesis and Thermal Stability of Poly (DL-Lactide)," *Polymer Bulletin*, **1992**, *27*, 623-629.

[19] Nalbandi A., "Kinetics of Thermal Degradation of Polylactic Acid under N_2 Atmosphere," *Iranian Polymer Journal*, **2001**, *6*, 371-376.

[20] Gupta M. C. and Deshmukh V. G., "Thermal Oxidative Degradation of Poly (Lactic Acid)," *Colloid & Polymer Science*, **1982**, *260*, 308-311.

[21] Gupta M. C. and Deshmukh V. G., "Thermal Oxidative Degradation of Poly (Lactic Acid)," *Colloid & Polymer Science*, **1982**, 260, 514-517.

[22] Wachsen O., Reichert K., Kruger R., Much H., et al., "Thermal Decomposition of Biodegradable Polyesters-III. Studies on the Mechanisms of Thermal Degradation of Oligo-L-Lactide Using SEC, LACCC and MALDI-TOF-MS," *Polymer Degradation and Stability*, **1997**, *55*, 225-231.

[23] Wachsen O., Platkowski K., and Reichert K., "Thermal Degradation of Poly-L-Lactide- Studies on Kinetics, Modeling and Melt Stabilization," *Polymer Degradation and Stability*, **1997**, *57*, 87-94.

[24] Omura M., Tsukegi T., Shirai Y., Nishida H., et al., "Thermal Degradation of Poly (Lactic Acid) in A Blend with Polyethylene," *Industrial & Engineering Chemistry Research*, **2006**, *45*, 2949-2953.

[25] Sivalingam G. and Madras G., "Thermal Degradation of Binary Physical Mixtures and Copolymers of Poly (E-Caprolactone), Poly (*D, L*-Lactide), Poly (Glycolide)," *Polymer Degradation and Stability*, **2004**, *84*, 393-398.

[26] Mohammadikhah R. and Mohammadi-Rovshandeh J., "Thermal Degradation and Kinetic Analysis of Pure Polyglycolic Acid in Presence of Humid Air," *Iranian Polymer Journal*, **2008**, *17*, 691-701.

[27] Drumright R., Gruber P., and Henton D., "Polylactic Acid Technology," *Advanced Materials*, **2000**, *12*, 1841-1846.

[28] Kopinke F. D. and Mackenzie K., "Mechanistic Aspects of the Thermal Degradation of Poly (Lactic Acid) and Poly (*B*-Hydroxybutyric Acid)," *Journal of Analytical and Applied Pyrolysis*, **1997**, *40-41*, 43-53.

[29] Shih Y. and Chieh Y., "Thermal Degradation Behavior and Kinetic Analysis of Biodegradable Polymers Using Various Comparative Models," *Macromolecular Theory and Simulations*, **2007**, *16*, 101-110.

[30] Chujo K., Kobayashi H., Suzuki J., and Tokuhara S., "Physical and Chemical Characteristics of

Polyglycolide," *Macromolecular Chemistry and Physics Journal*, **1967**, *100*, 267-270.

[31] Mohammadi-Rovshandeh J., Farnia S., and Sarbolouki M., "Synthesis and In-Vitro Degradation of D, L-Lactide-Glycolide Copolymers," 6th ISOC. Tabriz, Iran, **1997**.

[32] M. Rovshandeh J., Farnia S. M. F., and Sarbolouki M. N., "In-Vitro Degradation of Lactide-Glycolide Copolymers: PLA35GA30," *Journal of Applied Polymer Science*, **1998**, *69*, 627-630.

[33] Mohammadi-Rovshandeh J. and Sarbolouki M., "Synthesis and In-Vitro Hydrolytic Degradation of Polyglycolic Acid and Its L-Lactide Copolymer," *Iranian Polymer Journal*, **2001**, *10*, 53-58.

[34] Proikakis C. S., Mamouzelos N. J., Tarantili P. A., and Andreopoulos A. G., "Swelling and Hydrolytic Degradation of Poly (D, L-Lactic Acid) In Aqueous Solutions," *Polymer Degradation and Stability*, **2006**, *91*, 614-619.

[35] King E. and Cameron R., "Effect of Hydrolytic Degradation on the Microstructure of Poly (Glycolic Acid)," *Applied Polymer Science*, **1997**, *66*, 1681-1690.

[36] Mobedi H., Mashak A., Nekoomanesh M., and Orafai H., "L-Lactide Additive and In Vitro Degradation Performance of Poly (L-Lactide) Films," *Iranian Polymer Journal*, **2011**, *21*, 237-245.

[37] Dorgan J. R., Lehermeier H., and Mang M., "Thermal and Rheological Properties of Commercial Grade Poly (lactic acids)," *Journal of Polymers and the Environment*, **2000**, *8*, 1-9.

[38] Denq B., Chiu W., and Lin K., "Kinetic Model of Thermal Degradation of Polymers for Nonisothermal Process," *Journal of Applied Polymer Science*, **1997**, *66*, 1855-1868.

[39] Minying L., Lijun G., Qingxiang Z., Yudong W., et al., "Thermal Degradation Process and Kinetics of Poly (Dodecamethyleneisophthalamide)," *Chemical Journal on Internet*, **2003**, *5*, 43-53.

[40] Friedman H., "Kinetics of Thermal Degradation of Char-Forming Plastics from Thermogravimetry. Application to A Phenolic Plastic," *Journal of Polymer Science*, **1964**, *6*, 183-195.

[41] Li X., Huang M., Guan G., and Sun T., "Kinetics of Thermal Degradation of Thermotropic Poly (P-Oxybenzoate-Coethylene Terephthalate) by Single Heating Rate Methods," *Polymer International*, **1998**, *46*, 289-297.

[42] Li X. and Huang M., "Thermal Decomposition Kinetics of Thermotropic Poly (Oxybenzoate-Co-Oxynaphthoate) Vectra Copolyester," Polymer *Degradation and Stability*, **1999**, *64*, 81-90.

[43] Kelen T., "Polymer Degradation," Van Nostrand Company, New York, **1983**.

[44] McNeill I. and Leiper H., "Degradation Studies of Some Polyesters and Polycarbonates: 3-Polyglycolide," *Polymer Degradation and Stability*, **1985**, *12*, 373-385.

[45] Smith JM., Van Ness CH., and Abbott M. M., "Introduction to Chemical Engineering Thermodynamics," McGraw-Hill, New York, **2001**.

[46] Prausnitz J. M., Lichtenthaler N. R., and Azevedo G. E., "Molecular Thermodynamics of Fluid Phase Equilibria," New Jersey, Prentice-Hall Inc., **1999**.

[47] Balakrishnan R. and Guria C., "Thermal Degradation of Polystyrene In The Presence of Hydrogen by Catalyst in Solution," *Polymer Degradation and Stability*, **2007**, *92*, 1583-1591.

PERMISSIONS

All chapters in this book were first published in JPST, by Research Institute of Petroleum Industry; hereby published with permission under the Creative Commons Attribution License or equivalent. Every chapter published in this book has been scrutinized by our experts. Their significance has been extensively debated. The topics covered herein carry significant findings which will fuel the growth of the discipline. They may even be implemented as practical applications or may be referred to as a beginning point for another development.

The contributors of this book come from diverse backgrounds, making this book a truly international effort. This book will bring forth new frontiers with its revolutionizing research information and detailed analysis of the nascent developments around the world.

We would like to thank all the contributing authors for lending their expertise to make the book truly unique. They have played a crucial role in the development of this book. Without their invaluable contributions this book wouldn't have been possible. They have made vital efforts to compile up to date information on the varied aspects of this subject to make this book a valuable addition to the collection of many professionals and students.

This book was conceptualized with the vision of imparting up-to-date information and advanced data in this field. To ensure the same, a matchless editorial board was set up. Every individual on the board went through rigorous rounds of assessment to prove their worth. After which they invested a large part of their time researching and compiling the most relevant data for our readers.

The editorial board has been involved in producing this book since its inception. They have spent rigorous hours researching and exploring the diverse topics which have resulted in the successful publishing of this book. They have passed on their knowledge of decades through this book. To expedite this challenging task, the publisher supported the team at every step. A small team of assistant editors was also appointed to further simplify the editing procedure and attain best results for the readers.

Apart from the editorial board, the designing team has also invested a significant amount of their time in understanding the subject and creating the most relevant covers. They scrutinized every image to scout for the most suitable representation of the subject and create an appropriate cover for the book.

The publishing team has been an ardent support to the editorial, designing and production team. Their endless efforts to recruit the best for this project, has resulted in the accomplishment of this book. They are a veteran in the field of academics and their pool of knowledge is as vast as their experience in printing. Their expertise and guidance has proved useful at every step. Their uncompromising quality standards have made this book an exceptional effort. Their encouragement from time to time has been an inspiration for everyone.

The publisher and the editorial board hope that this book will prove to be a valuable piece of knowledge for researchers, students, practitioners and scholars across the globe.

LIST OF CONTRIBUTORS

Hamid Soltanian, Mohammad Ataei and Reza Khalokakaie
School of Mining, Petroleum and Geophysics Engineering, University of Shahrood, Iran

Ezzatollah Kazemzadeh
Petroleum Engineering Department, Research Institute of Petroleum Industry (RIPI), Tehran, Iran

Effat Behboudi
Department of Petroleum Engineering, Amirkabir University of Technology, Tehran, Iran

Nasser Keshavarz and Asghar Naderi
Exploration and Production Research Center, Research Institute of Petroleum Industry (RIPI), Tehran, Iran

Jian Xiaofei
China University of Geosciences, Beijing 100083, China
National Research and Development (Experimental) Center for Shale Gas, LangFang 065007, China

Tang Shuheng and Jia Lingxiao
China University of Geosciences, Beijing 100083, China

Lin Wen and Zhang Qin
Research Institute of Petroleum Exploration and Development-LangFang, LangFang 065007, China

Liu Renhe
National Research and Development (Experimental) Center for Shale Gas, LangFang 065007, China
Research Institute of Petroleum Exploration and Development-LangFang, LangFang 065007, China
China University of Petroleum, Beijing, 102249, China

Liu Jun and Liu Qingyou
School of Mechatronic Engineering, Southwest Petroleum University, Cheng Du Si Chuan 610500, China

Wang Wei and Sang Guo
School of Civil and Architecture Engineering, Southwest Petroleum University, Cheng Du Si Chuan 610500, China

Xiao Guohua and Zhang Jianzhong
Drilling Technology Research Institute of Jidong Oilfield Company, Tang Shan He Bei 063000, China

Fang Yu
Advanced Materials and Analysis Center, National Institute of Clean-and-Low-Carbon Energy, Beijing 102209, China

Sepideh Kashefi and Mohammad Nader Lotfollahi
Faculty of Chemical, Gas, and Petroleum Engineering, Semnan University, Semnan, Iran

Abbas Shahrabadi
Exploration and Production Division, Research Institute of Petroleum Industry (RIPI), Tehran, Iran

Seyed Saied Bahrainian, Amin Nabati and Ebrahim Hajidavalloo
Department of Mechanical Engineering, Shahid Chamran University of Ahvaz, Ahvaz, Iran

Wenjie Xu, Pengcheng Bian, Hongze Gang, Jinfeng Liu and Shizhong Yang
State Key Laboratory of Bioreactor Engineering and Institute of Applied Chemistry, East China University of Science and Technology, Shanghai 200237, China

Bozhong Mu
State Key Laboratory of Bioreactor Engineering and Institute of Applied Chemistry, East China University of Science and Technology, Shanghai 200237, China
Shanghai Collaborative Innovation Center for Biomanufacturing Technology, Shanghai 200237, China

Artyom E. Chemodanov, Vladislav A. Sudakov, Sergey A. Usmanov and Rinat K. Khayrtdinov
Institute of Geology and Petroleum Technology, Kazan (Volga Region) Federal University, Russia

Robert R. Akhmadullin
PAO "Tatneft", Russia

Pouran Nazarian Samani, Maryam Mirshahani and Navab Khodaei
Petroleum Geology Division, Research Institute of Petroleum Industry, Tehran, Iran

Badr Salem Bageri, Mohamed A. Mahmoud, Reyad A. Shawabkeh and Abdulazeez Abdulraheem
King Fahd University of Petroleum & Minerals, Dhahran, 31261, Kingdom of Saudi Arabia

Samuel Getnet Tsegaye, M. E. Nton and O. A. Boboye
Department of Petroleum Geosciences, Pan African University Life and Earth Sciences Institute (PAULESI), University of Ibadan, Ibadan, Nigeria
Department of Geology, University of Ibadan, Ibadan, Nigeria

Welela Ahmed
School of Earthsciences, Addis Ababa Science and Technology University, Addis Ababa, Ethiopia

Mohammad Reza Talaghat and Ahmad Reza Bahmani
Department of Chemical Engineering, Petroleum and Gas, Shiraz University of Technology, Shiraz, Iran

Marzieh Neyzan Hosseini
Department of Geophysics, College of Basic Science, Kermanshah Branch, Islamic Azad University, Kermanshah, Iran

Ezatallah Kazemzadeh
Department of Petroleum Engineering, Islamic Azad University, Science and Research Branch, Tehran, Iran

Ehsan Sobhani
Department of Geophysics, College of Basic Science, Kermanshah Branch, Islamic Azad University, Kermanshah, Iran

Bita Arbab
Department of Petrophysics, Iranian Offshore Oil Company, Tehran, Iran

Abolfazl Moradi and Behzad Tokhmechi
Mining, Geophysics and Petroleum Engineering Depatment, Shahrood University of Technology, Shahrood, Iran

Vamegh Rasouli
Petroleum Engineering Department, University of North Dakota, USA

Mohammad Fatehi Marji
Mining and Metallurgical Engineering Department, Yazd University, Yazd, Iran

Ghasem Khatinzadeh, Ahmad Roozbahani and Ali Mehdizadeh
Chemical, Polymeric & Petrochemical Technology Development Research Division, Research Institute of Petroleum Industry (RIPI), Tehran, Iran

Index

www.ingramcontent.com/pod-product-compliance
Lightning Source LLC
LaVergne TN
LVHW081030210926
848485LV00004B/986
9781639894109